The Worker at Work

The Worker at Work

A textbook concerned with men and women in the workplace

T. M. Fraser
Former Professor
Department of Systems Design Engineering
and Director
Centre for Occupational Health and Safety
University of Waterloo, Canada

Taylor & Francis
London · New York · Philadelphia
1989

UK	Taylor & Francis Ltd, 4 John St, London WC1N 2ET
USA	Taylor & Francis Inc., 1900 Frost Road, Suite 101, Bristol PA 19007

Copyright © T. M. Fraser 1989

All rights reserved. No part of this publication may be reproduced, stored in a retrieval system, or transmitted, in any form or by means, electronic, electrostatic, magnetic tape, mechanical, photocopying, recording, or otherwise, without the prior permission of the copyright owner and publisher.

British Library Cataloguing in Publication Data
Fraser, T. M.
 The worker at work: a textbook concerned with men and women in the workplace.
 1. Industrial health & industrial safety. Management aspects
 I. Title
 658.3'82

 ISBN 0-85066-476-4
 ISBN 0-85066-481-0 pbk

Library of Congress Cataloging-in-Publication Data

is available

Typeset in 11/12 point Bembo by
Bramley Typesetting Limited, 12 Campbell Court, Bramley, Basingstoke, Hants.

Printed in Great Britain by
Taylor & Francis (Printers) Ltd, Basingstoke, Hampshire

Contents

Preface ix

Part I **People at work: An interactive relationship** 1

Chapter 1 Origins 3
  Introduction 3
  The development of occupational hygiene and
    ergonomics 5

Chapter 2 A systems viewpoint 15
  The anatomy of a system 15
  The person–machine–environment system 17
  Human capacities in a person–machine–
    environment system 23
  Human limitations in a person–machine–
    environment system 27
  System function 28

Part II **The work environment** 33

Chapter 3 Work, skill and fatigue 35
  Definitions 35
  Energy metabolism 36
  Types of work 42
  Measurement of energy cost of work 50
  Operational factors in work fatigue measurement 55

Chapter 4 The industrial work station 62
  Work station analysis 62
  Work place dimensions 64
  Other design principles 74

Chapter 5	A guide to manual materials handling	80
Chapter 6	Repetitive strain injury	91
	Terminology and pathology	91
	Occurrence in industry	94
	Prevention and control	100
Chapter 7	Human stress, circadian rhythms and shift work	102
	The nature of stress	102
	Circadian rhythms and shift work	111
Chapter 8	Job satisfaction and work humanization	118
Part III	**Design for human use and function**	131
Chapter 9	Equipment design for human use	133
	Introduction	133
	Displays	134
	Controls	146
	Control and display stereotypes	151
	Task analysis for control layout	152
Chapter 10	Hand tools	153
	Guidelines for tool design	153
	Generic tools	165
	Power-driven tools	173
Chapter 11	The office work station	178
	The office system	185
	Environment	195
Part IV	**Physical agents in the work environment**	201
Chapter 12	Vibration	203
	Parameters of vibration	203
	Whole-body vibration	209
	Human response to segmental vibration	216
Chapter 13	Noise and sound	223
	Noise	224
	Noise hazard	236
	Hearing conservation	241
	Noise control at source	244

Chapter 14	Heat gain and heat tolerance	252
	Introduction	252
	Prevention and control of heat stress	270
Chapter 15	Heat loss and cold tolerance	275
	Introduction	275
	Cold protection	277
Chapter 16	Vision, illumination and visual hazards	280
	Vision	280
	Illumination	289
	Glare	295
	Visual hazards	296
Chapter 17	The atmosphere, respiration and barometric pressure	300
	Atmosphere	300
	Respiration	301
Chapter 18	The effects of barometric pressure	312
	Partial pressure effects	312
	Total pressure effects	321
Part V	**Chemical agents and aerosols in the work environment**	335
Chapter 19	A review of organic chemical terminology	337
Chapter 20	A review of the principles of industrial toxicology	344
Chapter 21	Toxicity in the work environment: Some selected chemicals	357
	Respiratory irritants	358
Chapter 22	Dusts and other solid particulates	375
	Classification of dusts by effect	378
Chapter 23	Control of chemical agents and aerosols in the work environment	384
	Hazard control	392
Chapter 24	A review of the principles of ventilation for contaminant control	399
	Dilution ventilation	400

Exhaust ventilation	402
Hood design	404
Ducts	408
Fans	411
Air cleaning devices	413
Types of collector	414

Appendix Walk-through survey: A checklist questionnaire   419

References   425

Index   438

Preface

As the title would imply, this book is concerned with the definition, organization, and solution of work-oriented problems besetting men and women in the workplace. Consequently it embraces elements of what are commonly known as ergonomics, occupational hygiene, and occupational health. Yet this is not a book specifically oriented to the needs of ergonomists or occupational hygienists, although I believe it could act as a text or reference in both of these fields. Instead it is intended more for those managers and others in the workplace whose background is far from ergonomics and occupational hygiene, but who are concerned with human problems, and human needs at work, as well as with the work environment, and the design of tools, equipment, and procedures which optimize human performance while maintaining health and comfort.

Thus, while I hope it will be of value to ergonomists and occupational hygienists, it is intended also for persons such as industrial engineers, industrial designers, production engineers, human resource managers, office managers, architects, industrial psychologists, industrial toxicologists, laboratory managers and technologists, occupational health physicians and nurses, trade union executives, members of occupational health and safety committees, and so on, along with students in these fields and in various community colleges, technological colleges and high schools.

No man can be a specialist in all fields. In consequence this book is not written for the specialist; it is written *by* a generalist, *for* a generalist. It is written for a person who has a little knowledge in many of the areas covered, and perhaps specialist knowledge in some, but wishes to see as best he or she can the totality of the inter-relationships occurring amongst the tangibles and intangibles of work. The specialist can seek more comprehensive and refined data in the specialist literature.

It is perhaps only when one completes a book that one feels competent to begin it, and as the author of this book I wonder where I found the temerity even to start. Consequently its faults are mine and its errors are mine. I can only hope that its strengths will outweigh its weaknesses.

T. M. Fraser
University of Waterloo, Canada
January 1989

Part I
People at work:
An interactive relationship

Chapter 1
Origins

Introduction

People have worked in one way or another since the beginnings of the human race. Work, indeed, for most is a prerequisite for survival. For some it is an arduous endeavour with physical demands that reach to the levels of human tolerance or beyond; for others it is an emotionally demanding exercise that exhausts the spirit, but for most it provides returns either intrinsic in the act of working or in the rewards that the work can bring. Indeed, in many of the great religions, and in the concepts of the great philosophers, the act of working is a virtue, and work is inherently ennobling. And yet, it is only in relatively recent history that attention has been paid to the sacrifices of life and health that have all too often been demanded of the worker, or to the frequently haphazard way in which a worker, sometimes with unsuitable tools and equipment, has been tossed into an inappropriate working environment. Indeed, the ancient Greek and Roman physicians, such as Hippocrates and Galen, may have observed some of the consequences of hazardous exposures in the working environment, but they did little or nothing to change either the work or the environment.

In addition to the medical and nursing professions, which are primarily concerned with the care and treatment of the sick and injured, and not so much with prevention and provision of safe, healthy, and comfortable environments conducive to high level production, two other professions in recent times have evolved to consider the problems of the worker in the workplace. These are the professions of occupational or industrial hygiene, and ergonomics. Each is a relatively new profession, and it might be argued which has the longer history. Each is transdisciplinary in origin, taking its knowledge, skills, and expertise from the arts, the sciences, technology and engineering, but each in its own way is dedicated to improving the lot of the worker in the workplace, and the manner in which the working tasks are performed.

To some extent these professions have gone their separate ways, each with its own required bank of knowledge, each with its professional institutions, and demands for official recognition, but a very large overlap exists in the requirements for knowledge and skills between the two professions. Indeed, many occupational hygienists have tended to subsume the concepts of ergonomics under some general rubric of other factors in occupational hygiene without being fully aware of what they were attempting to encompass, while the practising ergonomist expects to consider the problems of the working environment as part of his legitimate interest in studying the worker in the workplace. Indeed, in parts of Europe the concepts of occupational hygiene tend to be subsumed under the general heading of ergonomics.

Occupational hygiene and ergonomics: definitions

It is well to begin with some definitions. The American Industrial Hygiene Association, one of the most significant world bodies in occupational hygiene, has defined the discipline of occupational hygiene as: 'That science and art devoted to the recognition, evaluation, and control of those environment factors and stresses, arising in or from the workplace, which may cause sickness, impaired health and well being, or significant discomfort and inefficiency among workers or among the citizens of the community'. However, Berry (1983) points out that most occupational hygiene practitioners consider that this definition is not completely adequate, and in particular that it excludes recognition of the fact that a primary concern should be an anticipation of potential hazards rather than *post hoc* management, and that expansion into family and community aspects of occupational hygiene is a significant part of the hygienist's work.

Occupational hygiene takes its origins in chemical engineering, chemistry, physiology, toxicology, and industrial engineering, although it has spread beyond these bounds into whatever disciplines might be necessary for its practitioners to understand and control their problems. Clearly, like ergonomics, it is transdisciplinary in its background and in its practice, and equally clearly no practitioner can be an expert in all areas.

Ergonomics, on the other hand, has been defined as the scientific study of the relationship between man and his working environment (Murrell, 1965). To expand that definition, ergonomics can be considered as the study of the anatomical, physiological, and psychological aspects of workers in their working environment with the object of optimizing health, safety, comfort, and efficiency. It is concerned with ensuring that what has been called the work system (see later) is conducive to good performance and work effectiveness, and consequently that the work environment is compatible with the health, safety, and comfort of the worker.

Ergonomics takes its origins in human psychology, physiology, anthropology, biomechanics, and industrial design and engineering, but like occupational hygiene, it too is transdisciplinary and has expanded into whatever fields are necessary to its practice.

The development of occupational hygiene and ergonomics

Although not commonly recognized, a term similar to ergonomics was developed by the Polish scientist, Wojciech Jastrzebowski, in the early 19th century to describe activities involved in the study of human work (Rosner, 1979). The Japanese also undertook work in this area shortly after the first World War. The English term Ergonomics, however, was proposed by Professor K. F. H. Murrell at the founding of the Ergonomics Research Society of Great Britain in 1950 (Edholm and Murrell, 1973). The world itself is derived from the Greek *ergos*, meaning laws, and *nomos*, work; the term ergonometrics is also occasionally used.

While the term ergonomics is in common use in Britain, and also throughout Europe and Asia, in forms modified according to language, there are three other terms in use particularly in North America, namely human engineering, human factors, and human factors engineering. These three terms are very largely synonymous, although proponents of each can be found. Various attempts have been made to compare them with ergonomics and detect some gross or subtle differences. There are differences, discussed below, but they are subtle, and are becoming less and less important as more people throughout the world tend to use ergonomics as a blanket term.

Human factors engineering, and related terms, although involving all the human sciences, takes much of its origin from applied psychology. The term has been used particularly in North America for work needed in meeting the requirements of engineering design for human use. With a multidisciplinary team approach, human engineering developed as a form of science and technology applied particularly in the aeronautics and space industries, and in military usage.

Ergonomics, on the other hand, while involving all the human sciences, and also being concerned with design and development for human use, took more of its origin from human physiology, particularly in the areas of work physiology and biomechanics, and directed perhaps more of its activities to the problems of human work. As noted above, however, the distinctions, if they can be defined at all clearly, are largely academic, and for all practical purposes all the terms are synonymous, since indeed an ergonomist undergoes essentially the same training, and undertakes essentially the same work as a human factors engineer. An important consideration is that the concept is transdisciplinary in scope, and whether one approaches it from the background of psychology, physiology, engineering, or whatever, the ultimate objective is the same.

The development of industrial hygiene is not so chronologically clear cut, since although not called by that name it was an integral part of the humanitarian and reform activities of the 19th century. The first industrial hygienists, and for that matter the first safety inspectors, were indeed the factory inspectors whose activities were defined by the early British Factory Acts, (see later) and while factory or plant inspectors no longer serve the same function as industrial hygienists, it is not uncommon to find that they have undergone much the same training, and indeed some factory or plant inspectors may also be qualified industrial hygienists.

The industrial hygienist today commonly has a professional background in chemistry and/or chemical engineering, with further specialized training and experience in recognition, evaluation and control of physical and chemical hazards in the workplace that equip him to meet the requirements of higher academic degrees and professional certification from the appropriate professional bodies. With suitable training he may also go on to further specialization in, for example, noise management, radiation physics, air pollution control, toxicology, and so on.

To understand the contemporary role of occupational hygienists and ergonomists in the joint recognition, evaluation, and control of occupational hazards in the working environment, however, it is necessary first to go back into early human history and follow the development of human thinking in this regard.

Historical review

While the ancient Egyptians knew gross human anatomy from their embalming procedures and practised both the medical and surgical arts, and while ancient Egyptian ruins from the time of 2100–1700 B.C. showed evidence of sanitary facilities and sewerage systems, they recorded little or nothing with respect to the management and control of occupational problems. The history of occupational health and safety in fact begins in the fourth century B.C. when the Greek physician Hippocrates, considered to be the father of medicine, described the occurrence of lead poisoning among miners and metallurgists. The book from the Hippocratic school entitled *Airs, Waters, and Places* indicates the relationship between persons and their environment. Hippocrates, however, was concerned more with the arts of healing than with the prevention of illness and did nothing to alleviate the miners' lot.

From the time of Hippocrates until the first century B.C. there was little further development, although a body of knowledge had obviously accrued concerning the potential hazards of working conditions. The Romans built public hospitals and administered some far-reaching health laws, and indeed, Pliny, the Roman historian and naturalist, in the second half of the first century A.D., wrote in his encyclopedia on the dangers of dust from the ore cinnabar, or mercuric sulphide, and also on the dangers of lead fumes. He went so far as to devise a dust mask made from an animal bladder for the protection of miners and others working in dusts. Lead was widely used in Roman plumbing. Gibbon, the great eighteenth century English historian, suggests in the *Decline and Fall of the Roman Empire* that a primary factor in that fall was widespread lead poisoning that sapped the will and the strength of the Roman people.

The second century A.D. saw great developments in ancient Roman medicine, culminating in the work of the physician Galen, who actually was a Greek working in Rome. Galen wrote prolifically, and not always from deep knowledge or personal experience. He wrote on general medicine, but also described various occupational diseases. His work, which unfortunately was often more speculative than scientific, was later adopted into the teachings of the Christian Church, to be accepted without question. It became almost sacrosanct for thirteen hundred years, and remained virtually unquestioned until the end of the Renaissance.

The Renaissance led to a host of new ideas, and in particular permitted translation of some of the old Arabic medical manuscripts, which, while valuable for the development of medicine, were not concerned with occupational problems. Similarly the great medical schools at Pisa and Padua in Italy, and Montpellier in France were formed around that time but they too were not much concerned with the problems of the worker in the workplace, although in 1473 Ulrich Ellenborg published a treatise on toxic fumes and vapours and their control.

Not much more of significance happened, however, for nearly 50 years, and then at the beginning of the sixteenth century there appeared a new figure, Phillipus Aureolous Theophrastus Bombastus von Hohenheim, a Swiss physician, metallurgist, and naturalist, better known as Paracelsus, who strode a bombastic and arrogant path among the scientific and intellectual elite of his time, discarding cherished beliefs and practices, and drawing attention to new ways of looking at life, health, and sickness.

He emphasized the importance of the direct observation of nature, rather than theoretical speculations, and was the first to put forward the doctrine that life processes are chemical. In his chemical studies he discovered hydrogen. Certainly a mystic, and perhaps a charlatan, he was a practising alchemist who believed in the almost holy virtues of gold and mercury. As an alchemist he was devoted to turning lead into gold, and as a metallurgist he became well aware of the diseases of miners and metal workers associated with mercury, lead, copper, zinc, iron, silver, as well as sulphur and alum. He revered mercury and considered it almost a panacea for disease. Indeed he developed the use of mercury ointment in the treatment of syphilis which was becoming an epidemic at that time, a treatment which was used well into the twentieth century. He was also the first person, at least in Europe, to use the opium preparation laudanum in the treatment of disease. His exotic behaviour and intemperate habits, however, as well as his outspoken opposition to the teachings of Galen, led him into disrepute among his contemporaries.

Knowledge, nevertheless, was continuing to grow, and some ten years after the death of Paracelsus in 1541, a German physiologist and mineralogist, using the Latin name Agricola, wrote a textbook entitled *De Re Metallica (On Metallic Matters)* which also examined dust diseases.

Again there was a hiatus until the beginning of the eighteenth century when an Italian physician by the name of Ramazzini published a remarkable book on occupational disease called *De Morbis Artificum Diatriba (A Treatise on the Diseases of Workers)*. Ramazzini is considered to be the founder of occupational medicine and hygiene. The book, as well as being a medical text, is an extraordinary social document which describes the diseases associated with many trades, such as cobbling, tanning, felt-making, lead-working, and so on, along with a detailed account of the nature of the processes and the habits and activities of the workers. It acts as a social history of the times, and describes, for example, how a worker might enter a trade in his late childhood, knowing that, because of the hazards of the trade, his life expectancy might be no more than twenty years .

In another field and country, in the eighteenth century, an Englishman conducted what might be considered the first epidemiological study in occupational health and demonstrated that 'Devonshire Colic', a form of recurrent griping, constipation and/or diarrhoea, which had previously been considered a local and circumscribed disease of unknown origin, was in fact a form of lead poisoning derived from lead dissolved in the process of cider making and consumed as part of the daily beverage. He was instrumental in changing the process and thereby eliminating the disease.

Also in the eighteenth century another English physician, Percivall Pott, was the first to recognize that the constituents of soot could give rise to a form of cancer of the scrotum in chimney sweeps. These chimney sweeps started as young boys, small enough to climb up and down the fireplace chimneys of the time, who were constantly exposed to soot at a time when washing was perfunctory at best, and washing of the more intimate areas rare if it occurred at all. His activities, and the activities of other humanitarians then and later, notably Charles Kingsley, author of *The Water Babies*, led ultimately to the banning of the practice.

By the nineteenth century, however, the ravages of the Industrial Revolution were widespread; provisions for safety and health were negligible. Injury and illness from work exposure were commonplace; reform was essential.

Reform legislation

In the Middle Ages, in England and Europe, common law regulated public nuisance, and when an occupational craft created a public nuisance the citizen or the community could call upon the practices of common law to control it. While work procedures and working conditions were normally regulated by craft guilds, some regulatory ordinances and statutes began to appear as demands became greater. Some were propounded by Elizabeth 1 of England (1558–1603), and in France about the same time. Control in the United States remained with common law or local authorities until the nineteenth century.

One of the earliest pieces of reform legislation was the Health and Morals of Apprentices Act of 1802 in England, which, while making some attempt to improve the workplace, was even more concerned with church attendance and the provision of separate accommodation for the sexes. It was, however, a beginning, which led to more useful legislation both in England and elsewhere.

The first piece of social welfare legislation took place in England in the eighteenth century, by an act of 1757, in which a fund was founded to provide sick benefits, as well as support for invalids and for death. Some years later, in 1793, France initiated a national system of social assistance and medical care, but no definitive attempts at social reform really developed until the early nineteenth century.

Much of the early reform is associated with the name of the industrialist and politician Sir Robert Peel, known also for the introduction of a police force who were known as 'peelers' from his name. He, with the support of another industrialist Robert Owen, and the physicians Sir John Simon and Thomas Southwood Smith, introduced a series of Factory Acts in England. The first, in 1833, made provision for some factory inspectors in designated factories, introduced some safety clauses pertaining to the guarding of machinery, and reduced the working hours of children. The 1833 act was subsequently modified in 1864 and 1867 to include many industries, and places employing more than fifty persons. The eating of meals was prohibited in noxious places, and requirements were set up for more machine guarding than had previously been specified, as well as ventilation for the control of dust. In the United States, about the same time, the state of Massachusetts enacted a law regulating child labour, which was followed by another in 1848 in the state of Pennsylvania. In an area more strictly related to the topic of safety the United States federal government passed the Federal Safety Appliances Act for railroads in 1887, which was expanded four times between then and 1920.

The effects, however, were still minimal. One inspector, writing in 1841, describes a girl working within 10 inches of a revolving shaft who '... was caught by her apron, which lapped around the shaft, and being tight around her body, she was whirled round and repeatedly forced between the shaft and the carding machine. Her severed right leg was found some distance away'. Engels, the associate of Karl Marx, writing of England in 1844, wrote: 'Besides the deformed persons, a great number of maimed ones may be seen going about in Manchester: this one has lost an arm or part of one, that a foot, a third, half of a leg; it is like living in the midst of an army just returned from a campaign'.

Relevant legal doctrines

The 'maimed ones', however, had little recourse to compensation other than suit in the civil courts. Few workers had the education or wherewithal to pursue such a course. When they did so, at least in the English-speaking world, they were not uncommonly met with a group of legal doctrines based on English common law which did not always work to their advantage. Two of the more relevant are known as the Master-Servant Relationship, and the Master-Stranger Relationship.

The Master-Servant doctrine established that an employer had a legal obligation to provide the employee with a suitable place to work, safe tools to use, knowledge of hazards that are not immediately obvious, competent fellow servants (that is, fellow employees) and competent supervisors. In addition the employer had to provide rules for safe performance and means to implement those rules.

The Master-Stranger doctrine established that the employer had a greater obligation to a stranger than he had to an employee. Under common law, then, an employee was free to choose employment (although that freedom was often more theoretical than real), but he was expected to be knowledgable about hazards, as defined by his employer, which in turn he was theoretically free to accept or reject.

Should he be injured, an employee could sue on the basis that his employer had failed to meet the legal requirements, that in fact he was negligent. This, of course, was difficult, expensive, and time-consuming for an employee who was neither educated nor knowledgable, nor in fact could afford time to pursue a suit. The employer, in turn could present three forms of defence, namely, (a) Contributory Negligence, that is that the employer could not be held responsible if the employee himself was negligent, (b) Negligence of a Fellow Servant, that is, that the employer could not be held responsible if a fellow employee was negligent, and (c) Assumption of Risk, namely that an employee by accepting the job tacitly accepted the risks that went along with it. The result was that suits were seldom successful.

There were, in addition, other problems associated with suits. Commonly there was a great delay in resolution; a suit skilfully defended could be extended over several years. Even if it were found in favour of the plaintiff the compensation was often grossly inadequate, with inconsistencies in awards from court to court. And, of course, the mere existence of suits led to great antagonism between employee and employer. Thus there was a great need to develop a more effective recourse should compensation be necessary, and more effective protection for the worker to obviate the need for suit.

Social security and workers' compensation

Under pressure from the great humanitarians, as well as the legal profession and the trade unions, this need began to be recognized. In Germany, where reconstruction and reform were taking place under Bismarck, the great statesman of the time, the first compulsory social security system for wage-earners was set up in 1883. The scheme included sickness insurance, occupational accident insurance (1884), and disablement and old-age insurance (1889). The German example was followed by other countries which were facing serious social tensions and growing demands from the workers. In some instances this led to the development of state supported mutual benefit societies for sickness and accident, or the development of insurance schemes supported by both management and workers, but in others it led to the development of a new concept, namely that of a no-fault Workers' Compensation system jointly financed by the employers and the state.

The first piece of workers' compensation legislation was the Workmen's Compensation Act in Britain in 1897. This Act provided for automatic compensation to an injured worker except in the case of wilful misconduct. The Act was financed by a levy on employers, based on proportional payments arising from the intrinsic hazards in the industry, as modified by the specific record of individual companies within that industry. In return the worker gave up his right to sue, and employers yielded their traditional defences. The original Act was limited in scope to a few industries and situations, but was expanded in 1907 and ultimately became a world model. New Jersey and Wisconsin in 1911 were the first North American states to follow, and Ontario became the first in Canada in 1920. The concept spread over Europe, North America, Australia and the other English-speaking countries, as well as Japan, until workers' compensation, or state and company financed insurance, became worldwide in the developed countries.

The twentieth century saw great social advances throughout the world. In 1935 the United States adopted a social security law, although it was neither innovative nor very comprehensive, but in 1938 New Zealand set up a national health scheme with cash benefits paid out of income tax, and included compensation for occupational conditions. In 1942, Lord Beveridge, in Britain, published a far-reaching committee report which outlined the substance of the future health care system of the United Kingdom, and in 1944 an International Labour Conference adopted two important recommendations, which outlined the requirements for social security as follows:

(a) maintenance of income by granting replacement benefits in the event of loss of earnings either due to temporary or permanent incapacity for work (caused by sickness, accident, invalidity, old age) or due to loss of employment, or death of the breadwinner,

as well as benefits designed to help meet the exceptional expenses necessitated by the maintenance of children;
(b) access to medical care in the form of preventive care, curative care and rehabilitation (Tamburi, 1983).

While these recommendations were not immediately put into effect, and in fact are only beginning to be put into effect by some of the more advanced countries, they provide guidelines for future development.

International Labour Organization

In 1919, as part of the Treaty of Versailles following World War I, the International Labour Organization (ILO) was formed in Geneva, on the basis that 'universal and lasting peace can be established only if it is based on social justice'. The constitution of the ILO states that all human beings, irrespective of race, creed, or sex, have the right to pursue both their material well-being and their spiritual development in conditions of freedom and dignity, of economic security and equal opportunity. The organization attempts to ensure that the fundamental rights of workers are respected throughout the world and by supporting the efforts of the international community to achieve full employment, to raise living standards, to distribute the rewards of progress fairly, and to protect the life and health of the workers.

Through its studies and conferences, and through the promulgation of instruments known as International Conventions, which are subject to ratification by its members, and International Recommendations, which provide guidance in the various fields of interest, the ILO has encouraged its international members throughout the world to develop standards and procedures for occupational health maintenance and accident prevention, while through its International Occupational Safety and Health Centre (CIS) it disseminates information on matters of occupational health and safety throughout the world.

Tripartism in Occupational Health and Safety Legislation

A feature of the ILO structure is the concept of tripartism by which all its decisions and recommendations are made on the basis of a three-way discussion among representatives from governments, employers, and employees. Tripartism is encouraged by the ILO as a guiding principle in all health and safety matters among its many member states.

The concept of tripartite responsibility involving the state, the employer, and the employee is intrinsic in most contemporary occupational health and safety legislation. In Britain, the Robens Committee in 1972 made significant recommendations to Parliament in this regard in which they pointed out that the primary responsibility for control of occupational accidents and diseases lies with 'those who create the risks and those who work with them'. They recommended that the complexities of the existing legislation should be reduced, that it should not be concerned with circumstantial details, but should attempt to shape attitudes and create an infrastructure for a better organization under the management of industry itself. The result led to a flexible structure of Codes of Practice, aiming at reasonable practicability, but with the force of law as required.

In contrast to the British approach of reasonable pragmatism, the French and Belgians outline detailed provisions for management and control. Most countries follow the British approach, some the French. In Sweden, in particular, which has a very socially advanced occupational health and safety control system, tripartism is strongly emphasized. Their control is under the Worker Protection Act, revised and modified by the Work Environment Commission in 1978. In most countries responsibility lies with either a department or ministry of Labour, or of Health, with the former commonly being favoured since it brings under the same control both the technical and the medical services. Since public health, however, is normally a responsibility of the department or ministry of Health, there is a need for liaison between the two bodies.

The role of NIOSH and OSHA

Two of the most significant bodies in this regard, with an influence beyond their country of origin, are the Occupational Safety and Health Administration (OSHA) and the National Institute of Occupational Safety and Health (NIOSH), both in the United States. With the passage of the United States Occupational Health and Safety Act in 1970 the responsibility for regulation of occupational health and safety conditions was transferred from the control of individual states into the hands of the Federal government, under the Department of Labor. This led to the establishment of uniform standards of health and safety across the nation and, because of the significance of the United States, it influenced health and safety standards throughout the world.

NIOSH, on the other hand, is an arm of the Department of Health, Education, and Welfare, and deals primarily with the implementation of safety and health in the workplace. It conducts research and provides needed technical services for workers through prevention and control of occupational hazards and diseases. It also responds to requests for evaluation of hazards from either the employers or the employees.

Interactive role of ergonomics and occupational hygiene

In the light of the foregoing it is useful to re-examine the role of ergonomics and occupational hygiene in the workplace.

Fraser (1984) has drawn attention to the fact that ergonomics is not a branch or subdivision of occupational hygiene, nor is occupational hygiene a branch of ergonomics. Nor, for that matter, is one in competition with the other. Each discipline is looking at the same problems from a somewhat different point of view, with somewhat different objectives in mind. The work of the ergonomist is complementary to that of the occupational hygienist. The latter is primarily concerned with the habitability of the working environment and the safety and health of the worker within that environment; the ergonomist, on the other hand, is concerned with design for human use, and the optimization of human performance at work and in adverse environments, while recognizing that maintenance of health and safety is a prerequisite in achieving these ends. Each profession takes a different approach, but in fact they are working towards the same objectives, namely the promotion and maintenance of health, safety, comfort, and efficiency in the workplace.

Fraser goes on to suggest that what in fact is needed is some joint approach to the problems of the worker in the workplace that embodies the best of both professions, along with the appropriate consideration of physiology, psychology, toxicology, engineering and so on, as the needs demand. This book is an attempt to review some of the needs that have to be met to achieve these objectives.

Chapter 2
A systems viewpoint

The anatomy of a system

The term 'system' as a technical descriptor long predates the concepts of what has come to be known in high technology jargon as systems engineering. Indeed these very concepts have been used in medicine and the biological sciences for centuries. One speaks of the body as a system. One can define the respiratory system, the cardiovascular system and so on. What then is the definition of a system? A system can be defined as an aggregate of interactive components operating together to perform a function. It should be noted that there are several significant words and phrases in this definition. For example, a system is an 'aggregate', or an otherwise loose association of components that are not necessarily homogeneous, but nevertheless are brought together to serve a purpose. One of the most significant words is that the components are 'interactive'. In other words they have a mutual or reciprocal effect, one upon the other. At the same time, these interactive components are 'operating together', that is they are working in mutual harmony, and in so doing they 'perform a function'. Thus a system is a highly integrated complex that is ultimately oriented to the completion of some act or task.

The definition of where a system begins and ends is arbitrary, however, and a system can be defined at any level of complexity. Thus each component of a system may itself be a system (or, in reality, a subsystem) with its own components or subcomponents, and each subcomponent may be still further subdivided at whatever arbitrary level one chooses. To return to the human analogy, the whole body is a system which in turn is made up of subsystems, some of which, like the respiratory and cardiovascular systems, have already been mentioned. There are many more, however, such as the central nervous system, the endocrine system and the digestive system. Each of these in turn can be broken down into still more refined subsystems such as organs, tissues, and cells, until at least in theory, one finally reaches the molecular or even atomic universe itself.

On the other hand, a system might comprise a multiman spacecraft served by a gigantic logistic support operation of people, equipment, ships, aircraft, launch and landing facilities, or, for that matter it might be a complex of electronic and mechanical components operating with little or nothing in the way of human control. For the mathematician, indeed, a system can be the ultimate abstraction and take the form, for example, of a group of interactive differential equations — and so on as each discipline defines its own.

Since the components of a system are interactive then, by definition, each properly functioning system has a certain equilibrium state — and of course it has a function. The equilibrium state is determined by the interactions that occur among the components; the effectiveness of the function is determined by the stability of the equilibrium.

The fact that the components of a system are dynamic and mutually interactive determines the corollary that no component should be considered in isolation from the system in which it exists, since as soon as a component is isolated both the component and the system change.

The person as a system component

In the working world a person of necessity is a component of many systems. Consideration of a person as a component of a system, however, is a relatively new concept forced upon engineers, designers, and planners by a developing technology which failed to recognize not only human limitations and liabilities as part of a joint operation in a technological environment, but also human attributes and assets.

How then did this new concept arise? As technology advanced, and as the pace of production, transportation, and communication quickened, people were called upon to perform more and more demanding tasks under increasingly adverse and time pressured conditions. Engineers and physical scientists, little concerned with the limitations and capacities of the person for whose benefit they were ultimately working, continued development of their technological world in the somewhat naive belief that if something were humanly possible then it was humanly acceptable. By the beginning of World War II, in particular, the performance capability of military aircraft was beginning to outstrip the human capacity for control, with the inevitable result that failure ensued, bringing injury or death to the pilot and crew, and damage or destruction to property.

The person-machine system

Consideration of the causes underlying this type of situation during World War II led to the initial realization of a concept which, although it was new at the time, today may seem commonplace and self-evident, namely that the operator and the machines and devices that he or she operates, or is associated with, cannot be considered as independent entities. One cannot design and manufacture a 'machine', be it a hoe, a milling machine or an aircraft, without considering the limitations and capacities of those who are going to use, operate, or maintain it. And so the concept of a person-machine system was born—a person-machine system being defined as an aggregate of people and machines (or a person and a machine) operating as a unit to perform a function. Thus a worker on a shop floor may not only be part of a larger system embracing a complex operation, but he or she may also be the human component of a single-person, single-tool system.

A person-machine system, regardless of its nature, exists within an environment. And just as one cannot reasonably consider that the person and machine are independent entities, one cannot consider the person-machine system as being independent of the environment in which it operates. One must think in terms of an interactive person-machine-environment system in which the components are in dynamic equilibrium. The operator, machine, and environment interact with each other. The operator exists within the environment and interacts with it; he or she forms an artifact, device, machine, or technology; his or her activities and requirements define and modify the machine; the machine in turn modifies or determines his or her activities; the person and machine form a system which interacts with the environment, as does the machine itself, changing the nature of the environment and in turn becoming part of it, while the environment conditions the nature of the machine.

Thus, in studying the worker at work one must try to encompass the totality of a person-machine-environment system, analyzing not only the prime components but also the interactions between components. Unfortunately this approach, the *systems approach* as it is called, is not always feasible, and one must often be content with a piecewise analysis of components and their limited interactions while bearing in mind the effect of these interactions on the totality of the system.

The person-machine-environment system

There are various ways in which one might make a representation or a model of a person-machine-environment system, each no doubt valid. The schematic model presented here is simple and pertinent, but, although attempts have been made to make it comprehensive, no claim is made that it is either the only or the most complete way of representing a complex system. It is modified from a model originally developed by the author in another ergonomic context (Fraser, 1964).

As can be seen from Figure 2.1, the totality of the environment is defined, using the terms of the environmental geographer, as the *ecosphere*.

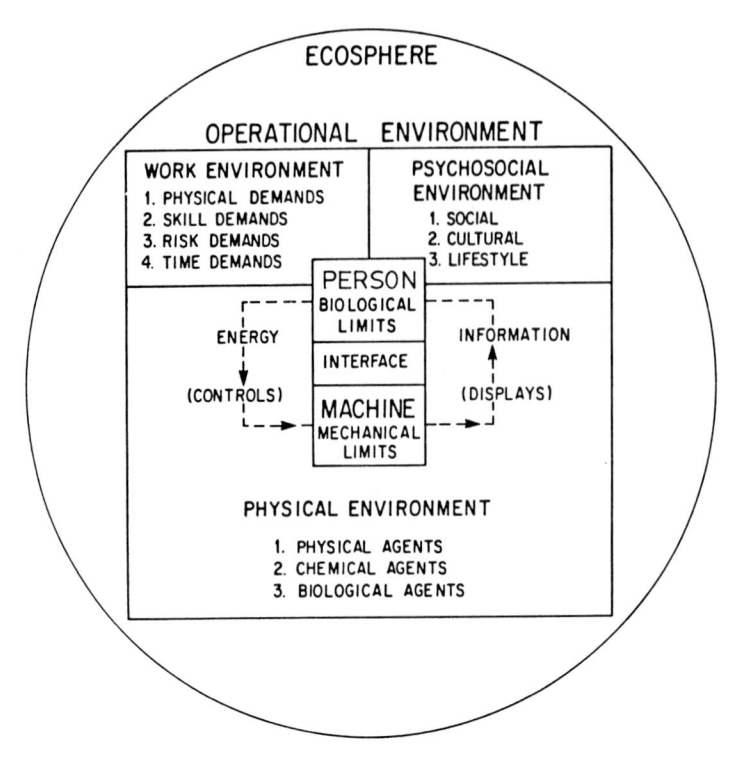

Figure 2.1. The person-machine-environment system.

People and all their artifacts exist within that total environment. From an anthropocentric point of view, the human role is to mould that environment to what extent it can to suit human purposes. In addition, the person interacts with the person-machine system at the all important *person-machine-interface* where, as shown in Figure 2.1, he or she provides energy to the machine by way of controls and receives information by way of displays.

The person-machine system also interacts with the environment. Clearly, however, it cannot interact with the totality of the environment, so that each person-machine system then carves out of the ecosphere its own *operational environment*, specific to that particular system. Since there is to all intents and purposes an infinite set of person-machine environments, there is also an infinite set of operational environments, although in practice many are more or less the same. Thus an operational environment might pertain to a person-tool relationship, a person-vehicle relationship, an industry-plant relationship, and so on.

Figure 2.1 also illustrates that the operational environment can be further classified into three basic divisions, namely the physical environment, the psychosocial environment, and the work environment.

Physical environment

The physical environment comprises that portion of the environment where, in addition to the land, the water, the air, and the natural and artificial structures of which it is comprised there are a variety of environmental agents which can be inimical to people, and hence ultimately destroy the system. These agents, can be defined as physical agents, chemical agents, and biological agents.

Physical agents

Physical agents (Table 2.1) include various manifestations of mechanical force, as well as the effects of heat and cold, exposure to ionizing and

Table 2.1 *Physical agents in the environment potentially harmful to people*

Kinetic
- ★ gravitational
  - ★ acceleration
  - ★ deceleration
  - ★ reduced (nulled) gravity
- ★ vibrational
  - ★ high frequency (>100 Hz)
  - ★ low frequency (<100 Hz)
- ★ acoustic
  - ★ sonic
  - ★ ultrasonic
  - ★ infrasonic

Thermal
- ★ heat gain
- ★ heat loss

Radiant
- ★ ionizing
  - ★ alpha and beta particles
  - ★ gamma and X-rays
  - ★ neutrons and others
- ★ nonionizing
  - ★ visible light
  - ★ ultraviolet light
  - ★ laser light
  - ★ electric fields
  - ★ magnetic fields
  - ★ extremely low frequency (ELF)
  - ★ radio frequency (RF)
  - ★ microwave

Barometric
- ★ hypobaric
- ★ hyperbaric

non-ionizing energy, and the untoward results of high or low barometric pressure. Many of these topics are examined in detail in Part 4, Physical Agents in the Work Environment.

Chemical agents

Chemical agents are so vast in number that they defy classification here. Their study comprises the field of occupational and/or environmental toxicology, and is the subject of Part 5 of this book, Chemical Agents and Aerosols in the Physical Environment. They include toxic or otherwise irritating vapours, mists, particulates and other aerosols.

Biological agents

Biological agents are living invasive or irritant entities which cause damage to, or destruction of, human cells, either directly or through noxious products of their metabolism. They include viruses, bacteria, spores, moulds, fungi, and others. Study of these agents belongs, more properly in biological and medical texts. They are not considered further here.

Psychosocial environment

In contrast with the physical environment the psychosocial environment, though significant and real, is intangible and not measurable by the metrics of the physical world. The model in Figure 2.1 shows the psychosocial environment embracing only the human portion of the person-machine system. This relationship, of course, represents reality, but since all the components of the system are interactive anything that affects the human portion of the system can ultimately influence the entire system. Thus one person's interactions with another can influence the function of any system of which he or she may be a part.

Since the psychosocial environment is intangible it is also difficult to define. Psychosocial stimuli can be considered as those which originate in social relations or arrangements in the environment which affect the organism through the mediation of higher nervous processes (that is, activities involving the higher levels of organization in the brain). While most of the resulting reactions are positive and desirable, adverse reactions can also occur.

The term psychosocial environment is comprehensive and is intended to embrace all aspects of psychological and social relationships. But, just as there are agents in the physical environment which can specially influence the function of a person-machine system, there are also components of the psychosocial environment which can influence the system. Perhaps somewhat arbitrarily, these latter can be defined in three categories, namely, those pertaining to social relationships, those pertaining to cultural and ethnic background and those pertaining to personal lifestyle, although of course these divisions are not mutually exclusive.

Social environment

The term social relationships refers to those relationships that occur with managerial authority figures, with peers and fellow workers, and with domestic and family figures and in particular with the results of these interactions as they affect the attitudes and behaviour of the person within the system.

Cultural environment

Sociologists will argue the extent to which attitudes are affected by the cultural background of the persons involved. Is there indeed a difference, for example, in the attitude towards regularized time-paced work between the middle-class city dweller and the peasant farmer and, if so, to what extent is it measurable? The answers, even to that kind of simple question, may be complex, and perhaps even unobtainable, but experience would seem to suggest that cultural, and perhaps ethnic, background play a part in influencing human behaviour and thereby human function within a person–machine system.

Lifestyle

Lifestyle, with respect to the effect on person–machine function, is a more readily definable factor. Although this is not the place to pursue the evidence, there is no doubt that life patterns involving the abuse of alcohol, tobacco, and other drugs can have an adverse effect on human behaviour, and that proper nutrition, sleep, housing, and use of leisure time are necessary for optimal function. Again, however, any direct causal relationships may be difficult to specify with clarity although they manifestly exist. Specific aspects of the psychosocial environment are dealt with in later chapters.

Work environment

In the sense that the term is used here, the work environment does not so much refer to the physical locale and ambiance of the work place as to the requirements of the work itself. Indeed the locale of the workplace is embraced by the term physical environment.

The work environment, in this respect, then, can be considered to have four components, namely those of physical demands, skill demands, risk demands, and time demands. The phrases are virtually self-explanatory. The term physical demands, for example, refers to the extent to which there is a requirement for manual or other physical labour: skill demands refers to the need for dexterous or intellectual expertise; risk demands refers to the safety quality of the working environment, and time demands refers to such matters as duration of work, distribution of shifts and the extent and type of rest pauses. All of these matters are dealt with later in detail.

Human limitations

As also illustrated in Figure 2.1, over and above any stress imposed from the exterior, intrinsic human constraints exist within a system since a person, of course, is a living being with biological limitations which are relatively insusceptible to modification. These contraints can be classified comprehensively, but not exclusively, as:

Physiological: limitations in power, strength, endurance, and capacity to maintain homeostasis under adverse conditions;
Psychological: limitations in learning capacity, skills, performance capability, tolerance of adverse conditions, and motivation;
Anthropometric: limitations derived from a fixed morphology, tissue structure, size and shape of work envelope, and postural requirements;
Nutritional: limitations occasioned by need for maintenance of appropriate food and water intake and requirements for elimination;
Clinical: limitations occasioned by a person's state of health, presence of disease and accompaniments of aging.

These various factors combine in diverse ways to affect the nature of the human response and modify the effective system. Their influence is discussed throughout.

Mechanical limitations

Just as there are constraints that affect the human portion of the person-machine system there are also factors that constrain the mechanical component. These in a sense are analogous to the human constraints, with, however, one major difference. Whereas human constraints are inherent, inescapable, and not significantly susceptible to modification, mechanical constraints exist in the system because they were placed therein wittingly or unwittingly by the system designer and are open to modification. They can be classified as constraints arising out of:

Unsuitability of design: a machine or device can do only what it is designed to do. If the design is inadequate the function will be imperfect;

Unsuitability of materials: the materials selected for the structure may be inappropriate to the purpose;
Unsuitability of construction: despite adequacy of design and choice of materials the construction may be unsatisfactory.

If any, or some combination of these, is at fault the effectiveness of the system will be compromised. This compromise may be manifest in system function, in its direct effect on the person, in the environment, or in any combination of these, but in the long run the compromise will affect human function and human capacity.

Human capacities in a person-machine-environment system

In an era of high technology, where robotic devices and quasi-intelligent machines invade the workplace, and where the worker is not infrequently expected to operate in actual or potentially adverse environments, it is well to consider what are the capacities that make his or her presence desirable in an operational situation, what can he or she do better than any artificial device, and also what are the limitations that might make him or her a liability within such a system.

Table 2.2 presents a comparison of human and machine capacities. The list is comprehensive but cannot be considered to be exhaustive; in particular, while human capacities remain more or less unchanging, machine capacities increase as technology advances. The principles, however, remain the same. In Table 2.2, in each case where possible, human capacities are contrasted with related machine capacities. It will be noted, however, that certain human and machine capacities have no equivalent.

In aviation and space operations, where interface conditions impose demands that can meet or even exceed all human capacities, it has been shown on more than one occasion that the presence of an operator in the system was essential to save the operation from abject failure. At the same time, that presence imposes engineering demands on the design and construction of the system that make it much more complex than would otherwise be necessary.

What then are the capacities or assets that make a person desirable within a functioning system? These might be considered broadly in four categories, namely the capacity for *management*, the capacity for *information acquisition and processing*, the capacity for *learning*, and the capacity for *design and creativity*, along with a few ancillary attributes.

Management

The term management is employed here in its broadest sense and is used to describe the control of people, machines (in their broadest sense), or

Table 2.2: Comparison of Human and Machine Capacities

Human Capacities	Machine Capacities
Detection of wide variety of defined signals.	Monitoring of wide variety of pre-specified signals.
Detection of significant signals in noise.	Detection of signals outside human experience
Detection of unusual or unexpected signals.	Very simple pattern recognition
Detection of incomplete or distorted signals.	Slow, comprehensive, retrieval of pre-specified, detailed, information
Very complex pattern recognition	Very rapid, consistent, response to expected events in the face of distraction
Rapid, but erratic, retrieval of information, relevant or related	
Flexible response to unexpected low probability events	Deductive reasoning-anlaysis and classification
	Decision making on the basis of pre-specified logic
Inductive reasoning, and the exercise of informed judgment	
Decision making in the absence of full information	Very limited learning capacity
	Limited intercommunication
Initiation of control strategies	Checking (counting) presence or absence of pre-specified requirements
Learning—profiting from experience and instruction	
Sophisticated intercommunication	Graphic or other detailed implementation of designer concepts
Subjective estimation and evaluation	Rapid, accurate, complex computation and data processing
Definition and application of principles and strategies	
Design and creativity	Pre-specified, limited, precise, repetitive movement of moderate force
Interpretation and conceptualization of numerical and other data	Exertion of great strength, smoothly, as required
Prioritization and optimization of task requirements under overload	Wide ranging, controllable, mobility
Dextrous, flexible, manipulation, of low level force	Consistent, reliable, performance of routine, repetitive tasks
Wide variety of unplanned low strength applications	Capacity to operate in environments beyond human tolerance
Limited flexible mobility	Simultaneous performance of multiple activities
Limited self-maintenance	Maintenance of extended performance without fatigue
Limited self-repair	
Reproduction	Indifference to emotional demands

both. It involves the capacity to evaluate a situation, to initiate action, to maintain control once action is initiated, to monitor the feedback, and to revise subsequent action as necessary.

The process of evaluation involves a number of sequential and parallel steps beginning with *perception*, which is the process of developing awareness through stimulation of the senses, followed by *cognition* or development of knowledgable appreciation. These activities occur in the brain by integration of sensory stimuli from the environment along with input from the higher centres of the brain and associations derived from memory. The process is one of *inductive reasoning*, or in other words a chain of logical thought, based on perception and leading to a conclusion on which a judgment or decision is made.

Decision-making is in fact the capacity to render a verdict or instruction on the basis of the judgmental process. A computer of course can, and does, make decisions. There is, however, an important distinction between the decision-making process of a computer, and that of a person. A properly programmed computer will come to a conclusion on which a decision can be based. It will do so only if it possesses all the necessary information in logical sequence, in which case the conclusion will be inevitable. A great human attribute on the other hand, lies in the fact that a human decision-maker can make a decision (although not always correctly!) on the basis of inadequate information, distorted information, or illogically presented information. This *flexibility of response*, or the capacity for unprogrammed decision-making, is probably one of the greatest human assets in a person-machine system, in that it allows an operator to make sudden unprepared responses to unexpected, changed or changing situations.

Information acquisition and processing

The process of sensory perception is, in machine terms, the process of information acquisition and processing, part of which is *signal detection*, or sensing. In comparison with machine instrumentation people have certain advantageous, and also disadvantageous, characteristics. Their sensing capacity, or capacity to detect signals, using auditory, visual, tactile, olfactory (smell), gustatory (taste), kinaesthetic (motion) and acceleration senses is very wide. In general it is much greater than can be built into any one machine. On the other hand, the signals are open to misinterpretation and confusion, and are curiously lacking in some areas. For example, a person can sense acceleration directly, but not velocity; a person has no sense of potential energy, for example height, and no direct sense of proximity, to name a few, each of which can be machine sensed. The portion of the sensory spectrum to which a person is sensitive is in fact rather narrow, but within that portion the sensitivity is great. People have a highly developed capacity to detect signals in the presence of noise, that is, to detect a signal of significance in the presence of closely related signals — the so-called 'cocktail party' effect where one can hear one's name mentioned across the room in a noisy cocktail party. Even more remarkable is the capacity to infer a signal from information presented in part, or distorted; in other words, to interpolate or fill in the missing parts, for example, in an auditory or visual signal and to make it complete, a task not yet successfully accomplished in detail by the most sophisticated of computer mediated systems. A related property is the capacity to recognize a signal out of context.

Part of the skill of sophisticated information processing lies in the art of *pattern interpretation*, or the capacity to take apparently unrelated parts and unite them into a composite whole. Again people are much superior to any machine in this capacity. The whole art of seeing, for example, lies in pattern interpretation derived from the integration of nerve signals as generated in the retina. It is a learned function, but with experience a pattern can be interpreted virtually regardless of its presentation; for example a representation of an object, say a drawing that is distorted or incomplete, can commonly be recognized with relatively little difficulty. A pattern can even be recognized where it exists only with the knowledge of the underlying meaning; for example the apparently random portrayal on a radarscope, or the symbolic lines on a display representing a railroad operation.

Information once acquired, however, is only of value when it can be stored and reliably retrieved. *Storage or retrieval*, of course, occur by way of memory, and human memory is remarkable in its scope. Unfortunately, it is also extremely erratic in its accessibility, and unlike computer memory it has no controlled erase function so that all too often it is absent when required and present when unnecessary. It is normally extraordinarily rapid in its accessibility and serves one of its most useful requirements by being able to provide unprogrammed information for use in unexpected situations.

Learning

The ability to learn is almost uniquely biological. One is learning probably from the moment of birth, if not before. The random motions of a baby become transformed, as it unconsciously monitors the motion feedback, into purposeful movements; the primitive efforts of the apprentice become the skills of the master; the painfully acquired knowledge of the student becomes the wisdom of the philosopher and, in practical terms, the human operator within a system can learn, consciously or unconsciously, the nuances of system behaviour as it functions, and put into practice the results of experience. Of recent years, 'thinking' computers, and robots that monitor their own actions, have been developed but the extent of learning even in these advanced machines is as yet still very primitive, and does not approach the learning capacity of even the least endowed worker.

Design and creativity

Design and creativity are probably the most unique of all human attributes. They involve the capacity to take from the old and develop something new; the capacity to innovate and invent, the capacity to think of something in a different manner, to develop a new idea, to generate a new concept. Not all persons have these capacities to the same degree, but even the most advanced computer does not possess the capacity to design or create anything other than in a random manner. This is not to say that computers or other similar devices are not of value in design. Indeed they are, but a computer can only do what it is instructed to do; it does not have the creative spark.

Ancillary attributes

People have certain other capacities that assist them in their tasks. They are both mobile and dexterous, in that they can move of their own volition and can manipulate objects. Each of these capacities is limited; people are neither so mobile nor so dexterous as special purpose machines, but combined with their other attributes these capacities are of great value. They are also capable of a limited amount of self-maintenance and self-repair, although readily susceptible to damage. And not only are they capable of reproduction, they seek the opportunity whenever feasible.

Human limitations in a person-machine-environment system

Structural

Human limitations arise largely out of the nature of human structure and function. Inherent in this structure are limitations in strength and power which are normally far exceeded by appropriate machines.

In addition, because of the mechanisms of their physiological function, human activities and environmental demands are grossly influenced by the need for water, food, waste disposal, and provision for exchange of oxygen and carbon monoxide.

Behavioural

Unlike machines, people both experience and need social relationships at work, at recreation, and at rest. Unfortunately, appropriate social relationships are not always available, and some that are inevitable are not always desirable.

They are also subject to a variety of intangible conditions such as fatigue, anxiety, fear, prejudice, boredom, and a wealth of other complex emotions which can obstruct or distort their perceptions, their decision-making, their management capacity, and their endurance. And indeed, unlike machines, which are at their most efficient when continuously functioning, people have a mandatory need for rest and recreation, which in total consume a large part of their day.

Operational

Operationally the range of human tasks is somewhat limited in its scope, even if broad in its application. As already noted, the nature of human structure and function dictate the physical tasks that one can do and the environment in which one can operate. In addition, however, people have difficulty in conducting several demanding tasks at the same time, the more so if one or more is complex. Either one task will be completed to the detriment of the others or all will ultimately suffer.

Work capacity, of course, whether physical or intellectual, varies from individual to individual, and even from time to time in the same individual. In intellectual work, however, a person must be recognized as a poor handler of data, particularly numerical data, both in handling quantities of data and in computation. In these respects a computer is far superior. On the other hand people are capable of an intuitive reasoning and conceptualization of data patterns far beyond the capacity of a computer.

Human reliability, however is neither assured or assurable. Humans are unreliable in the quality and consistency of their performance. They are also unreliable in their capacity to tolerate adversity.

Environmental

Again arising out of the nature of their structure and function, humans are sensitive to the environment in which they live, and specifically to heat, cold, acceleration and motion, radiant energy, toxic materials, and of course to infection, other biological agents and the inevitability of aging.

System function

From the foregoing it can be seen that a person-machine-environment system functions in a dynamic equilibrium. Within a range of tolerance, determined by the components and their limitations, some acceptable variation from the optimum is permissible. When the balance is disturbed beyond the permissible limits (which may vary from time to time and circumstance) an overload or stress will exist. The source of the overload normally lies in the physical, psychosocial, or work environment, or at the person-machine interface. Ultimately, since it is an interactive system, the effects of that stress will show as a strain somewhere in the system. Since the machine portion of the system can be designed to meet all foreseeable strain then the most significant effects, from an anthropocentric viewpoint will be manifest in the weakest component of the system, namely the operator.

Figure 2.2, which uses the term man-machine system rather than person-machine system, is a model of the functional operation of a person-machine-environment system, from the viewpoint of the system operator. Function begins with some situational demand made upon the system. The demand might be a requirement to complete a task, or it might be some internal change within the system parameters. As a result of the demand the operator influences the system to take some action. On the basis of qualities within the operator, such as experience, memory, and judgment, he or she expects a certain outcome from that action. Because of some disrupting element of human inadequacy, machine failure, or environmental stress, the actual outcome of that action may not be the same as the expected outcome. Furthermore, the apparent outcome, because of such further disruptions occasioned by human error (misperception) or some inadequacy in display and communication of information, may not be the same as the actual outcome. The operator, indeed, may observe something different not only from what he or she expected but also from what in fact actually occurred, and may not be aware of the reality.

If, indeed, the expected outcome occurs, and is so perceived by the operator, the operator is now in a state of potential homeostasis, which is the term for physiological equilibrium. If that state is maintained and no other disruptions occur the whole system will remain in equilibrium during, and subsequent to, the original action; the system is now prepared for a new action which in turn will be subjected to the same sequence.

Now, if the apparent outcome is expected but not perceived, if it is perceived but not expected, or if it is neither perceived nor expected, then a potential disruption exists, which, while mediated by the operator, can ultimately affect the whole system. The extent to which a potential disruption may become actual is determined by the capacity of the system to adapt or compensate. These capacities, of course, both human and machine, are limited. If tolerance is adequate, homeostasis will be retained, and the system function will be restored to normal, but if tolerance is exceeded there will be a failure of homeostasis, in which case there will be a disruption of the entire system. This disruption will manifest itsef at various points of the system function, as indicated in the figure, and tend to aggravate still further the pre-existing problems, until ultimately there is some catastrophic failure.

Use of people in a system

In the face of their limitations, then, and in the light of their assets, how should people best be used and how not used in a person-machine-environment system? A person is best used as a manager: of machines, of other persons, or of total systems; as an evaluator and decision maker; as a diagnostician and trouble shooter; in a position where he or she can use their capacities for design and creativity and where they can receive meaningful feedback on their activities. A person is poorly used as a monitor or observer of

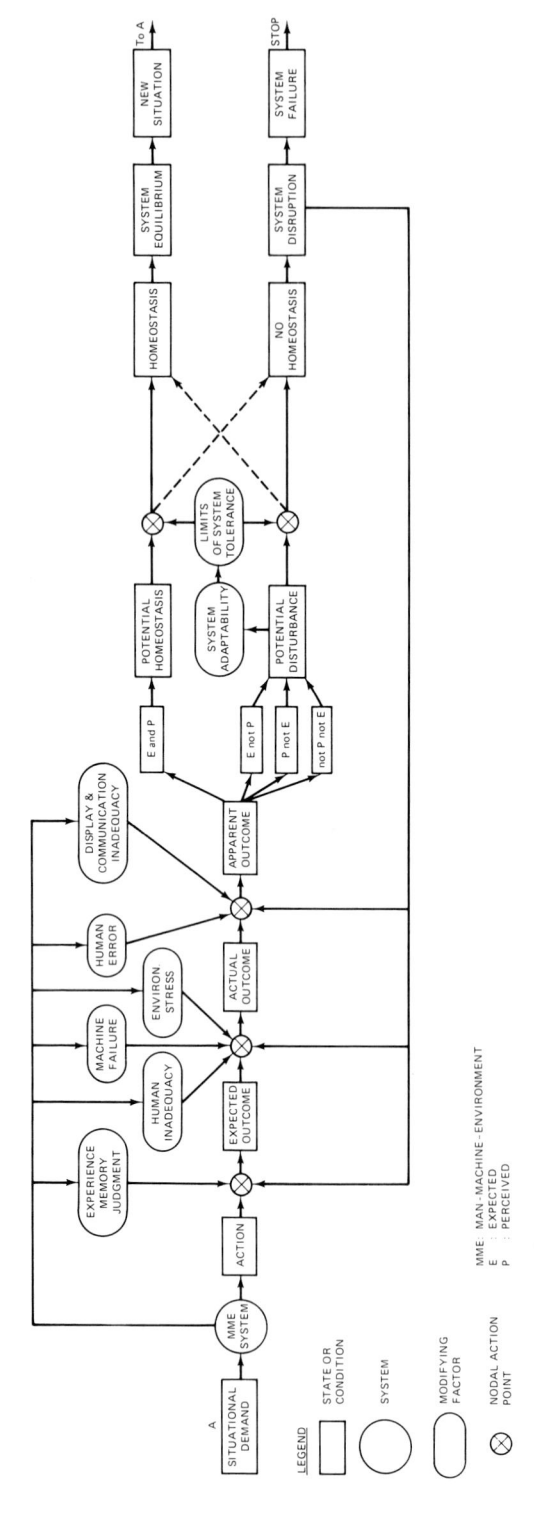

Figure 2.2. *Human function in a person-machine-environment system.*

detail, as an 'engineering component' in a production system, as an information processor or data handler or in a position of routine, repetitive, motor activity. Unfortunately, for economic and other reasons, it is in these latter type of occupations that the vast majority of workers are employed.

It will be apparent that while tasks and machines, where conditions permit, can be so designed as to obviate the requirement for a human operator, and environments can be selected to suit his or her convenience, there are still many situations where a human operator is desirable. Under these conditions it must be the objective of the occupational hygienist, the ergonomist, the safety engineer, or for that matter the primary designer, to so design or modify the system as to minimize the probability of adverse reactions.

The remaining chapters in this book, while specifying in detail the topics already mentioned, will indicate some of the approaches that can be made to this end; and while these chapters are independent in themselves they illustrate and exemplify the concepts of, and interactions within, the person-machine-environment system.

Thus, the whole concept of human work, as presented in Part 2, The Work Environment, is treated as occurring within the type of operational environment already described, in full realization of the human-environment interactions that take place therein. Indeed, Chapter 8, Job Satisfaction and Work Humanization, specifically examines some of the significant psychosocial interactions. The interactions that occur between humans and machines, which indeed determine the design, use, limitations, and hazards of these machines are examined in Part 3, Design for Human Use. To complete the picture the nature, hazards, and management of the operational environment are considered in two further parts, namely Part 4, which examines the nature and management of physical agents in the environment, and Part 5 which is concerned with chemical agents. Biological agents are not considered here.

Part II
The work environment

Chapter 3
Work, skill and fatigue

In his monograph on human work and stress Fraser (1983) points out that the concept of work as an entity, independent of the needs of day-to-day living, is one that is perhaps unique to post-Industrial Revolution societies. Primitive man was concerned with survival—hunting, fishing, fundamental agriculture and the social and religious rituals that accompanied these aspects of his life. Even in the pre-industrial age, except for the few who could afford leisure, work on the land or among the crafts and trades was still for most a dawn-to-dusk way of life.

With the advent of industrialization, and specifically with the Industrial Revolution, came the realization that work could in fact be distinguished from other aspects of living, although the distinction is not always clear cut. One man's work can be another man's play. The athlete on the football field or the advertising agent entertaining stars of the stage and screen may get very well paid for work which is the envy of others engaged in more mundane activities. In fact, the situation in which the distinction between work and leisure is the least clear-cut is commonly that in which the greatest enjoyment is found by the worker. There are of course exceptions, and one can soon get tired of entertaining celebrities. Most of what is regarded as work, in fact, tends to have connotations of compulsion, either self-induced or applied from the exterior, and involves expenditure of time and effort on activities other than those of one's personal desire. In its simplest form, then, work is what one gets paid for, in currency or other consideration.

Definition

To place matters in a more scientific perspective, however, it is necessary to consider some more rigorous definitions. The following are modified from the International Standard ISO/DIS 6385 developed by the International Standards Organization, Geneva, and are intended as basic guidelines for the design of work systems. They apply to the design of optimal working conditions with regard to human well-being, safety, and health, taking into account technological and economic efficiency and are presented as follows:

Work system
  The work system comprises a combination of persons and equipment, acting together in a work process to achieve a particular outcome, in the work place, and in the work environment, under the conditions imposed by the work task.

Work task
  The intended outcome of the work system.

Work equipment
  Tools, machines, devices, installations, and/or other components or items used in the work system.

Work process
  The sequence in time and space of the interaction of persons, work equipment, materials, energy, and information within a work system.

Work place
  The area allocated to persons in the work system.

Work environment
  The physical, chemical, biological, social, and cultural factors surrounding a person in his/her work place.

Work stress
  The sum of those conditions and requirements in the work system which act together to disturb a worker's homeostasis.

Work strain
  The effect of work stress on a person in relation to his/her individual characteristics and abilities. The consequences may be physiological and/or psychological.

Energy metabolism

Sources of energy

All human energy ultimately comes from the sun, but for practical purposes it is derived from the ingestion, subsequent breakdown and reorganization of carbohydrates, fats, and occasionally proteins, from meat and vegetable sources, with the release of energy stored in molecular bonds. The total process is known as *energy metabolism*. Metabolism means change, and in this context refers to what otherwise might be called energy conversion. In this process of metabolism carbohydrates are broken down into carbon dioxide and water with the release of energy. Fats are broken down in a further metabolic process into carbohydrates, and in situations approaching starvation even proteins will be similarly broken down.

Carbohydrates are composed of carbon, hydrogen, and oxygen, united to form complex molecules. The simplest of these molecules are known generically as *monosaccharides* ('simple sugars') and can take several forms. The most common form to be found in food is *glucose* which can be represented by the following formula:

$$\begin{array}{c} CHO \\ | \\ H-C-OH \\ | \\ HO-C-H \\ | \\ H-C-OH \\ | \\ H-C-OH \\ | \\ CH_2OH \end{array}$$

It should be recognized that the molecule of glucose does not have the actual shape shown above. The formula merely demonstrates schematically the relationships of the carbon, hydrogen, and oxygen atoms. Two other monosaccharides commonly found in food are *fructose* or fruit sugar, and *galactose*, a variety of milk sugar. These have somewhat similar formulae to glucose with different arrangements of the '—H' and '—OH' radicals. Normally each of these sugar molecules combines with itself to form a chain in which the molecules are linked together by a process of *condensation*. In condensation one molecule loses an '—H' radical and the next loses an '—OH' radical at the point of linkage. These radicals unite to form water. The polymers in turn become more complex chemical compounds, of which the most common are the *starches*. Ordinary sugar is still another form of saccharide known as a *disaccharide* since it exists as two linked molecules.

Digestion of starches

When starches and/or sugars are ingested they are broken down by enzymes (chemical catalysts) in the gastro-intestinal tract to form the basic monosaccharides glucose, fructose, and galactose, of which by far the most common is glucose. The glucose is absorbed into the blood whence it is carried to the liver and converted for storage into another polymerized form known as glycogen. The glucose concentration in the blood then remains fairly stable, and the supply is maintained when necessary by further breakdown of glycogen in the liver to its constituent molecules, and subsequent re-distribution by the blood.

Cellular energy

Ultimately energy is required by all body cells, but although the energy is derived from glucose it is not glucose that is used as the primary cellular source. The chemical used is *adenosine triphosphate,* or ATP, which is found in all cells and provides an almost explosive form of available energy. Its structure is shown schematically below:

$$\text{Adenosine triphosphate structure:}$$
$$CH_2-O-P(O^-)(O^-)-O \sim P(O^-)(O^-)-O \sim P(O^-)(O^-)-O^-$$

Each molecule of ATP contains a disproportionately large amount of energy locked into *high energy phosphate bonds,* shown with the cursive link in the formula above. When energy is required a phosphate bond is broken away by enzyme action and releases the energy held therein. The ATP now becomes *adenosine diphosphate,* or ADP. The reaction can be represented as follows:

$$\text{ATP} \xrightarrow{\text{(enzymes)}} \text{ADP} + \text{energy}$$

To restore the energy balance, however, the ATP must be regenerated. To achieve this, at least in part, advantage is taken of another chemical system existing within the cell. This system is in the form of two interrelated chemicals known as *creatine* and *phosphocreatine*. Phosphocreatine also contains a high energy bond. During muscle contraction, when energy is required, phosphocreatine is changed by enzyme action into creatine, releasing the energy contained in the bond. The phosphorus atom combines with hydrogen and oxygen to form *phosphoric acid* which is taken up in other metabolic processes. The energy, however, is used to re-establish the ATP from ADP. The complete reaction can be shown as follows:

$$\text{Phosphocreatine} \xrightarrow{\text{(enzymes)}} \text{creatine} + \text{phosphoric acid} + \text{energy}$$

$$\text{ADP} + \text{phosphoric acid} + \text{energy} \xrightarrow{\text{(enzymes)}} \text{ATP}$$

Meantime chemical activity has also been taking place in the glucose which has been retained within the cell. The initial part of this activity occurs in the absence of oxygen and consequently is known as *anaerobic* metabolism. The second part involves an oxidative process and is known as *aerobic* metabolism.

Anaerobic metabolism

The process of anaerobic metabolism of glucose is also known as the *glycolytic cycle*. It occurs within the general cell body, or *cytoplasm,* and involves the breakdown of glucose, through a series of chemical steps, from a six-carbon molecule of glucose to two three-carbon molecules of *pyruvic acid*. Pyruvic acid is chemically very similar to *lactic acid,* and is commonly stored in that form for short periods in the muscles before being removed to the liver for further metabolism. The two together are known as *acid metabolites* and are ultimately broken down to form carbon dioxide and water. The carbon dioxide is exhaled via the lung, while the water is passed to the kidneys for excretion.

During this process of glucose metabolism some 5 per cent of the available energy held in glucose is released for use and provides energy for the regeneration of phosphocreatine.

All this biochemical activity takes place without the necessity for oxygen, and indeed a significant amount of energy can rapidly be made available in this manner. For instance, a trained athlete can run 100 metres without taking a breath, and us lesser mortals can run up a short flight of stairs and only start breathing heavily when we are resting after the activity is over. This capacity is sometimes known as generating an *oxygen debt*. Like all debts, however, it eventually has to be paid. Payment takes place by way of the second part of the process of energy metabolism, namely *aerobic metabolism*.

Aerobic metabolism

The biochemical activities of aerobic metabolism are sometimes known as the *Krebs* or *citric acid cycle,* from the name of the Nobel prize winner who elucidated it, and the requirement for citric acid in the process. The biochemistry is immensely complicated, involving removal of the carbon dioxide molecule and hydrogen atoms from the pyruvic acid molecule by enzymes known as *decarboxylase* and *dehydrogenase* respectively, in a process referred to as the *tricarboxylic cycle*. The free hydrogen atoms are subsequently transferred into hydrogen ions by way of an *oxidative cycle* which involves systems referred to as *co-enzymes, flavoproteins,* and *cytochromes*. The hydrogen ions unite with oxygen ions to form water.

During this total process of glycolysis, for each molecule of glucose that is metabolized, two molecules of ATP are formed in glycolysis, two in the Krebs cycle, and thirty-four during the oxidative cycle. The significance of oxygen thus becomes very apparent. The process is very efficient. Of the 686 000 calories available in glucose, some 266 000 are stored for use in ATP. The remainder are dissipated in heat, some of which is used to maintain body temperature, and some dispersed. As will be noted, the dispersal of excess heat during conditions of heavy exercise can be a physiologically limiting factor. Not all of the energy in the glucose molecule is used at any given time. Of the pyruvic acid formed only one-fifth is converted; the remainder is restored to glycogen.

Muscle function

Energy is required for all physiological functions. The greatest part of that energy, however, is used in performing muscular activity or exercise. A muscle begins in an attachment to bone, swells to form a muscle body, and ends in a tendon (gristle) attached to another bone. Muscles are located across joints so that when the muscle body contracts the two ends of the muscle are brought closer together and thus cause movement of the joint. Several muscles, with different origins and insertions, are commonly attached across a joint. In particular, one group of muscles, the *agonists*, act to cause a joint to bend or flex, while another, the *antagonists,* acting on the same joint, cause it to straighten or extend. The agonists and antagonists both act together in a coordinated fashion to produce a smooth movement.

Muscle belly

The muscle belly comprises bundles of muscle *fibres* of diameter ranging between 10 and 100 microns (1 micron equals 1/1000 mm) and of length from 1 to 500 mm. When a skeletal muscle fibre is looked at under a magnifying glass it appears to be striped, or *striated*. These striations are caused by the microscopic structure. When looked at under a microscope each fibre will be seen to be made up of 100–1000 *fibrils* called *myofibrils*. An electron microscope shows their structure.

Myofibril

Each myofibril in turn is made up of submicroscopic filaments which, in fact, are complex molecules. One of these molecules is called *myosin,* another *actin*, arranged so as to overlap like interlaced fingers.

The junction of the actin and myosin filaments, or as it is sometimes called, the *Z-line*, causes the apparent striations. When contraction occurs the actin filaments slide in between the myosin filaments and the molecule complex becomes shorter. The mechanism that allows this sliding to occur is still speculative. In the so-called 'Ratchet' theory it is assumed that the actin and myosin molecules are linked by multiple *cross-bridges*. At rest the linkage is potential only, any action being inhibited by the presence of an inhibiting chemical system. On stimulus from the appropriate motor nerve, or other form of stimulus, calcium ions are released which block the inhibitory system. With energy derived from ATP the cross-bridges become activated and complete the linkage. The cross-bridges then pull the actin filament into the myosin, producing contraction.

Muscle stimulus

The stimulus to contract normally comes from a nerve signal derived from nerve cells in the brain or spinal cord and transmitted to the muscle by way of nerves and their nerve fibres. The nerve fibres coming to the muscle terminate in *muscle end-plates* which are microscopic structures acting as a junction between the nerve and the muscle. When a nerve signal is received at the muscle end-plate a minute quantity of a chemical, *acetyl choline*, is excreted into the submicroscopic space between the end-plate and the nerve fibre. The presence of this chemical initiates a chain of biochemical events, culminating in the release of calcium ions which eventually cause the muscle to contract. As long as acetyl choline is present the fibrils affected will remain in contraction, and indeed the muscle in these circumstances would go into violent spasm. This is prevented by the presence of still another chemical, *acetyl choline esterase*, which is an enzyme that destroys acetyl choline. (The suffix *-ase* refers to an enzyme.) Thus, with each pulse of acetyl choline there is a concomitant pulse of the destroying enzyme. During World War II certain war gases, known as *Tabun* and *Sarin*, were developed which had the capacity to destroy the acetyl choline esterase. Inhalation of the gas in suitable concentration caused spasm of muscles and respiratory failure. Fortunately they were never used. Some time after the war the gases were modified and made into commercial insecticides.

Each microscopic nerve fibre is connected to some 150–200 muscle fibres dispersed throughout the muscle belly. The nerve fibre/muscle fibre complex is referred to as a *motor unit*. Stimulation of one motor unit produces a weak muscle twitch. Stimulation of many, by a stronger nerve signal, causes *multiple motor unit summation* and produces a strong contraction. Repeated frequent stimulation, with up to 10 or more twitches per second, produces *wave summation* and maximal contraction until it is terminated by muscle fatigue. The strength of contraction, however, is also a function of the length of the contracting muscle fibre, and hence the length of a muscle, before contraction. The greater the pre-contraction length the greater is the strength of the contraction.

Types of muscle

Three different types of muscle can be identified. These are skeletal muscle, smooth muscle, and cardiac muscle. While in general they are similar, and function in a similar manner, each has its own characteristics both with respect to structure and function. These will not be defined here except to note that skeletal muscle is attached to the bony skeleton and is the muscle which permits movement to occur, while smooth muscle is the muscle found in internal body organs such as the arteries or the intestines. It gets its name from its smooth, as opposed to striated, appearance under the microscope. Cardiac muscle, as the name would imply, is the muscle found in the heart. It has the remarkable capacity to continue repeated contraction for the lifetime of its owner (and occasionally beyond!). The type of muscle that we are most concerned with in discussion of the nature of physical work is of course skeletal muscle. It is because of its function that we are capable of doing work, although of course all types of muscle are involved when we work.

Types of work

Work is considered by the physicist or engineer to be the product of the intensity of the applied force and the distance of the resulting movement. While this concept may be useful in physics it bears a very limited application to the real world of the industrial worker or the needs of the ergonomist or occupational hygienist. Clearly there is real and physically measurable work involved in moving shovelfuls of sand from one location to another, a task which gives rise to equally evident fatigue; but anyone who has held that same shovel at arm's length for a relatively short time without doing any apparent physical work rapidly becomes aware of an equally evident fatigue. Thus, from a physiological or ergonomic point of view work can exist even where there is no movement. This concept will be examined further in consideration of the difference between *static* and *dynamic* work.

Another distinction, however, must also be made, namely that between *physical* and *skilled* or *intellectual* work. Physical work largely comprises those activities involving physical strength, while skilled or intellectual work comprises those involving the exercise of skills or judgment. In industry, tasks may commonly involve some element of both.

As we have seen, all human function, including that concerned with the basic processes of life, requires the consumption of energy. Over and above the basic energy needs, physical work is characterized by the expenditure of relatively large quantities of energy in the contraction of torso and limb muscles against an imposed load, e.g. running, lifting, carrying, pushing, pulling, striking, or support against gravity and so on. In skilled work the expenditure of energy in muscle contraction may be less, since the activity may require relatively minor physical movements. Nevertheless, considerable energy may be expended in simply maintaining the posture, or the position of hands and arms, as for example in typing, or writing. In addition, skilled movements have to be made with precision, requiring coordination of muscles in some complex learned pattern, the maintenance of which requires evaluation, judgment, feedback of information, and continued or repeated experience. As well, there is a significant, although ill-defined, consumption of energy in the processes of thought, evaluation, and decision-making.

Physical work

Whether physical or skilled, or some combination of both, each form of work ends in the same outcome, namely that if persisted in for a long enough period with inadequate breaks the output of the worker will diminish and he will undergo the subjective experience of fatigue.

Man, however, is not physically very active. Even the hardest working coal miner or lumberjack spends three-quarters of his working day either in some form of rest or asleep, while a sedentary worker spends even more. Much of that activity, as has been already noted, is not work in the normal physical sense, but nevertheless requires significant energy expenditure. Both static and dynamic work therefore have to be considered.

Dynamic, or moving, work occurs when the task requires repeated, commonly rhythmic, contraction and relaxation of muscle, as for example in operating a lever, or some other form of flexion and extension of joints. In these circumstances the work is in fact measurable in terms of the physical dimensions of force times distance. Physiologically, dynamic work requires what is referred to as *isotonic* muscle contraction, that is, contraction of the muscle with physical shortening of the muscle fibres and resultant movement of a joint.

Static work

Muscles, however, can contract without any shortening of the muscle fibres. When no shortening can or should take place the fibres contract under tension. This occurs during static, or holding, work when contraction is maintained without rhythmic relaxation. There is therefore no visible or directly measurable work. Physiologically this is called *isometric* contraction. No joint movement occurs.

Fatigue from dynamic and static work

Static work is much more fatiguing than dynamic, as is exemplified by the previously noted effort required to hold a shovel at arm's length. There are, of course, sound physiological reasons for this phenomenon. The application of muscle force, in addition to performing the function, compresses other soft structures either in the muscle body or between the muscle and underlying bone. Where this application is rhythmic, or repetitive, as in dynamic work, it allows, for example, continued provision of blood and nerve supply. The blood, in particular, supplies nutrients (carbohydrate) and removes waste products of acid metabolism (acid metabolites).

The rhythmic action also has another significant function, namely to assist in returning the venous blood to the heart and lung where gas exchange can take place, allowing the carbon dioxide formed during metabolism to be exhaled and replaced by oxygen. Normally, of course the physical pumping action of the heart creates a pressure which drives the blood around the body through the arteries, veins, and their smaller derivatives. Some of this action takes place against the flow of gravity, particularly if the body is in the standing position. One-way flow is ensured by the presence of valves at short intervals within the veins. These valves can be visualized as swing doors which open in the direction of flow towards the heart and close to flow in the opposite direction. If the circulation is impeded, or if there is a prolonged requirement for continued circulation against gravity, as in prolonged standing with muscle tension like a soldier on parade, there will be a reduced blood return to the heart. The heart can only transmit onward the blood it receives. If it receives an inadequate flow then, in particular, the blood supply to the brain will be reduced. This can lead to collapse and fainting, as again may be found on the parade ground. Rhythmic muscle action assists the blood flow by massaging the blood along the veins past the venous valves. And indeed soldiers on parade are encouraged to rock slightly as they stand erect in order to maintain a slight pulsing action of the big calf muscles.

In static contraction, of course, the blood flow, and particularly the venous blood flow, is obstructed, as muscles squash the veins either in the muscle substance or against underlying bone. This impedes the blood flow, reduces the nutrient supply, and allows a build-up of toxic acid metabolites which, in turn, interfere with the normal metabolic processes. In addition, compression of the nerve fibres which provide the nerve supply to the functioning muscles ultimately leads to reduced force of contraction and inability to maintain contraction.

The effect of this is that while static force can be maintained at about 15–20 per cent of maximum for prolonged periods, when that force is increased to 50 per cent of maximum it can be applied for only a few minutes. At less than 20 per cent of maximum the blood supply is maintained at adequate levels (Rohmert, 1960). A comparison between the effects of dynamic and static work is shown in Table 3.1.

Significance of static/dynamic work in the work place

It is easily understood, and intuitively obvious, that the activities of dynamic work are physiologically demanding and subjectively fatiguing. The contribution to overall physiological demand and fatigue deriving from static work is much less obvious. All work, however, whether it be the strenuous exercise of heavy lifting or the sedentary activity of the scholar, has a large although varying element of static work in addition to the obvious dynamic. Thus — to return to our sand shoveller — as he stoops to load his shovel his arms are held outstretched in isometric contraction, his back and legs are stabilized by muscles contracting in tension without shortening, his head is held steady, and so on, while he bends at the hips and perhaps slightly flexes his spine. It is of course possible to analyze all the activity, static and dynamic, that is taking place during a complete shovelling cycle. This analysis belongs in the field of biomechanics and will not be considered further here. Suffice it to say that while much of his fatigue results from the dynamic work that he is performing a large amount also derives from the static supporting work that is being done to maintain his posture and stabilize his limbs and head.

Similar types of situations occur in all manner of work wherever there is a requirement to maintain posture of body and limbs against gravity. The worker who holds a hand drill with both hands in front of him is not only stabilizing his posture but is also required to sustain the weight of the drill with hands outstretched. The press operator who works with both hands manipulating controls at shoulder level is supporting these arms against the gravitational pull. The miner crouching in a seam, or the construction worker operating a screwdriver above his head, are both attempting to maintain very awkward postures. Literally hundreds of similar examples could be presented. But the problems of static work are not confined to the manual worker. The secretary, typing with hands outstretched over the keyboard, the violinist with his arms at shoulder level, the scholar studying his books, or the student writing at his desk, are all the time attempting to maintain some supporting limb position or body posture. And all too often, unfortunately, the design of the

Table 3.1. *Effects of dynamic and static work*

Physiological function	Static work	Dynamic work
nerve activity	continuous (reduced)	intermittent
blood supply	reduced	assisted pump
waste products	accumulated	removed
nutrients	reduced	maintained

equipment, tools, other devices, and procedures used in the tasks takes little account of normal human capacities and limitations, and thereby tends to aggravate these pre-existing problems. It is one of the main objectives of the art and science of ergonomics to minimize by appropriate design of hardware and procedures those problems that arise because of a demand for otherwise unnecessary static work.

Physical fatigue

Much of the foregoing discussion has centred on the occurrence, and the need for reduction, of what we know as fatigue. While the concept of fatigue is intuitively well understood, it is a term that is very difficult to define.

It can be considered as an operational, but undesirable end-result of work. It may be acute (short term) or chronic (long term). Acute fatigue occurs as a result of a short term overload and may be a generalized whole-body response, as at the end of a hard day's work, or it may be localized to a certain region in response to a certain task, as arm and shoulder fatigue from intensive use of a screwdriver. Both may occur at the same time. It is characterized by general body weariness, perhaps sleepiness, pain, aching, and perhaps swelling of the affected muscles, with immediate inability to continue the task, followed, where severe, by persistent aching and stiffness.

The causes are inherent in the physiological mechanisms of muscle function, including accumulation of waste products in the muscle body which interfere with that function, and other more complex factors to do with interference with the circulation of blood and tissue fluids as well as inhibition of the controlling and coordinating activities of nerve cells and fibres in the brain and nervous system. Indeed, perhaps, there is also interference with the capacity of cell mechanisms to transform energy, and in the capacity to acquire and transport oxygen and effect gas exchange between oxygen and carbon dioxide. Acute fatigue responds readily to rest, as will be discussed later.

Chronic fatigue is a less tangible condition characterized by dullness, apathy, and loss of drive, perhaps depression, perhaps irritability, and loss of productivity. It is associated with prolonged experience for weeks or months of physical overload, prolonged hours, inadequate rest periods, inadequate sleep, aggravated by poor working conditions. Like acute fatigue, it too is relieved by rest and is modified by change, although the rest and change need to be of much longer duration.

Skill and intellectual fatigue

In any skilled job utilizing tools and/or equipment there is always an element of physical work which may contribute more or less significantly to the total pattern of fatigue. Similarly during intellectual work there are physical factors relating to maintenance of posture and limb position. Performance of a skill, or intellectual activity, is itself fatiguing even with minimal additional physical work. Subjectively the fatigue is experienced as weariness, staleness, loss of motivation, and vague intangible discomforts dependent on the nature of the task. Inevitably, however, with continuation of the task there is deterioration of performance.

This problem was first systematically examined in World War II when it became apparent that the skills of Royal Air Force bomber pilots began to deteriorate after some hours of stressful flying. Investigation was put into the hands of Sir Frederick Bartlett of Cambridge.

Bartlett's studies became known as the Cambridge Cockpit studies (Bartlett, 1953). For these studies he built and operated an aircraft simulator with functioning instrumentation and controls. Using this simulator he observed the performance of pilots over flights of several hours' duration, and measured their ability to maintain their desired position.

In flight, one's attention is concentrated on three instruments, namely, the airspeed indicator, the altimeter (for altitude) and the compass (for heading). Information from these three has to be continuously integrated mentally by the pilot during flight, along with information of lesser immediate significance from other instruments, for example, oil pressure, temperature, aircraft attitude, and so on. It is critical to maintain the altitude, airspeed, and heading within a very narrow range.

It was observed that this ability deteriorated with time in the cockpit. On analysis of data Bartlett concluded that several factors contributed to the outcome. Firstly, with time, there developed an acceptance of lowered personal standards of accuracy and performance by the pilot. He began his flight with great attention to detail, particularly with respect to the critical instrumentation. Over a period of time, however, he would become satisfied with less. In addition, again over time, there would be a reduction in the range of his attention. At the beginning of the flight he would encompass the whole complex array of instrumentation; towards the end attention would be concentrated only on the essentials. This would be accompanied by a progressive failure to integrate the information from his instruments, and particularly the totality of the available information in terms of his previous established priorities. Thus eventually his actions would take the form of responding or reacting to independent pieces of information, 'putting out fires', instead of smoothly coordinating the whole. With increasing fatigue he would also tend to lose the short-term memory of information derived from peripheral instruments which would require repeated refreshing. As an added complication it was observed that as he approached the end of the flight, perhaps as he recognized a relief from stress, there was a sudden further let-down of concentration, with sometimes disastrous results.

How does this apply to a generalized theory of skill fatigue? It is argued that in performing a skilled task the subject perceives the task and its environment as a whole. This whole is constantly undergoing change but remains completely identifiable, just as a waterfall is constantly changing but maintains a consistent identifiable shape. The skilled operator perceives the task and its environment as a total pattern of stimuli which form a frame of reference. The arrangement and values of the components of this perception may change without degrading the total pattern, in the same manner as the waterfall changes. A skilled operator learns to recognize the totality of the pattern. He recognizes changes in the pattern which do not upset the whole, and sets up subjective interactive standards of deviation which are not permitted to be exceeded. He takes action when some perturbation threatens the integrity of the whole.

In skill fatigue his adherence to his pre-set standards begins to deteriorate and he becomes satisfied with less. His timing becomes disorganized; he may still take the right actions but not in smoothly coordinated time. The totality of his stimulus field begins to disintegrate and he no longer sees the task as a whole. The unconscious pattern that he has identified becomes a collection of unrelated signals, each of which may require a greater or less degree of action. On top of this, as in the cockpit studies, the peripheral demands, that is those that are not closely organized, become overlooked. Ultimately the entire task proficiency will suffer until performance ceases.

Initially the phenomena of skill fatigue are momentary. There is a brief, almost instantaneous, lapse in judgment, followed by further lapses. These lapses, over time, tend to increase in duration until the operator becomes aware of his condition and relieves it by extra effort to the extent possible. Initially the fatigue tends to be specific to particular skills and can be relieved by changing the activity or the demands. But ultimately it will develop in the new activity and will progress to other skills. Performance, however, can be maintained at a relatively high level in spite of fatigue, particularly where survival is concerned.

Arousal and boredom

The capacity to perform continued skilled work, and conversely the onset of skill fatigue, are closely related to the presence or absence of the phenomenon termed *arousal,* a concept attributed to the work of a neurophysiologist named Magoun in the 1940's. Arousal is the term given to a state of physical and mental alertness which is under the control of a special network of nerve cells and communicating fibres in the brain. This network is referred to as the *reticular activating system* (RAS), which extends from a region in the middle of the brain known as the *thalamus,* where the nature and general topographic origin of sensation is determined, to the *cerebral cortex,* or outer layer of the brain where, amongst other activites, refinements of sensation and implementation of motor activity are determined. The network also extends downwards into the spinal cord and peripheral nerves.

The RAS was at one time thought to be a supporting structure, or 'scaffold', on which the rest of the brain was organized until Moruzzi and Magoun (1949) demonstrated that it had a very significant physiological function, namely to vary the level of alertness or sensitivity to stimuli, from a level of full arousal to a level of sleep. The activity is coordinated in the thalamus which transmits and receives stimuli to and from the cerebral cortex, and to and from the periphery, or in other words the external environment. The cerebral cortex, of course, is ultimately responsible for perception, cognition, and the initiation of voluntary activity. The peripheral nerve endings are responsible for receiving sensory stimuli and implementing muscle action.

The system then, coordinated by the thalamus, can be stimulated by outgoing signals from the cerebral cortext or by incoming signals from the sensory nerves. Similarly, outgoing signals from the RAS centre in the thalamus can raise the level of alertness in the cortex and the sensitivity of the peripheral nerves.

Thus a feedback loop is formed. For whatever reason, whether it is a self-generated thought, or a perception, the cortex becomes aware of a threat and alerts the RAS. The RAS in turn increases the sensitivity of the sensory input. Thus a loop is formed which reinforces the cortical arousal. Alternatively, a sensory input is received by the sensory nerves, is transmitted to the thalamus, which in turn alerts the cortex, and a loop again is formed. Thus, for example, if one is on the point of sleep and suddenly hears a strange perhaps faint but unexpected noise, one becomes instantly alert; similarly, if one again is approaching sleep and suddenly recalls an overlooked important task, or a future unwelcome demand, one again is immediately alert.

Once initiated, arousal is maintained at a greater or less intensity depending on the stimulus, and diminishing over time, for a period of up to 15–20 minutes, if not reinforced, following which it will diminish to a relaxed state, and ultimately, should there be no other demands, to a state of sleep. To maintain even a relaxed arousal, however, some sensory input is required. In the absence of adequate stimuli, boredom, loss of alertness, and ultimately loss of proficiency will occur which, under appropriate circumstances will lead to sleep.

There would, indeed, appear to be a related inhibitory system which, when stimulated, acts to inhibit arousal. The two systems work in antagonism, the degree of alertness depending on the balance. When activation is dominant, there is subjective alertness, high muscle tone, increased stimulation in that portion of the nervous system responsible for control and mediation of bodily function and involuntary activity (namely, the *autonomic nervous system*), as well as readiness for motor action.

The person feels subjectively refreshed, and is said to be in *ergotropic* adjustment.

In complete absence of stimuli, which does not occur naturally, but can be partially achieved in experimental conditions, there is gross emotional disturbance, with varying degrees of anxiety, fear, hallucinations, and disintegration of the personality. Some of the elements of reduced stimulus environments can be found in certain types of conveyor belt work where the constant noise, relatively unvarying environment, and sheer boredom of the work produce a similar kind of situation. An even more unvarying environment is found in the all-white 'clean-rooms' used in electronic chip manufacture. Other low stimulus environments may be found on some occasions in underwater diving, in solo flight with no visual input, and in sailing on extremely calm waters in conditions where the horizon cannot be distinguished from the sea and the sky.

Measurement of energy cost of work

From a thermodynamic viewpoint the human body can be considered as a heat engine. In other words, it utilizes energy to perform work and in so doing generates heat. Thus by measuring the heat generated it is possible to determine the energy required. In practice this is almost never done, but it can be accomplished by placing the body in a special chamber called a whole-body calorimeter. Under highly controlled circumstances, with complex equipment and allowance for interfering factors, precise measurements can be made of the change in temperature in the calorimeter resulting from the presence of the body at work or rest. From these measures an accurate estimate can be made of the energy consumption during the period under consideration.

Since this type of approach is normally not feasible, advantage is taken of the fact that under controlled dietetic conditions (in particular with no excess of protein) the consumption of 1 litre of inhaled oxygen is approximately equal to the production of 5 kilocalories (kcal) of heat. One calorie, also called small calorie, is the amount of heat required at a pressure of one atmosphere to raise the temperature of one gram of water through one degree Celsius at 15° degrees Celsius. One kilocalorie, also called one large calorie, equals 1000 small calories.

Measurement of oxygen consumption

The Douglas bag is a large airtight bag which is carried strapped to the back of the subject. Air is breathed through a mask and hose assembly with appropriate inhalation and exhalation valves. Expired air is collected in the bag for 10 minutes or more while the subject is at rest, or during a task. The bag is cumbersome and interferes to some extent in the subject's capacity to perform a task, but virtually all tasks can be completed. The total volume of air is measured by expelling it from the bag through a gas meter, and the volume per minute (minute volume) calculated. The oxygen concentration in that expired air is then determined by the use of one of several sensitive chemical devices, ranging from portable oxygen electrodes made by the Beckman company, to complex and sophisticated gas chromatographs. Since atmospheric air at sea level contains 21 per cent of oxygen it becomes a simple matter to calculate the oxygen used.

To overcome the clumsy awkwardness of the Douglas bag, and the time limitations imposed by the size of it, various other more portable devices have been deloped, such as the Max Planck respirometer which is a lightweight dry gas meter carried on the back and used to measure the total volume of expired air during physical activity. The subject breathes through a face-mask with a low resistance one-way valve. Inserted into the valve at the entrance of the respirometer is an oxygen sensor which gives a continuous reading of the oxygen concentration.

Still another device is the integrating motor pneumotachograph (IMP) developed by Wolff (1970). In this device expired air passes from a full face mask through a flow meter in which there is a transducer which translates the instantaneous flow into a varying voltage output. This in turn is fed into an integrating device which produces the time integral and so provides a reading of the total volume of air that has passed through the meter. At regular intervals during this process a sample of air proportional to the volume that has passed through the flowmeter is withdrawn by an electrically driven sampling pump for future analysis.

Other techniques

It can be shown experimentally that the volume of air breathed (ventilation volume) over a given time is roughly proportional to the energy expenditure during that time. The relationship is shown in Figure 3.1 below, which illustrates that a given total ventilation is associated with a certain energy expenditure. According to Wolff (1970) the measurement is accurate within 10–20 per cent.

The Wright anemometer which can be used for this purpose is a miniature turbine incorporated into a face mask. The inspired air passes through slits to a rotor which in turn drives a revolution counter and displays the results on a dial calibrated in volume units.

Energy utilization is also proportional to heart rate. Heart rate, of course, is easily measurable by palpation of the pulse at the wrist. Obviously, of course, this cannot be done during work, and so various other more sophisticated devices have been developed which can be worn by a working subject. These include the cardiotachometer, which is a modified electrocardiograph that converts electrical heart pulses into

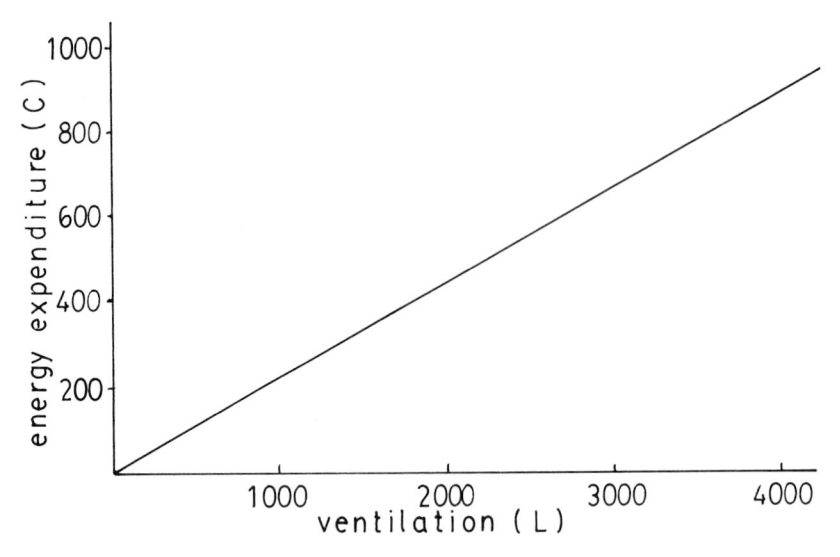

Figure 3.1. Relationship between total ventilation (litres) and energy expenditure (kcal) (after Wolff, 1970).

voltages which it then displays in graphic form as instantaneous heart rate, as well as the photoelectric ear pulse meter which uses a miniature photoelectric device attached to the ear lobe to achieve the same purpose, and the plethysmograph which senses pulsed changes in volume of, for example, a finger, to provide a measure of heart rate. All of these devices, unfortunately, are very sensitive to motion, and have to be located on the body with great care to avoid extraneous movement and resulting spurious information or electronic 'noise' in the signal.

Pulse rate measures during and post-exercise have been very valuable in evaluating work stress (Brouha, 1960). From these measures a heart rate recovery curve can be derived. The effect of two different periods of work and rest on the pulse rate is shown in Figure 3.2.

In the same context Christensen (1953) developed a table for grading work in terms of oxygen consumption, as seen below (Table 3.2).

Basal Metabolic Rate (BMR)

Using the techniques of oxygen consumption measurement, a measure can be made of the energy required for the multiplicity of physiological activities required in merely living, such as heart rate and lung function, maintenance of muscle tone and neural activity. This is referred to as the *basal metabolic rate*. In standard conditions, that is, at total rest, fasting, in a thermoneutral environment, it can be shown that the BMR for a healthy 70 kg Caucasian man is 1750 kcal per day. The equivalent number

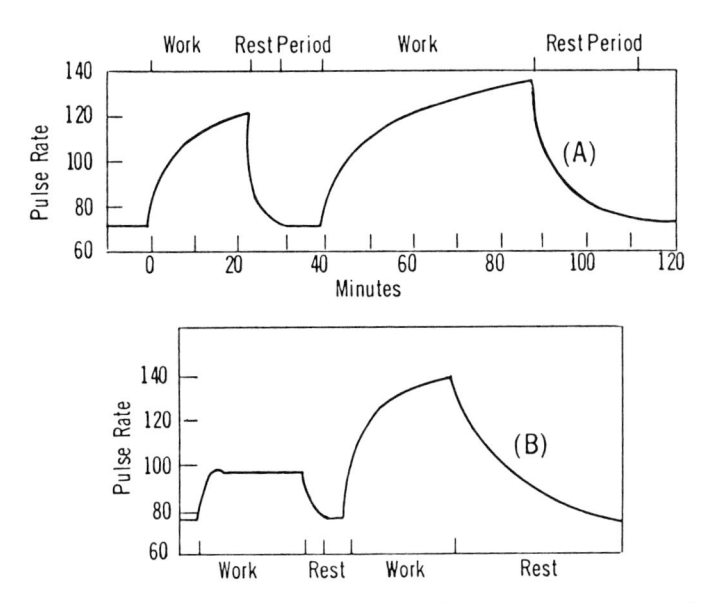

Figure 3.2. Effect of work and rest on the pulse rate (after Brouha, 1960).

for a woman is 1450 kcal. This requirement must be met by the ingestion of foodstuffs which can be metabolized to produce that amount of heat energy. Any additional requirements, imposed for example by work, must be met by an increased input. Conversely, any input over and above requirements has to be dissipated as heat or stored as fat.

Similarly it can be shown that for sedentary work an additional 1000 kcal might be required, to a total in a standard male of 2750 kcal. For light industry a worker will need about 1500 kcal extra, to a total of 3250 kcal, while for heavy industry, such as coalmining, there might be a demand for a total of 3750 kcal.

Normally the maximum expenditure for continued daily work should be maintained in the range of 3750 to 4000 kcal per day. It will be shown later in discussion of the human response to hot environments that at

Table 3.2. Energy costs of work. (after Christensen, 1953)

Grade of work	Energy expenditure kcal/min	kcal/8hr	Approximate oxygen consumption, litres/min
Unduly heavy	over 12.5	over 6000	over 2.5
Very heavy	10.0–12.5	4800–6000	2.0–2.5
Heavy	7.5–10	3600–4800	1.5–2.0
Moderate	5.0–7.5	2400–3600	1.0–1.5
Light	2.5–7.5	1200–2400	0.5–1.0
Very light	Under 2.5	Under 1200	Under 0.5

about that level the body becomes limited in its capacity to dissipate the generated heat.

For short periods of a few weeks to a few months, 5000–6000 kcal can be tolerated, and for very short periods of a day or less (such as in marathon running) up to 10000 kcal of expenditure can be achieved. Expenditure beyond these levels, or expenditure beyond the levels of one's intake, gives rise to depletion of resources and becomes a primary cause of chronic fatigue.

Many studies have been conducted to determine the energy cost of different activities. One of these was undertaken by Passmore and Durnin (1967), from which representative extracts are shown below. They indicate the range of energy cost in kcal/min.

Other similar studies were conducted earlier by Lehmann (1953) and provided the following findings:

Table 3.3. *Energy expenditure of persons engaged in various activities. (derived from Passmore and Durnin, 1967)*

Occupation	Energy expenditure, kcal/day Mean	Minimum	Maximum
Men			
Elderly retired	2330	1750	2810
Office workers	2520	1820	3270
Laboratory technicians	2840	2240	3820
University students	2930	2270	4410
Building workers	3000	2440	3730
Steel workers	3280	2600	3960
Coal miners	3660	2970	4560
Women			
Elderly housewives	1990	1490	2410
Middle-aged housewives	2090	1760	2320
Department store clerks	2250	1820	2850
University students	2290	2090	2500
Factory workers	2320	1970	2980

Table 3.4. *Energy expenditure in various occupations (derived from Lehmann, 1953)*

Occupation	Type of work	Required energy (kcal/day)
Watchmaking	Light, sedentary, manual	2700
Mechanic	Walking, light, manual	3000
Shoemaker	Sedentary, heavy, manual	3300
Wood sawing	Standing, very heavy manual	3900
Carpenter	Climbing, medium heavy	3900
Coal miner	Standing, stooping, very heavy	4200
Lumber jack	Standing, climbing, very heavy	4200

Operational factors in work fatigue management

Physical and ergonomic

1. Work load

The activities of physical work involve standing, sitting, lying, crouching, kneeling, moving at varying rates on the level or on slopes, lifting, carrying, and utilizing the limbs and body for the exertion of varying forms of force. As already noted, to ensure optimum efficiency and health the load that a worker might be regularly expected to bear should not exceed, except for short periods, the load that permits him to maintain an energy balance over a prolonged working life.

Although physiological limitations are the ultimate criterion, there are other motivational and cultural factors which may modify the load tolerance to less than the ultimate. Consequently, in addition to variations imposed by physical size and sex, the tolerable load may vary quite widely from one cultural group to another, and from individual to individual. These possibilities will have to be borne in mind in establishing load levels. Training may be required. Unless the individual worker is considered to be expendable, however, shortsighted attempts to maintain high overloads for periods of more than a few weeks are doomed to failure because of the inevitable reduction of proficiency that will ensue.

In terms of lifting, carrying, pushing, pulling, and so forth, the recommended maximal loads are discussed in Chapter 7.

2. Working hours

The question of working hours is closely related to that of work load and has been examined generally in looking at fatigue. It is possible to maintain a consistent high output with a regular work duration of 16 hours or more a day for a few weeks. Beyond that, the attempt again becomes self-defeating. It has been shown by accumulated experience and by experiment that in a Western industrialized population, at least, the optimum work routine is found with an 8-hour day in a 40–50 hour week. Figures for undeveloped countries are not available. For short periods of several weeks at a stretch, with appropriate motivation in the form of a dedicated attitude, or appropriate reward, a 12–16 hour day can be maintained within a 100-hour week. Other approaches are considered in a later discussion of shift work.

Prolonged hours can have an effect both on proficiency and health. The effect on proficiency has been studied on numerous occasions, although some of the most definitive work was conducted by the Industrial Fatigue Research Board in Britain as long ago as 1920. Essentially they showed that, within limits, although these might be hard to define, the longer the working day the lower the average hourly output.

Not surprisingly there is also a measurable effect on health, measurable at least in terms of sickness absenteeism. Following much labour unrest in the 1920's and early 1930's, working hours were widely reduced to about 48 hours per week. With the onset of World War II there was a great demand for productivity. Forgetting the lessons learned in World War I and after, working hours were again increased, but it soon became evident that the increase could not be sustained without a significant increase in sickness absenteeism. While other factors were no doubt also involved, there is nevertheless a clear relationship between reduction in working hours and reduction in sickness absenteeism.

Many companies today are moving to extended working days with more subsequent time off. While this may have more appeal to the workers than the traditional 8-hour day, 5-day week that has come to be commonplace, there are intrinsic disadvantages. Mackie and Miller (1978) studied performance and alertness changes in bus and heavy truck drivers working regular and irregular schedules. They found that drivers on regular schedules showed poorer performance after 8½ hours on late night trips while drivers on irregular schedules with continually changing start and stop times, but the same total hours, showed poorer performance after only 5 hours on late night trips. In addition, relay truck drivers expressed significantly greater feelings of fatigue during the second half of all 9½ hour trips irrespective of the time of day on both regular and irregular schedules. For bus drivers there were significant increases in subjective ratings of fatigue after 7½ hours of regularly scheduled driving and after 6 hours of irregularly scheduled driving. Performance changes occurred after 6½ to 7½ hours on regular trips and after 4½ to 5 hours on irregular trips.

Grandjean (1982) in a discussion of working hours rejects a 10 hour or more day as being unacceptable on medical and physiological grounds, noting that even a 9-hour day will lead to excessive fatigue.

3. Role of monotony

As noted in consideration of arousal, where conditions are intrinsically lacking in stimulus, arousal diminishes, and if conditions are appropriate (and sometimes even where they are not!) the reduced arousal will lead to sleep, as many a boring lecture has demonstrated. Studies of Japanese train drivers have shown a steady decline in alertness over an 8-hour shift (Endo and Kogi, 1975). Studies of German railway engineers have shown it can be difficult to maintain alertness while driving trains. Recordings were made from a reset alerting device which produced a signal on the average once a minute. If the signal were not cancelled a light would flash, then a horn would sound, and finally after 30 seconds of no response the automatic brakes would be applied. Hildebrand *et al.*, (1974), investigated 2238 such automatic brakings, representing a rate of one automatic braking for every 15 drivers. In addition, this study detailed 20 000 instances of acoustic warning signals being given when the warning light on the alerting device were not heeded. These 20 000 instances occurred on relatively few locomotives, namely 10, over the short period of one month. This suggests that inattentiveness and possibly drowsing are serious problems for train crews.

4. Work breaks

The need for work breaks or rest pauses has been long established. As shown earlier in discussion of the physiology of work, rest is essential to allow restoration of physiological imbalances imposed by work, as well as to generate some change in the environment of monotonous work. Indeed, when formal work breaks are not provided spontaneous breaks will be introduced by the worker on one pretext or another.

Much of the knowledge of work breaks goes back to the studies of the Industrial Fatigue Research Board previously mentioned. It has been shown, in these and other studies, that after an activity has continued for a short time a worker requires a short break, even if highly motivated. Even before the worker has realized that his concentration is slipping, or his muscles are failing, there is evidence of deterioration in his performance, and if the activity is continued beyond this time the deterioration may become serious to the extent that output will begin to fall (Murrell, 1969). The problem for the supervisor becomes one of determining how long should be the period of work without a break, how long should be the break, and what form should it take.

In point of fact the desirable duration of work has never been clearly established on scientific grounds, and in practice work breaks have been determined more by negotiation than by empirical research. Several generalizations, however, can be enunciated. Firstly, work breaks should be introduced when performance is at a maximum, just before a reduction in productivity. The timing of the break is more important than its length, although optimal periods can be determined for individual jobs.

Secondly, work breaks may be more useful for relatively ineffective workers; better workers seems to develop more efficient procedures and therefore have less need for rest. Thirdly, work breaks are more effective with work requiring mental concentration (vigilance) than on jobs that are more or less automatic.

Several types of breaks exist. Spontaneous breaks occur following some intensive, perhaps short-term, effort where the worker halts his activity and rests for a short period to recover. This rest re-establishes, at least in part, the physiological steady state of homeostasis that was disturbed by his work. In some cases the break will not be obvious. The worker may in fact stop the activity that is causing the fatigue but will busy himself for a short period doing some other activity related to his job that is different and less fatiguing, for example sharpening a tool, cleaning the surrounds, and so on, in such a manner that a supervisor cannot find fault. Formal breaks, however, have become an established pattern in most industrial work and generally are permitted to occur over and above spontaneous breaks. Studies of output, sickness, and absenteeism have shown that where formal rest periods are permitted the overall proficiency is higher, an even greater difference being observed in the older worker doing heavy work, as is seen in Table 3.5 below:

In practice, it is commonly accepted that there will be a mid-shift break of about half-an-hour, along with a 5–15 minute break during each of the half shifts. In addition, where the work is physically or mentally demanding most authorities recommend a five minute break every hour. The National Institute for Occupational Safety and Health (United States) indeed recommends that for extremely vigilant work, as for example intensive use of a video display terminal, there should be a work break of 15 minutes every hour.

The term *work break,* of course, is open to interpretation. Commonly it is interpreted as meaning a physical rest from the task, and as such it is very effective, although perhaps expensive. Various authorities, however, confirm that a work break is not necessarily a physical rest from labour. Murrell (1969) points out that as an alternative to a rest pause, 'actual changes in occupation may be introduced at regular intervals and this may be as effective as actual rest.' This view is confirmed, for example, by Cakir *et al.,* (1980), the video display terminal authorities, who state . . . 'rest pause may be interpreted as an interruption from work, as a specific rest period, as a period of inactivity, as an interruption from specific phases of a task, etc.'.

5. Shift work

The capacity to work at a high level of performance is a function of the time of day, as well as the habits and what have come to be known as

Table 3.5. *The effect of rest periods on absenteeism (quoted in Murrell, 1965)*

Nature of Pause	Heavy	Absenteeism (loss in shifts %) Medium	Light
Under 45 years			
None	6.8	3.7	2.5
Some	4.4	3.4	2.7
Frequent	3.6	3.9	2.7
Over 45 years			
None	7.6	6.0	4.6
Some	5.2	4.2	3.9
Frequent	3.5	4.7	3.4

the *circadian rhythms* of the worker. Circadian rhythm and its associated problems will be considered in Chapter 5. Suffice it to say at this time that fatigue is grossly increased and performance reduced during changing shift work.

6. Design of work, work stations, tools and equipment

Design for human use and human work is the major thrust of ergonomics. Where work, work stations, tools, and equipment are inappropriate fatigue will be increased and proficiency reduced. It is the objective of the ergonomist to ensure wherever possible that task and tools fit the man. Various aspects of these problems and their solutions will be detailed in subsequent chapters, and in particular in Part III.

Personal and administrative

1. Health

It goes without saying that a healthy person is more capable of sustained work of whatever kind than an unhealthy person. Equally obviously a person of strong physique is more capable of heavy work than a person of poor physique.

Natural good posture tends to derive from physique. A person with poor postural habits imposes on himself or herself an unnecessary static muscle load which adds to the fatigue of the task being undertaken. This is particularly significant in sedentary tasks at, for example, a conveyor belt or a desk. A significant part of the job of an ergonomist is to ensure that faulty postures are minimized by ensuring that the intrinsic design of the work station, tools or equipment does not encourage the development of faulty posture.

2. Nutrition

In early discussion of energy metabolism it was noted that the amount of work that can be achieved is a function of the available energy. Thus to match work output in kilocalories there must be an equivalent input in kilocalories. Indeed, it is necessary not only to maintain a balance commensurate with the type of work being done, but also to provide the necessary protein, fat, minerals, vitamins and other essential substances over and above the carbohydrates necessary for energy production. These are normally available naturally in a well balanced diet. Synthetic dietary supplements, including additional vitamins, are not necessary for the normal healthy person eating a balanced diet. The requirements for a balanced diet, the type of which may vary from culture to culture, are not considered here, but can be found in other references.

3. Sleep

The question of sleep is, of course, related to that of rest. The physiological need of muscles for the recovery time of sleep is not at all clear. The heart muscle and the chest muscles used in breathing continue to function without a break from life to death. Yet for physical and emotional well-being sleep is necessary, although when motivation is high one can in fact go for several days without sleep with minimal decrement in performance and little in the way of physiological manifestation. While there is a great range of tolerance, from a minimum requirement of about 4 hours daily to a maximum of 9 or 10, most persons would seem to require 6–8 hours of continuous sleep per day for continued well-being.

The question of sleep will be examined further in the chapter dealing with circadian rhythms.

4. Motivation and employee/management relations

Although it is a very difficult, if not impossible matter to measure, there is a general understanding to the effect that a well-motivated person will perform better and be more resistant to fatigue than a poorly motivated person. Motivation may be derived internally, as found in a dedicated person who wishes to accomplish a personal goal, or it may be imposed externally by way of reward of whatever kind. The reward is not necessarily financial. It could be in the form of prestige or other social compensation.

One interesting and often quoted set of studies that bear on this matter is known as the Hawthorne study because it was conducted among employees in the Hawthorne division of Western Electric in the United States by Mayo and his colleagues (reported by Roethlisberger and Dickson, 1939). The studies were conducted over 12 years and intially were concerned with determining the relationship between lighting and production. Although a variety of different conditions of brightness and dimness was imposed on the working environment, the workers tended to respond favourably regardless of the nature or direction of the change. It was concluded that the interest in the workers that was implied by management action in pursuing the studies was at least as important as the fact of the changes. In other words, where the employees were regarded as persons, unique and important, and where some corrective action was seen to be taken, they responded with an increase in motivation. The response is often known as the Hawthorne effect.

Environmental

The physical conditions of the working environment, such as noise, vibration, heat, lighting, and the quality of breathing air, are vitally important in influencing the health, well-being, and performance capacity of the woker in that environment. They are considered in detail in Chapters 12–18.

Chapter 4
The industrial work station

The work place can be considered as the location where a person or persons perform a task for a prolonged period. The work station is one of a group of work places which may be used sequentially by the same person to perform a job.

Work station analysis

Work station analysis is the tool by which actual and potential ergonomic hazards are recognized, identified, and evaluated. The process will be considered in greater generality, but also more detail, in examination later of problems in the category of occupational hygiene.

Although it is not a difficult task it is one that requires preparation and preliminary analysis. Familiarity with the operations is essential before any observations or measurements can be undertaken.

The analysis comprises a visual observation backed up by photographic or video record and measurement. A preliminary discussion with management, supervisors, and where feasible and useful, workers from the tasks involved, is essential to determine the reason or reasons for an evaluation and the nature and extent of the related problems. The discussion should include an outline of the demographics of the work force, any shift system in use, the number and duration of rest pauses, the involvement of trade unions and health and safety committees in plant working conditions, the existence and acceptance of incentive systems, and the extent and frequency of known work-related problems. Wherever possible the preliminary approach should include an analysis of absenteeism records, including workers' compensation or insurance reports, and discussion with human resource personnel regarding the extent and nature of complaints and injuries. Discussion with health personnel and an intepretation of health records within the limits of confidentiality is of value.

Observation begins with a walk-through overview of the whole work area of concern, (see Chapter 23) which might be the entire plant, a production line, or some section of a production line. The region chosen, however, should be adequate to allow understanding of the material flow and the processes involved. In the course of this preliminary walk-through the investigator should note areas of concern or specific tasks to which he will return when he has completed his initial survey. On completion of the preliminary walk-through he will return to each of the noted tasks and examine them in detail, observing any excessive workload placed upon the worker by reason, for example, of improper working heights, poorly organized work surfaces, improper placement of storage bins, unnecessary or exaggerated requirements for lifting, bending, stretching, twisting, excessive static load, maintenance of undesirable postures, use of poorly designed tools or equipment, and so on, in terms of the topics already outlined.

Various checklists, for example as found in Grandjean (1980), have been designed for use in this observational process. A checklist, however, is essentially an *aide-memoire*, or reminder, rather than a device for checking off particulars which may very well not apply under the circumstances of a specific investigation. It is often much more useful to conduct the survey, taking careful notes, with sketches where applicable, and then use the checklist to ensure that nothing of value has been missed.

Measurements are essential, but they are simple, and commonly require nothing more than an ordinary tape measure with which to measure heights, widths, depths, reaches, and so on. Weights of loads, tools, components, and so forth, are also required, many of these are obtainable from persons within the plant.

Records are also essential, whether in the form of notations or the completion of appropriate forms. It is extremely useful for later reference and consultation to have these records accompanied by photographic or video evidence. Video records, particularly when taken with a small video camera, are particularly useful.

One of the difficulties facing the investigator analyzing the motions of workers in completing a task is to record in detail the movements the worker makes. This problem, of course, has been of interest to the industrial engineer in determining the motions involved in completing a particular task, with the ultimate intent of developing standard tasks with standard pay rates. The same principles have been applied in ergonomic analysis, for example by Salvendy and Seymour (1973). Others, such as Corlett and his colleagues (1979), have developed more ergonomically oriented systems, while Baleshta and Fraser (1986) devised a recording system based on choreographic notation which lends itself to direct computer mediation. All of these systems, however, are as yet intrinsically cumbersome, and unless applied by the dedicated inventors themselves are probably no more efficient than equally cumbersome detailed note taking.

Work place dimensions

The dimensions of the work place are of necessity determined by the dimensions of the worker. The study of human dimensions is known as *anthropometry*. Anthropometry is concerned with the definition of more or less fixed human dimensions such as stature or height, mass, which under normal terrestrial gravity is the same as weight, limb length, width and circumference as well as functional dimensions such as seated height, arm reach, and the capacity for such activities as lifting, pushing, pulling, gripping, and so on.

Engineering anthropometry is a term coined by Roebuck and his colleagues (1975) to describe a special aspect of anthropometry which deals with the application of scientific physical measurement methods to human subjects for the development of engineering design standards and/or specifications. It includes static or fixed measurements (see later) as well as dynamic or functional measurements of dimensions and physical characteristics of the human body as it occupies space, moves, and applies energy to physical objects. These are determined as a function of age, sex, occupation, ethnic origin, and other demographic variables.

Body dimensions vary from individual to individual and even from time to time in the same individual. The latter occurs chiefly with age. In addition to the changes that occur with growth and development, there may be, even in adults, for example, a signficant difference in the stature of an individual of 30 years and the stature of the same individual at, say, 65 years. With age there is loss of elasticity of tissues, and in particular loss of some of the supporting tissue in the spine. The result is to shorten the spine and consequently the stature, a phenomenon which is aggravated by the almost inevitable development of a slight stoop. The resulting difference may amount to one centimetre or more.

Variations in dimensions can even occur on a day-to-day basis, depending for example on the state of hydration (that is, water content,) of the tissues. Water content can vary considerably with intake, and also with excessive output as in heavy sweating. After excessive consumption of alcohol the tissues can become grossly over hydrated. Change in hydration can cause measurable changes for example in weight, or in the diameter and circumference of limbs, and so on.

Static and dynamic dimensions

Mention has already been made of static and dynamic dimensions. Static dimensions are normal in classical anthropometry. They are measured at or from clearly defined anatomical points and indicate the precise length, width, diameter and circumference of the body part under consideration. Such dimensions are valuable for categorizing individuals or populations, but are of limited usefulness in engineering design.

In most working situations, or even in non-working situations, people, or parts of their bodies, are in motion. In performing physical functions the body operates as a whole. The practical limit of arm reach, for example, is not determined only by the length of the arm. It is determined in part by shoulder movement, bending and/or rotation of the back and perhaps even hip motion. Thus a static dimension, or dimensions, might be inadequate to describe the needed space. Of more significance is the extent of functional motion one might be able to undertake, say in the seated position. Thus the designer is not merely concerned with the length of the arm, but with the length of reach from some fixed point, say on the seat.

One such point is indeed the Seat Reference Point (SRP). When the back and bottom surfaces of a seat are hard flat planes the SRP occurs at the intersection of these two planes and the lateral midline of the seat. When cushions are employed the point is referenced to the occupied cushion surfaces. Obviously the thickness and compressibility of the cushions can introduce unwanted variables and consequently still another reference point, the H-point, was defined by the Society of Automobile Engineers for use in automotive seat design. The H-point is at the mean of distribution of the centres of an imaginary line between the hips. While no such real point can be measured, a model known as the H-point machine is maintained by the Society.

Use of anthropometric data

In using anthropometric data for engineering purposes it is, of course, important to use the data that are relevant to the topic. A multiplicity of data has been generated, covering a great variety of human dimensions and a large number of functional requirements. One must realize, however, that most of these data have been derived from measurements taken on young healthy males, the majority of whom were Caucasian and in the military services. There is, for example, a dearth of data on children, the elderly, the handicapped, specific industrial groups (e.g. transport drivers), and most ethnic populations other than Caucasian. Thus it is useless to use data derived from, say, U.S. Air Force personnel to provide design criteria for a tractor or machinery to be used primarily by Chinese females.

It has already been noted that body dimensions vary from person to person and from time to time in the same person, and that consequently one cannot use for design purposes dimensions derived from one person to appy to a group of persons. In selecting an appropriate dimension, then, it is desirable to choose some representative value which has application to a large group of persons. Such a value would represent some central point common to the group, and would be a measure of central tendency.

Measures of central tendency

The most common measure of central tendency is the *average or mean*, which is the sum of a group of values divided by the number of values in the group. Another such measure is the *median*, which is the middle value of a set. The median is not necessarily the same value as the mean. Still another measure is the *mode*, or modal value, which is the most common of a group of values in a set.

Biological characteristics are distributed randomly across a population. However, in a sufficiently large statistical sample these characteristics will assume a *Normal Distribution*, which is defined by the bell-shaped curve shown in Figure 4.1. Not all characteristics follow that curve; in some instances the peak of the curve is pushed, or skewed, to the left or the right. It will be noted, however, that with a normal distribution the mean, the median and the mode all fall on the same peak point.

The fact that with a normal distribution the mean is also a midpoint implies that half the values that are summed to calculate the mean lie below the average. Because of this, even an average value has to be used with great care. For example, if one were to specify an escape hatch to meet the average appropriate body dimensions half the population would not be able to exit; or if one were to specify seat width on the basis of an average value half the population could not sit on it in comfort, if at all. This is not to say that there is no place for average data in design. A fixed facility, not susceptible to adjustment, and used by a very large population, for example a supermarket check out counter, would probably

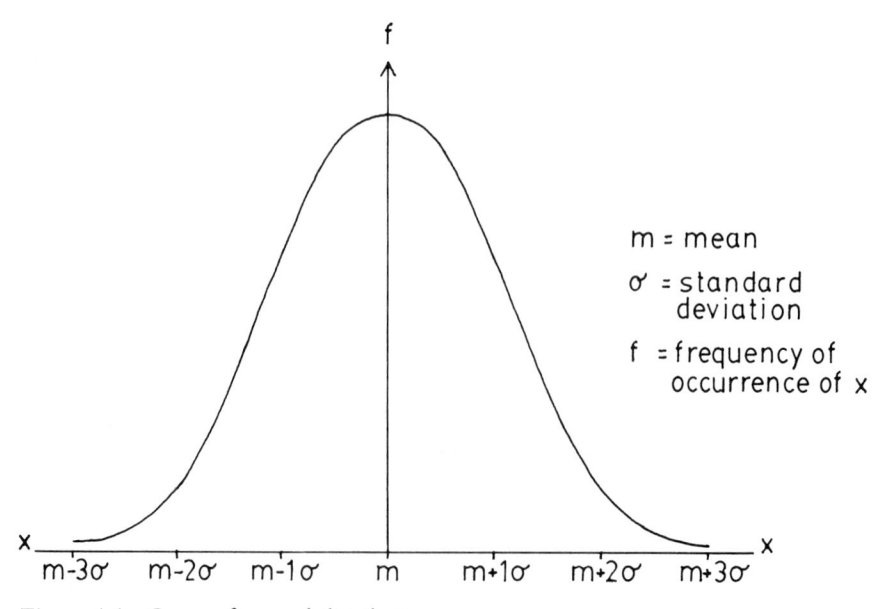

Figure 4.1. Curve of normal distribution.

discommode fewer persons at lesser cost if built to average specifications than if designed to meet specific user specifications. The use of averages, indeed, can meet a wide variety of applications when applied with care.

Where the use of an average is not appropriate the use of a range of values may be considered. One such range encompasses percentile values. A *percentile* is a value on a per cent scale that indicates the extent of a distribution in terms of per cent that is equal to or below it. Thus an escape hatch designed to meet a 95th percentile would also accommodate all persons of lesser dimensions. Similarly a requirement for arm reach in vehicle control that would meet the limitations of the 5th percentile dimensions would also meet all the larger dimensions. For design purposes it is common, wherever feasible, to encompass at least the 5th to the 95th percentile. Extension beyond that range may incur costs or difficulties out of proportion to the benefits.

Designing to meet a set of percentiles may of course still not make for an adequate design in terms of dimensions. Working activities are conducted under many differnt conditions, standing, seated or in motion, and the work station may be used by a number of different persons. It is therefore mandatory to allow for adjustability in the various work space dimensions, for example bench height, seat height, work surface angle, seat angle and so on. Various requirements for work stations are considered below.

Normal work area

The normal work surface is considered to be the area that can be swept by the forearm as the arm moves in an outward direction from the front of the body to full abduction as shown by the desk area in Figure 4.2. Such a movement includes full extension at the elbow along with some

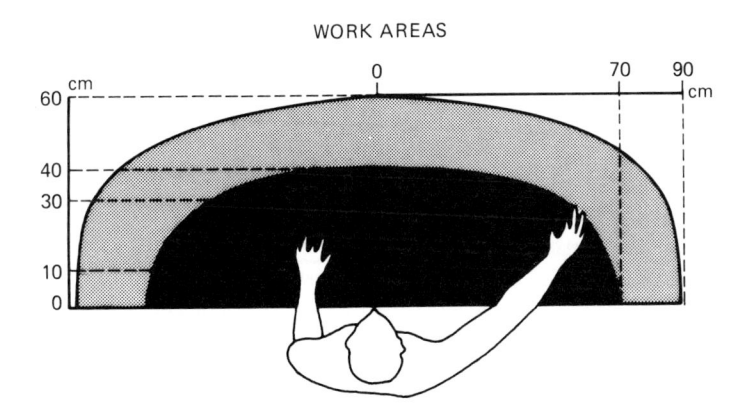

Figure 4.2. Work surface areas (after Clifford M. Gross, Principles of Ergonomics, Biomechanics Corporation of America, Deer Park, N.Y., 1986, with permission).

abduction at the shoulder. The work station, however, includes more than the work surface, and requires consideration of working activities, reaches, storage requirements, and so on, as the worker does his task.

Three basic types of work station can be defined, namely the sitting work station, the standing work station, and the sit/stand work station.

Sitting work station

All tasks that can be conducted reasonably and efficiently from a seated position should be so organized. This requires that all task items can be easily supplied and handled from a seated position and that the work can be manipulated at no more than 15 cm (6 in) above the work surface.

In addition, the seated position demands that forces do not exceed a 4.5 kg (10 lb) lift, and that most of the shift is occupied by such tasks as fine assembly, disassembly, writing, wrapping, packing, and so on.

Seated workers generally work in the space above the working surface which essentially provides space the shape of the interior of a segment of a sphere.

The volume of the workspace for seated and standing operators has been studied in detail and defined by the Human Factors Group of the Eastman Kodak Company (Kodak,1983). The data governing seated and standing reach capacities are illustrated in Figures 4.3, 4.4, and 4.5. For the seated worker they point out in their text that for example if the space is going to be used to pack small items into a kit, the distances from the centre of the workplace to each supply bin should be within this seated arm-reach work space. Although the same principles would apply to many other types of work, the distances from the centre of the workplace to each supply bin should be designed to be within this seated, arm reach, workspace. If eight items were to be clustered around the kit assembly area then at least a 25 cm (10 in × 10 in) work area would be needed in front of the operator. Supply bins would then be more than 25 cm (10 in) in front of the operator near the work surface. For the most efficient work motions the bins should be placed within 41 cm (16 in) to the right or left of the centre of the workplace and not more than 50 cm (20 in) above the surface, preferably lower. To avoid fatiguing the shoulder muscles, the procurement of items from supply bins must be kept to 25 cm (10 in) above the work surface. This technique would limit the comfortable reach distance to the left or the right of the body's centreline to 41 cm (16 in) and to about 36 cm (14 in) in front of the operator. Although more extended reaches can be made, they should not be incorporated into a highly repetitive assembly task.

Any object that is to be frequently grasped or procured should be located within 15–36 cm (6–14 in) of the front of the work surface. These ranges are the distances from which small objects can be procured without requiring the operator to bend forward. Large or heavy objects should

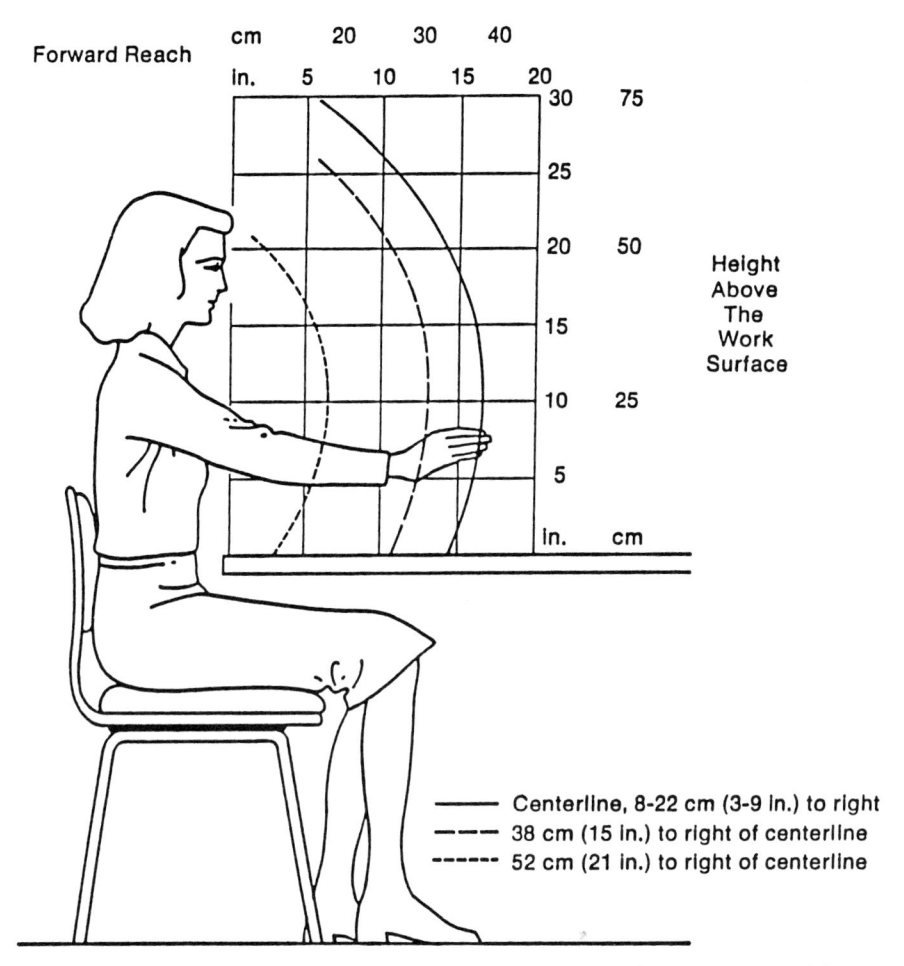

Figure 4.3. Forward reach capability for a small operator (after Eastman Kodak Company, Human Factors Section, 1983, with permission).

be located closer to the work space. It is permissible to have an operator occasionally (a few times an hour) lean forward to procure something outside the work area, but such reaches should not be made a regularly occurring part of a brief work cycle.

The recommended dimensions for seated work places are shown in Figure 4.6.

Standing work place

A standing work place is needed where the task cannot be done seated. This would include situations where there is a requirement for handling objects of greater than 4.5 kg (10 lb), where there is a need for high, low,

70   *The work environment*

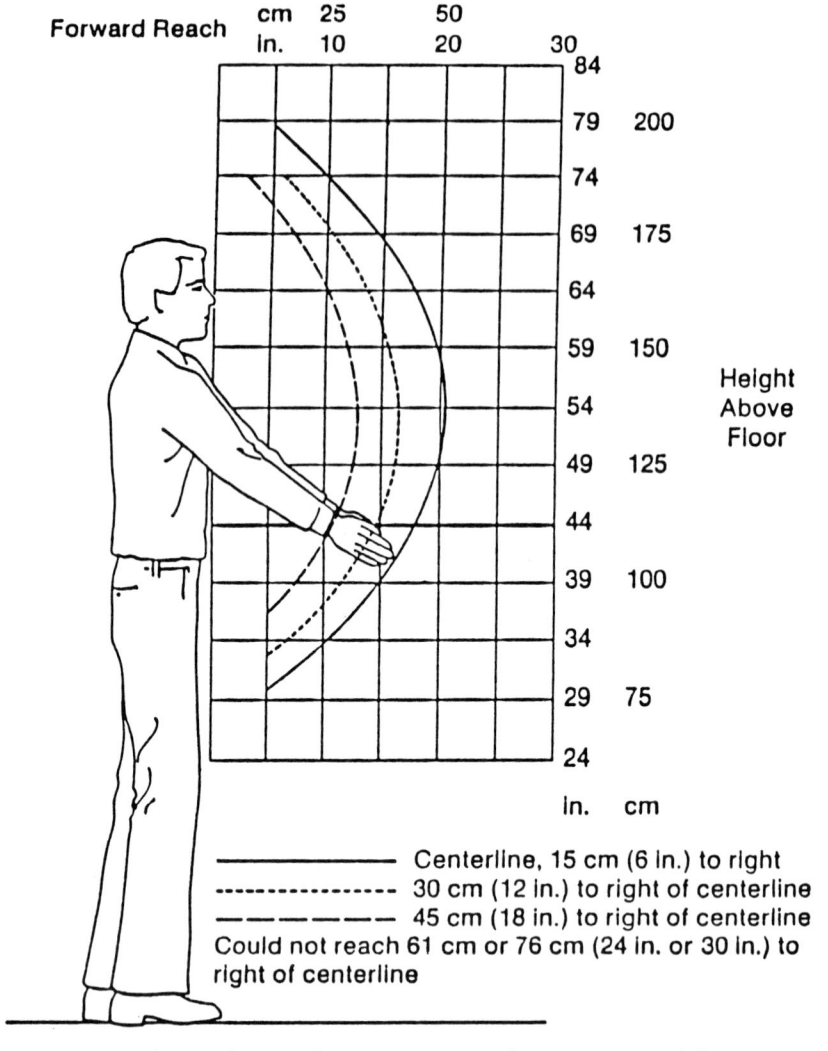

Figure 4.4. The standing reach area, one arm (after Eastman Kodak Company, Human Factors Section, 1983, with permission).

or extended reaches, where there is a requirement for downward application of forces, for example, packing, use of certain hand tools, or where the work station does not have room for knee clearance.

For standing work places the Eastman Kodak Group note that, without stretching or leaning forward very much, most people can reach about 46 cm (18 in) in front of the body as long as the object to be addressed is 110 to 165 cm (43 to 65 in) above the floor and not more than 46 cm (18 in) to the side of the body's centreline. At farther distances to the

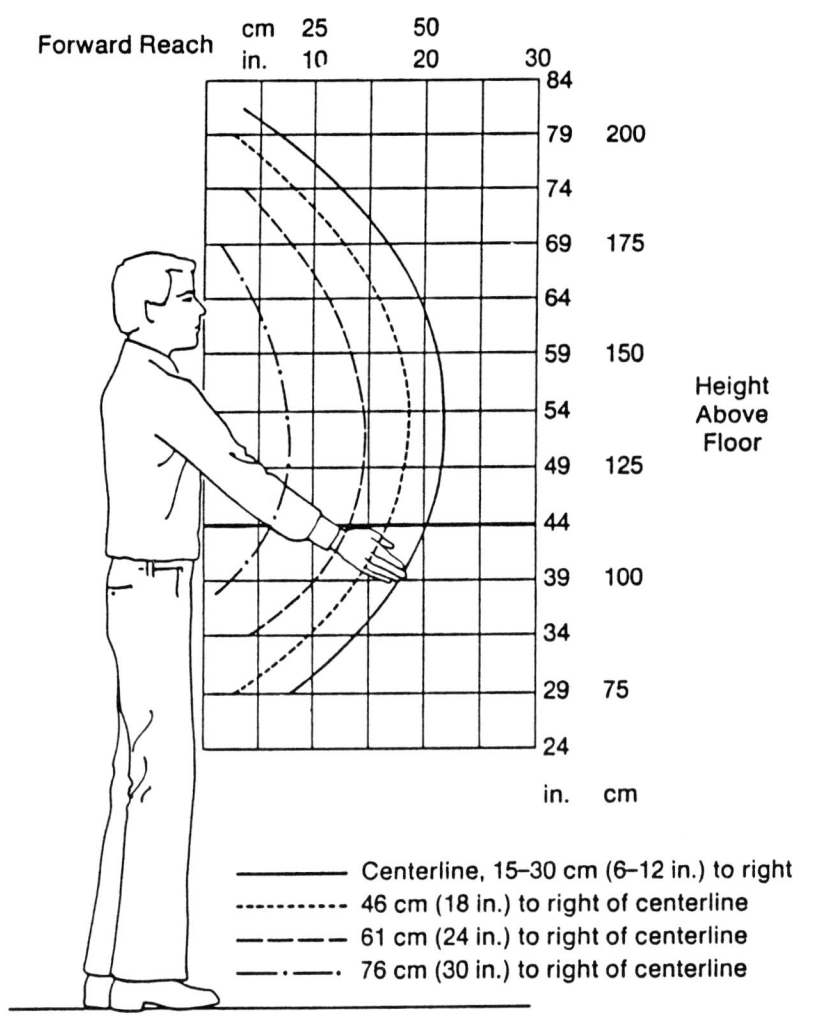

Figure 4.5. The standing reach area, two arms (after Eastman Kodak Company, Human Factors Section, 1983, with permission).

side, or height less than or greater than the above range, forward capability falls off. The operator can achieve an extended reach only by leaning, stretching, stooping, or crouching. These postures can all produce fatigue if they have to be assumed frequently or maintained for periods longer than a minute or so.

They also note that standing work place height should be designed according to the dimensions indicated in Figure 4.7. For light assembly, writing, and packing tasks, the optimal working height of the hands should be about 107 cm (42 in). For tasks requiring large downwards

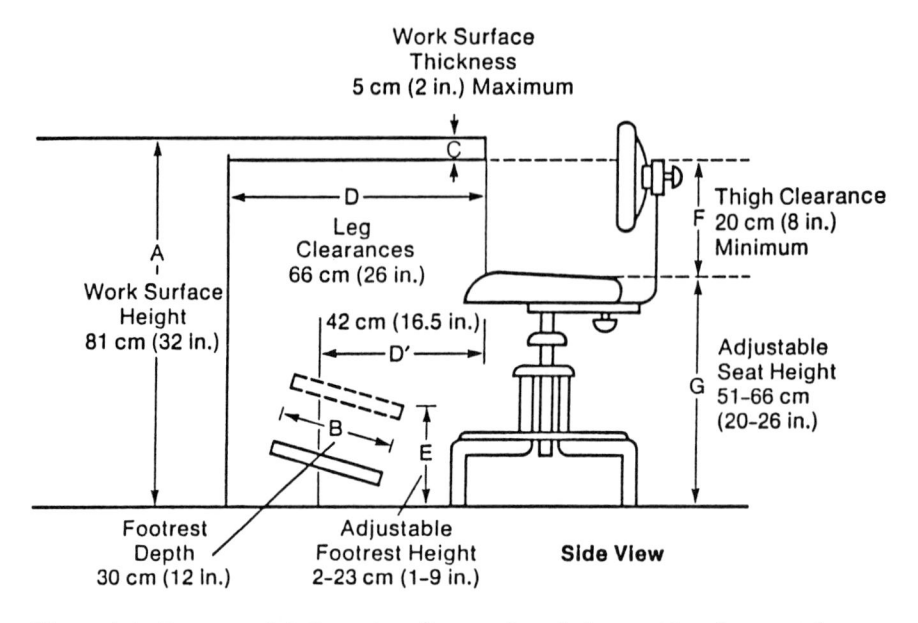

Figure 4.6. Recommended dimensions for seated workplaces with a footrest (after Eastman Kodak Company, Human Factors Section, 1983, with permission.

or sidewards forces, such as casing operations or planing, the working height of the hands should be about 91 cm (36 in). Even lower heights, to about 76 cm (30 in) may be appropriate for very heavy force exertions. For tasks requiring large upward forces, as in clearing machine jams and removing components, the optimal working height of the hands should be about 81 cm (32 in).

Standing operators should, where feasible, be provided with support stools for use while working, or at least a jump seat which they can use during machine time or slack time.

Sit/stand work place

A sit/stand work place is used where the task has dual seated and standing operations, where, for example, most of the task can be accomplished in the seated position but the operator has to stand occasionally to access some element of the task, or to apply forces not effective from the seated position. This would include an operation with frequent reaches greater than 41 cm (16 in) forward or 15 cm (6 in) above the work surface.

The height of the work surface in these circumstances is essentially a compromise that requires the operator to stoop slighty in the standing position and to have undesirable high chairs in the seated position.

For tasks where the operator is normally seated with occasional

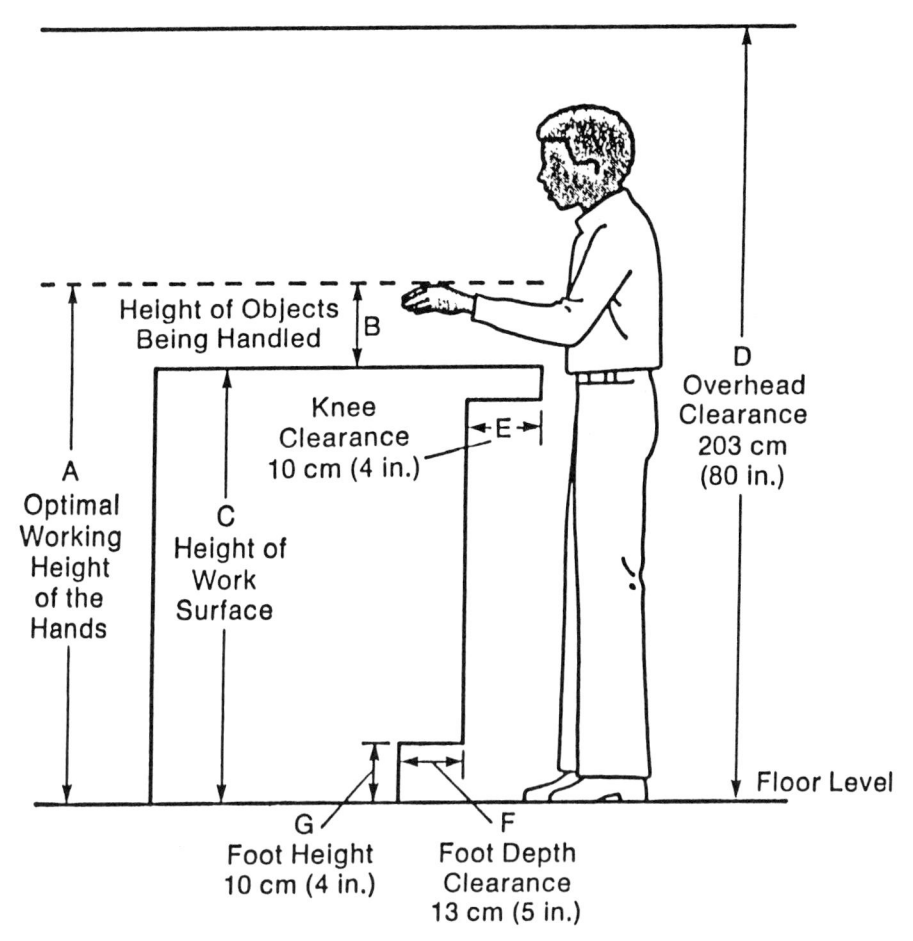

Figure 4.7. Standing work place dimensions (after Eastman Kodak Company, Human Factors Section, 1983, with permission).

standing, Eastman Kodak recommended that the height of the work surface should be 102 cm (40 in). Where the operations involve different tasks with different heights the basic work surface should be 91 cm (36 in) above the floor along with the use of movable platforms. For drawing surfaces, or light-tables, the work surface should be higher, at 112 cm (44 in), to allow close visual inspection.

General working height

As a general rule the height of the work surface should be such that the work can be conducted with the forearms approximately horizontal or sloping slightly downwards. In other words the work surface height should be about 50 mm (2 in) below the level of the elbow. Although work height is commonly referred to floor level it is better to refer it to elbow level to emphasize and account for the great variation in the heights of workers.

Proper work height can be achieved in three ways, namely, changing the surface height, changing the elbow height, or changing the height of the work on the work surface. The first can be accomplished either by designing a machine, device or surface of adjustable height, or by providing alternate work surfaces such as work tables of different heights on which to conduct different jobs or different aspects of the same job.

The second can be achieved, within limits, by having a chair or stool adjustable in height, or a floor platform to stand on. In this connection it should be noted that a seat higher than 90 cm (approximately 36 in) can be unstable and potentially dangerous, as well as unacceptable to the worker. The floor platform should range in height between 10 and 30 cm (4–12 in). The third approach, namely changing the height of the work on the work surface, can be accomplished where necessary and feasible by propping up the object of the work on supports.

Other design principles

Certain other principles of work station design can be isolated. Some of these are presented below, although not in any order of priority.

Static loads and fixed postures

The nature of the work activity, which in turn determines the nature of the work place, should be so organized and/or designed as to avoid or minimize static loads and fixed postures.

For example, tasks, machines, and equipment should be designed, selected, or organized to minimize holding work, as in operating hand tools at arms length, working with the arms overhead, and so on. Where necessary, aids for holding or balancing should be provided, such as weight compensated slings for heavy air wrenches. Tasks, machines, and equipment should also be designed, selected, or organized, to minimize awkward postures, for example, the need to perform a task in the stooped position because the work surface is too low, or the need to work with the arms up and forward, and indeed the need to work bent within a shaft or tunnel.

Load carrying

Over and above the need for restriction on the actual weight of the load, it must be recognized that undesirable static loading can also occur in carrying, particularly in one-sided carrying. Some element of carrying may commonly be involved in any industrial task. As noted in discussion of anthropometric factors, wherever feasible loads should be small and carried close to the body rather than bulky, with a need for outstretched arms. Prolonged carrying is not normally a feature of industrial work. It has been shown, however, that the technique of carrying can grossly influence the acceptable load. For example, carrying 15 kg in each hand is much less efficient and more fatiguing, even with proper handles, than carrying the same weight suspended equally at each end of a yoke across the shoulders, which in turn is less efficient and more fatiguing than carrying the load suspended in two packs, one across the chest and one across the back (Datta and Ramanathan, 1971). Maximum loads for one hand held by the side, according to Rohmert (1960), are as follows. It should be noted that these numbers refer to maxima and should not be considered as normally acceptable loads.

Table 4.1 Maximum loads for one hand held by the side

Conditions	Load
lightly built women:	27 kg
average women:	41 kg
powerfully built women:	54 kg
lightly built men:	63 kg
average men:	81 kg
powerfully built men:	94 kg

Seating

The requirements of industrial seat design are considered in detail in Chapter 11. Suffice it to say at this point that where a task can be done seated it is normally preferable to do so, provided the height and angle of the work surface are appropriate to the task and that the tools and equipment lend themselves to seated use. For example, seated work which causes the worker to work with arms forward and elbows upraised is more fatiguing and less efficient than it would be if the task were done from a standing position. Some times it may be possible, or even desirable, to use a so-called sit/stand position where the weight of the body is distributed between the feet and an appropriately designed seat. It is also desirable, again wherever feasible, to so organize the seating that the worker has the choice of working at any time in either the standing or the seated position, taking care in this case that the seat itself does not become an obstruction but can be moved, rotated, or folded out of the way of the standing worker.

Support of body parts

It is not infrequently overlooked that the head, which weighs about 6.5 kg has to be supported on a relatively small neck. This need is not normally particularly fatiguing, but when the head has to be frequently rotated on the neck, for example to view various displays, or has to be held out of line of the supporting spine as in tilt forwards or backwards, the resulting muscle strain can become limiting. Anatomically, backward tilt of the head is more difficult to maintain than forward tilt. Thus, any overhead display design requiring the head to be held at, or to make repeated motions through, an angle greater than 15 degrees above a horizontal line forwards through the eyes, (that is, out of the axis of the spine), should be avoided. Repeated sideways motions are also undesirable for the same reasons.

An average arm weighs somewhat less than 4.5 kg, while the hand and forearm weigh about 2 kg. In this connection, attention has already been directed to the need to avoid static holding or stretching.

A leg is much heavier (about 14.5 kg); consequently wherever feasible leg support should be provided. This can be in the form of an adjustable foot rest. Preferably this foot rest should be separate from a seat, or the equipment at which the worker is working. If it is not feasible to provide separate and movable foot rests, a foot rail under a work bench or around a seat is useful although less effective. While these requirements pertain to seated workers, it is also desirable to provide a foot rest, like the rail in a saloon bar, for the standing worker. Use of a rail of this type redistributes the load, and allows occasional tilt of the spine which is effective in temporarily relieving back fatigue. In this connection, the use of pedal controls by a standing worker should be avoided. Pedal controls throw all the weight on to one leg while the other is required for coordinated work.

Arm movement

Special consideration should be given to the radius of pivoting movements of the upper limb. It is desirable that movements should pivot around the elbow joint rather than the shoulder joint. Such motions improve movement time, movement accuracy and movement cost. A number of experiments quoted by Konz (1979) demonstrate that the number of repeated movements per unit time is significantly increased, with greater accuracy, and lowered physiological cost in terms of increase in heart rate during work, when the movements are pivoted around the elbow rather than the shoulder. Maximum accuracy, however, is found for cross-body movements, rather than for those pivoted around the elbow.

Preferred hand

Approximately 90 per cent of the population is right-handed. Tasks and equipment should preferably be designed to eliminate the need for right or left handedness, but where this is not possible they should favour the preferred hand. Konz and his colleagues (1969) found that the preferred hand is 10 per cent faster for reach-type motions, and is also more accurate; (7 per cent of subjects missed targets with the preferred hand as against 12 per cent for the non-preferred). The preferred hand is also 5–10 per cent stronger (Kroemer, 1974).

In this connection also, work flow should be designed such that work is passed to a work station from the operator's preferred side and leaves from the non-preferred side.

Decelerative movements

Sudden decelerative movements of the upper limbs are not uncommon in industrial work, as for example when a worker tosses a completed part into a bin, or flips a lever, or jerks something along a mandrel. These actions should be avoided wherever possible and replaced with smooth non-ballistic actions. Not only is the deceleration stressful in itself since it throws a sudden strain on muscles, ligaments, and joints, but in addition it commonly involves overextension at one or more joints with repeated minute damage to joint surfaces, which in turn may lead eventually to arthritic changes in the joint itself.

Conveyors

Conveyors may act as work stations in themselves, as in certain types of small assembly, or as links between work stations. While many are powered belts, some operate as a sequence of rollers along which materials, often in containers, move by gravity.

The dimensions of the conveyor are determined largely by the size of the units handled and the tasks they feed. In some cases the conveyors are used strictly for conveying materials to or from a task, in which case the materials are then transferred to the work station. The work station height then determines the desirable height of the conveyor. In other cases work is performed on materials still on the conveyor, in which case the dimensions are determined according to the general guidelines of seated or standing operations. Particular care must be taken, however, to allow suitable knee and leg room clearance for operators working at a conveyor since otherwise the supporting structure of the conveyor may act as an obstruction.

For efficiency, a conveyor should be accessible from both sides. Crossovers or gates should be provided where necessary. Walking on conveyors should be prohibited. The surfaces, particularly of rollers, are unstable for walkers. Since injuries are three times greater at the unloading end than elsewhere, conveyor storage space should be provided at the off-loading site to allow the operator better pacing of the work.

The rate of operation is a factor of significance. Clearly too slow a rate is inefficient. If the rate is too high then not only will some elements of a task be missed, or components stacked up, but there is also the possibility of inducing among operators a form of dizziness and disorientation sometimes called 'conveyor sickness'. This phenomenon arises from a rapid and continued stimulation across the line of vision which in turn disturbs the body balance mechanism in the labyrinth of the internal ear and gives rise to the dizziness and nausea. Consequently the optimal rate becomes a compromise between the least skilled and the most skilled.

Aisles and passages

Woodson and Conover (1954) note that proper design of aisles and passageways is essential to ensure the expeditious flow of human traffic. They point out that aisles and passages should be designed into a work place, taking into account such factors as the approximate traffic load at any one time, the number and size of entrances and exits, the number and location of doors opening into the passageway, and the illumination of the passageway and the proper identification of exits. The minimum dimensions for a two-person passageway are suggested as being 137 cm (54 in), while for three persons they should be 183 cm (72 in). For a two-wheeled hand truck the minimum dimension should be 76 cm (30 in), while for a fork truck there should be a clearance of 30 cm (12 in) on each side.

Stairs and ladders

Falls from ladders and slips on stairs are still a leading cause of injury. Although in many situations the requirements for stair and ladder design are detailed by regulation from the appropriate jurisdiction, it is well to outline some general standards. While various recommendations have been made, the concensus suggests that, the slope of a staircase should be about 30–35 degrees from the floor, with a maximum of 20–50 degrees. For a typical stair a riser or step height should be 116–18 cm (6.5–7.5 in) with a tread depth of 28–30 cm (11–12 in), and a tread overlap on the riser of 2.5–4 cm (1–1.5 in).

Ladders tend to have a slope in excess of 75 degrees from the floor, with a range from 50–90 degrees (Woodson and Conover, 1964). The distance beteen treads or rungs may vary from 18–41 cm (7–16 in), with 30 cm (12 in) being considered the most suitable. Openings above ladders should be designed to have adequate clearance for body and knees while climbing. Ladders mounted vertically on a wall should have 15 cm (6 in) clearance from the wall to allow for the feet.

General safety

Certain other guidelines can be considered on the grounds of general safety. These include the following:

1. Provide handles on objects to be lifted that weigh more than 4.5 kg (10 lbs).
2. Avoid or guard pinch-points.
3. Round off sharp edges and corners.
4. Protect against accidental activation of switches.
5. Provide lock-outs on machine controls for use during maintenance.
6. Provide lock-ins on ladders, stands, and so on, to prevent collapse.
7. Provide guardrails around platforms.
8. Keep aisles clear of projections.
9. Provide aids to identify emergency equipment, such as colour, lighting, location, and so on.
10. Provide a habitable environment.

Chapter 5
A guide to manual materials handling

Many jobs, and generally the lowest paid jobs, require the handling or movement of objects. Manual handling tasks are in fact the principle source of compensable work injuries. Of these, most occur in the lower back and are derived from improper lifting, lifting excessive loads, lifting otherwise acceptable loads too frequently, or lifting with a twisting movement or other improper posture.

Musculo-skeletal anatomy

The body framework consists of the skeletal system, comprising a sub-system of bones linked by joints and strengthened by ligaments, and also the muscular system, comprising muscles attached across the joints and capable of moving the joints. The whole is surrounded by various body organs. The soft tissues are surrounded by a varying thickness of fat and ultimately covered by skin. The various dimensions measured are a function of the size of bones and muscles, which in turn is determined largely genetically and partly by nutrition and training. They also depend on the thickness of various layers.

As noted, the bones are linked by joints. There are various kind of joints, some of which, such as the knee, permit motion in only one plane of space; others, such as the shoulder, permit free movement in all planes; still others, such as the small joints of the back, permit limited movement in all planes. Joint movement is restricted by tough fibrous bindings, called ligaments, the variety, number, and attachment of which varies from joint to joint.

The bones are made of a fibrous material heavily infiltrated with calcium and other minerals. Their structure is not static like that of the beam of a bridge, but varies continuously as calcium is deposited and re-absorbed and as the fibrous material is destroyed and restored.

The metabolism of muscles has already been considered (Chapter 3). Briefly, muscle fibres, and ultimately the entire muscle, contract in response to an electro-chemically mediated nerve stimulus which releases acetyl choline into the substance of the muscle fibrils. Since the muscle is attached across a joint, contraction causes movement at the joint. Coordinated control of the motion is directed by the brain through the peripheral nervous system, such that required motions may take place. It is, of course, through the contraction of muscles and the movement of joints that the handling and movement of objects can be undertaken. In the process of what is referred to as manual materials handling.

Much study has been given to the requirements for manual lifting. The details would, and do, encompass a text in themselves. All that can be done here is to outline the principles. The reader seeking comprehensive information should consult the sources from which much of this material is derived, such as *The Biomechanical Basis of Ergonomics* by E. R. Tichauer (John Wiley and Sons), or, what is probably the most definitive work, the *Work Practices Guide for Manual Lifting*, prepared and published by the National Institute for Occupational and Safety (NIOSH) in the United States.

The NIOSH Guide recognizes that there are four approaches to the development of guidelines for manual lifting, namely epidemiological, biomechanical, physiological, and psychophysical.

Epidemiological approach

The epidemiological approach is based on the science of epidemiology, which is concerned with identification of the incidence, distribution, and potential control of disease and injury, including the identification of factors which serve to modify risks of overexertion injury, with particular emphasis on the incidence and severity of low-back pain in industry.

Factors pertaining both to the job and to the person can be identified. Job risk factors include the weight and size of the object and the force required to lift it, as well as the location of the object, and its centre of gravity with respect to the worker; the further away the load centre of gravity is from the body, either from bulk of the object or layout of the workplace, the greater is the frequency and severity of problems. The frequency of the lift is also significant, as are the spatial aspects of the task in terms of movement distance, direction, the presence of obstacles and constraints derived from postural demands. Coupling between the worker and the object, for example the type and location of handholds, if any, can greatly modify the difficulty of the task; properly designed handholds placed in the line of the centre of gravity and, if possible, no wider than the natural hanging position of the arms can greatly improve the manoeuvrability of a load.

Personal factors include that of gender, not so much because of sex differences *per se*, but because females statistically speaking are less strong than males. It is also suggested that because of the structure of the female pelvis the centre of the hip socket in females is located in front of a line through the centre of mass of the body, while that of males is in the same plane; hence a greater bending moment exists on the spine of women to the extent that the lifting stress on the back muscles of women for the same object is some 15 per cent higher than for men (Tichaeur, 1978).

Age is another personal factor that requires consideration. Many studies have shown that the greatest incidence of low-back pain and back strain occur among workers in the 30 to 50 year old group. This, however, does not necessarily mean that the group is more susceptible; it is more likely that either workers over age 50 have selected themselves out of tasks that require heavy manual lifting, or that with experience they have learned techniques of avoidance of injury.

Some studies quoted in the NIOSH Manual suggest that increased body weight is favourably correlated with the ability to lift, probably because the heavier person is usually stronger than his lighter peer, and also that he has the mass required to counter-balance the handling of large objects. There is, however, controversy over whether stature in itself plays any part in the ability of a person to lift. Consensus suggests that there is little or no significance.

The style, or technique, of lifting can be significant. Despite protestations by various safety oriented organizations, there is no single ideal technique that will meet all circumstances. It is not always appropriate to adopt the commonly taught squatting method with the back straight while the knees are used to provide the lift. This might be appropriate in some circumstances, but in others, for example, it might be better to bend at the back and use the arms in a swinging ballistic movement, say, to complete several sequential actions at a low level. In general it is desirable where feasible, to separate the feet in order to maintain a balanced distribution of weight, then bend at the knees and hips with the back reasonably straight, keeping the arms as near to the body as possible with the load as close to the body as possible, and then lift smoothly with the legs.

The question of strength has been mentioned in connection with discussion of the significance of gender. It is generally considered that strength, and particularly back strength, is significant in protecting the back. Studies by Chaffin and his colleagues (1973, 1977) showed that the incidence of back pain in workers engaged in frequent lifting tasks was three times greater among persons whose strength as measured by isometric strength tests was not equal to or exceeding that demanded by their jobs. It was concluded that overstressing a person beyond his/her demonstrated strength cannot be tolerated by a person's musculoskeletal system, especially when such exertions are performed more often than about 100 times per week.

Biomechanical approach

In the biomechanical approach a person's physical capabilities are assessed along with the physical demands of the job, with specific reference to those physical attributes of the individual and the jobs that have been found to produce potential harm to the musculoskeletal system.

In this regard it will be recognized that a load held in the hands, as well as the weights (that is, masses × gravity) of the relevant body parts, create rotational movements or torques at the relevant body joints. Muscles are so positioned that they act through relatively small moment arms, and produce large motions of body segments with small degrees of shortening, along with high muscle and joint forces. The torques are found not only in the joints of the limbs that are specifically engaged in the task, but also in the supporting structures of the back. Thus, for example, when a 20 kg load is held at arm's length it produces a large torque at the lumbosacral joint of the back. For an 'average man', such a load would produce more than 1200 kg–cm torque, which in turn would produce a compression force within the lumbosacral joint equivalent to that developed when holding 40 kg load at a level between the knees. It is considered that jobs which place more than 650 kg of compressive force on the low back are hazardous to all but the healthiest of workers. In terms of a specification for design a much lower level of 350 kg or lower should be viewed as the upper limit (NIOSH, 1981).

Over and above the foregoing, specific recommendations have been made by NIOSH with respect to the biomechanics of lifting. These can be summarised as follows:

(a) Lifting a 5 kg compact load, that is where the mass centre of gravity of the load is within 50 cm of the ankles, can create compressive forces sufficient to cause damage to older lumbar vertebral discs. As the load mass centre of gravity is moved horizontally away from the body, a proportional increase in the compressive force on the low back is created. Thus even light loads need to be handled close to the body.

(b) When a load is lifted from the floor, additional stresses are exerted on the low back due to the body weight moment when stooping to pick up the load. Thus heavy loads should not be stored on the floor, but should be raised to about standing knuckle height (minimum 50 cm) to avoid the necessity for stooping over and lifting.

(c) The postures used to lift loads from the floor can exert complex and relatively unknown stresses on the low back when lifting. Specific instructions as to the safe posture to be used will necessarily be complex, reflecting such factors as leg strengths, load, and load size. Until such complexities are better researched it is recommended that instructions as to lifting postures be avoided.

(d) Lifting loads asymmetrically (by one hand or at the side with the torso twisted) can impart complex and potentially hazardous stresses to the lumbar column. Such acts should be avoided by instructions and workplace layouts which permit the worker to address the load in a symmetric manner.

(e) The dynamic forces imparted by rapid or jerking motions can greatly multiply the effect of a load. Instructions to handle even moderate loads in a smooth and deliberate manner are recommended.

Physiological approach

The physiological approach to the establishment of manual lifting guidelines uses measures of oxygen consumption, metabolic rate and heart rate as indices of the energy involved in conducting various tasks. As noted in Chapter 3, two types of work can be considered, namely dynamic and static. In either case, when the muscles are active there is a demand for an increased metabolism. Muscles can function either aerobically, that is utilizing oxygen, or anaerobically, that is without the use of oxygen. There is a limit to the amount of oxygen that can be supplied. This limit is the maximum aerobic capacity. At this point the muscles are fatigued and no longer able to do effective work, although they can still contract anaerobically. Anaerobic metabolism generally begins at about 50 per cent of aerobic capacity.

Results from laboratory experiments (NIOSH, 1981) suggest that lifting exercise in industry seldom reaches or exceeds the average male worker's 50 per cent aerobic capacity. Indeed, it seems that there are few practical situations where the average energy cost of living in industry is likely to exceed the 35 per cent level, a level of work which is considered by most research workers to be compatible with daily levels of work in 'heavy' industries, irrespective of age, sex, and other factors which are considered to limit the permissible daily energy expenditure in industry.

As previously noted, energy expenditure is commonly measured in kilocalories per minute (kcal/min). An oxygen consumption of 1 litre of oxygen per minute is equivalent to an energy expenditure of approximately 5 kcal/min. Some industrial tasks, particularly in heavy industries such as mining and steelworking require expenditures of as much as 15 kcal/min for short periods. To compensate for this expenditure, then, these work periods must be offset by rest periods and work at lower energy levels to provide an average expenditure throughout the day of 5 kcal/min.

Over and above general energy expenditure, various factors can be shown to affect the metabolic cost of lifting. These include the body posture and technique used in the lift procedure, the weight of the load, the frequency of lifting, the vertical distance of the lift, the vertical location at the beginning of the lift, as well as the temperature and humidity. Taking these various factors into account, and assuming an operational aerobic capacity of 15 kcal/min for men and 10.5 kcal/min for women, the NIOSH guidelines recommend that:

1. For occasional lifting (for one hour or less) metabolic energy expenditure rates should not exceed 9 kcal/min for physically fit males or 6.5 kcal for physically fit females.
2. Likewise, continuous (8 hour) limits should not exceed 35 per cent of aerobic capacity of 5.0 kcal/min and 3.5 kcal respectively.
3. Personal attributes of age, gender, body weight, etc., are insufficient to accurately predict work capacity for any particular individual, although such data are sufficient for making predictions of group averages.
4. The primary task variables which influence metabolic rate during lifting are:
    (a) the weight of the load handled
    (b) the vertical location at the beginning of the lift, and consequently the lifting posture
    (c) the frequency of lift.

Psychophysical approach

The psychophysical approach is concerned with establishing limits of human tolerance for load lifting in the light of human strength and endurance, and the will to proceed without undue fatigue.

Psychophysics describes the relationships between human sensations and their physical stimuli. In this connection it is concerned with defining human strength and the corresponding tolerance loads. Strength can be static or dynamic. Static strength is the maximal force muscles can exert isometrically in a single voluntary effort (Roebuck *et al.*, 1975). In testing for static strength subjects are invited to assume a specific body posture and to exert force against a stable resistance. Dynamic strength involves body motion, and is measured in terms of velocity and acceleration of the body members involved.

Strength varies consideraby with gender, female strength being only 35–84 per cent of male strength (Laubach, 1976). Other anthropometric variables such as height, weight, and age are not good predictors of strength (Keyserling *et al.*, 1978), although increased body weight is detrimental to strength with increasing age.

Figure 5.1 (Chaffin, 1974) is based on a predictive model which treats the body as a series of linked segments, for example arms, leg, torso, for each of which forces can be calculated. The figure shows the predicted lifting strength for a large, strong, 2.5 percentile male, using different loads lifted through different vertical distances from points at different distances in front of the ankles.

In measures of static strength all the factors are controlled. In dynamic

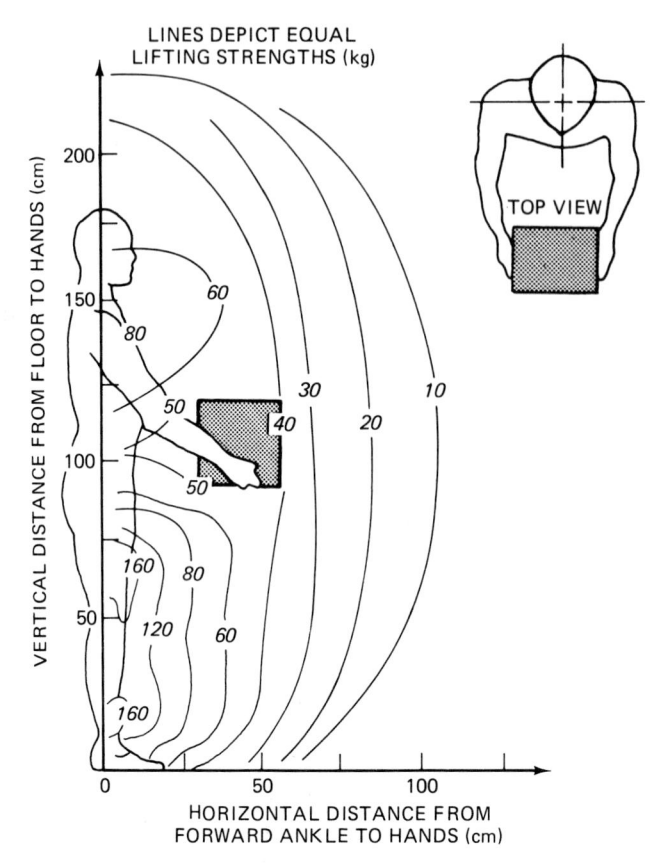

Figure 5.1. Predicted lifting strength of large, strong male (after Chaffin, 1974).

strength measurement the subjects are given control of one of the factors, normally the weight of the load. The subjects are then encouraged to do a prescribed lifting task, working as hard as they can without straining themselves, or without becoming unusually tired, weakened, overheated, or out of breath. Although numerous studies of this nature have been undertaken, much of this work has been done by Snook and his colleagues at the Liberty Mutual Insurance Company (Snook, 1978). The results are presented in comprehensive tables which outline the permissible rate of load lifting for males and females of three different percentiles, through three different vertical heights, (for example, floor to knuckle, knuckle to shoulder, shoulder to extended arm height), starting at varying distances in front of the ankles. The tables are large, and should be consulted in the original source or from the NIOSH Guide. Other tables are also presented.

General recommendations for lifting tasks

On the basis of their consideration of the epidemiological, psychological, physiological, and biophysical studies of lifting, reviewed briefly above, NIOSH developed a model for use in determining lifting standards. It is emphasized that the limits presented in these guidelines do not apply to all kinds of lift. They are intended to apply to a situation where there is smooth two-handed symmetrical lifting directly in front of the body, with no twisting during the lift. The object lifted should be no wider than 75 cm (30 in). The lifting posture should be unrestricted with good couplings between the hands and the object, and the shoes and the floor. The ambient environment should be favourable.

Various lifting task variables are incorporated. These include:

1. The object weight (L), (kg)
2. Horizontal location (H) of the hands at the origin of the lift, measured forward of the body centre line or midpoint between the ankles
3. Vertical location (V) of the hands at the origin of the lift measured from the floor level in centimetres
4. Vertical travel distance (D) from the origin to the destination of the lift, measured in centimetres
5. Frequency of lifting (F), average number of lifts per minute
6. Duration of period, assumed to be occasional (less than one hour), or continuous

For the purposes of the model, also, jobs are classified as infrequent, namely either occasional or continuous lifting less than once every three minutes; occasional high frequency, namely lifting one or more times per three minutes for a period of up to one hour; or continuous high frequency, namely lifting one or more times per three minutes continuously for 8 hours. These conditions are used in determining the following criteria.

Criteria for guidelines

A large individual variability exists among workers in lifting performance capacity and in the risk of injury. In consequence, any control measures instituted must be of both an administrative and an engineering nature. In other words some tasks require major modification and redesign, while some can be handled by selection and training of workers along with modification of exposures. To meet these requirements two lifting limits are defined, one being the Maximum Permissible Limit, and the other the Action Limit.

Maximum permissible limit (MPL)

This limit is based on four criteria, namely that:

(a) musculoskeletal injury incidence and severity rates increase significantly in populations when work is performed above the MPL;
(b) biomechanical compression forces on the lumbosacral disc are not tolerable in most workers at lifting levels above the MPL, that is, over 650 kg (1430 lb);
(c) metabolic rates among most individuals working above the MPL will exceed 5.0 kcal/min;
(d) only 25 per cent of men and less than 1 per cent of women have the strength to perform work above the MPL.

Action limit (AL)

The Action Limit is based on different levels of the same criteria, namely, that:

(a) musculoskeletal incidence and severity rates increase moderately in populations exposed to lifting conditions described by the AL;
(b) compression force on the lumbosacral disk within the AL (that is, up to 350 kg (770 lb)) can be tolerated by most young, healthy workers;
(c) metabolic rates for workers within the AL will not exceed 3.5 kcal/min;
(d) over 99 per cent of men and 75 per cent of women can lift loads described by the AL.

Three different levels are thus defined as follows:

1. Those above the MPL are unacceptable and require engineering controls for prevention of injury;
2. Those between the AL and the MPL are unacceptable without either engineering or administrative controls or both;
3. Those below the AL are considered to represent nominal risk to most industrial workers.

Guideline limit formula

To meet the above requirements a formula has been defined for the Action Limit which is then presented as a maximum load in kg for application in the criterion conditions described earlier. The Maximum Permissible Limit is considered to be three times the value of the Action Limit, that is:

$$MPL = 3 \, (AL)$$

The actual formula is as follows:

$$AL (kg) = 40(15/H)(1-0.004[V-75])(0.7+7.5/D)(1-F/F_{max})$$

where,

H = horizontal location (cm) forward of midpoint between ankles at origin of lift;
V = vertical location (cm) at origin of lift;
D = vertical travel distance (cm) between origin and destination of lift;
F = average frequency of lift (lifts/minute);
Fmax = maximum frequency which can be sustained Table 5.1, as follows:

Table 5.1 Maximum sustainable lifting frequency for use with NIOSH lifting guidelines (NIOSH, 1981)

Duration	V > 75 cm Standing	V ≤ 75 cm Stooped
1 hour	18	15
8 hours	15	12

In the foregoing formula the varying factors act as modifiers to the basic 40 kg. Thus, if the H factor, namely 15/H, is 15 cm no adjustment is necessary. Similarly, when V is 75 cm then the V factor equals unity and no adjustment is required. The D factor ranges from 1 to 0.74 as D varies from 0 to the maximum of 200 cm. A vertical travel distance of 25 cm is considered to be the minimal value allowed, and D is never set less than 25.

With respect to the F factor, when, for example, the lift originates below 75 cm and is performed continuously throughout the 8-hour day then, according to the table, Fmax would be 12. Thus if the frequency of lift (F) were 6 lifts per minute the F factor would be (1−6/12)=0.5, thereby halving the amount of load lifted.

The NIOSH approach, using the above formula, or a formula adapted for lb/in units, is by far the most comprehensive, but even it is useful only under the criterion circumstances defined. Other empirical approaches take little or no account of such critical variables as height of lift, height of starting point, frequency of lift, and so on. One such guideline is presented by the International Occupational Safety and Health Centre of the ILO. It defines by specified age groups the maximum permissible lift in kilograms for males and females, and is shown as follows (Table 5.2);

Table 5.2 *Permissible weight of lift (kg) for specified age groups (ILO, 1962)*

Age	14–16	16–18	18–20	20–35	35–50	over 50
Male	15	19	23	25	20	16
Female	10	12	14	15	13	10

Chapter 6
Repetitive strain injury

Repetitive strain injury (RSI) is one of a variety of names for a group of musculo-skeletal and tendinous injuries the majority of which are found in the upper limb, the shoulder girdle, and the neck. In industry these conditions tend to be the result of repetitive motion at joints, often to extremes of movement, and under load, or by a requirement to maintain constrained or awkward postures. Pathologically similar conditions can occur not infrequently in association with osteoarthritis, pregnancy, and in post-menopausal women. In such cases, of course, the spontaneously occurring condition is aggravated by overuse.

Terminology and pathology

RSI is also known as Repetition Injury and as Cumulative Trauma Disorder (CTD). To understand the pathogenesis of the conditions it should be recognized that, as noted before, muscles take origin in a bone, pass across a joint and terminate in a tendon which is inserted into another bone. Some tendons, notably those around the wrist, are encased in a smooth lubricating sheath through which they run. This sheath is known as the *synovial sheath.* At the base of the wrist, on the palmar surface, the tendons to the fingers are held together in a ligamentous tunnel called the *carpal tunnel* which prevents them from splaying out on contraction of the flexor muscles of the forearm. Also passing thorough this tunnel are small blood vessels, and fibres of the *median nerve* which provides nerve supply to the thumb, the first and second finger, and half of the third finger.

The joints themselves are lined with a lubricating membrane called the *synovial membrane.* Other anatomincal elements involved are the muscles and their attachments to bone. In particular, there are muscles responsible for *pronation,* or inward turning of the forearm, and *supination,* or outward turning of the forearm, which attach to protuberances on the outer corners of the humerus bone at the elbow, known as the *medial* and *lateral epicondyles,* respectively. Of interest also are the protective cushions or *bursae,* found in the shoulder joint, and elsewhere. Inflammation, with pain, swelling, and tenderness can occur in any of these structures, and although the resulting conditions are fundamentally the same, each condition goes by a different name according to anatomical site. The suffix, . . . *itis,* describes inflammation, and thus, for example, RSI can occur as tendinitis when the inflammation is in a tendon, synovitis when it is in a synovial membrane, tenosynovitis when it is in the tendon and tendon sheat, epicondylitis when it affects the epicondyles and bursitis when it affects a bursa. A separate but related condition is *carpal tunnel syndrome,* which occurs by reason of swelling of structures in the carpal tunnel.

While these various conditions can be recognized clinically, they can occur when there is little or no obvious pathology. Indeed, McDermot (1986), in a comprehensive review, states: 'The term repetition strain injury is unsatisfactory. First it implies that an injury has been caused by repetitive movement. However, the problem in many workers with RSI is induced by a static muscle load, for example among those who develop pain in the shoulders and arms after maintaining their upper extremities in the raised position that is necessary for their working activities. Secondly, the term "strain" implies a pathogenesis for RSI which is by no means certain. Thirdly, there is often no apparent injury present and there may be no discernible abnormality on clinical examination.' Accordingly, still another term, namely *overuse injury,* with no connotation of either repetition or strain, has come into usage. A corollary deriving from this viewpoint is that all forms of this type of injury, whether they be tenosynovitis, epicondylitis, carpal tunnel syndrome, occupational cervico-brachial syndrome, or whatever, arise from the same origin, namely overuse, and that the terminological subdivisions merely define the site or sites, and that what we are dealing within RSI, or whatever other name, is not a multiplicity of conditions but different aspects of the same condition. Some of the more common aspects are enlarged on below.

Tenosynovitis

Tenosynovitis is a general name given to an affliction that affects tendons and their sheaths at the wrist. The term is often misused to apply to other forms of overuse injury around the wrist, even carpal tunnel syndrome. It occurs as a dull ache, worsened by movement of the tendons concerned. There may be, in established cases, some swelling over the area, and when the wrist or fingers are moved a rasping sensation may even be felt. There is commonly weakness of the muscles involved.

With continued exposure the condition will worsen, and cause increased weakness and pain. There may be extension to other related tendons. Ultimately, if the worker persists, there could be permanent disability. If the worker is removed from the causative task however, before permanent damage is done, the condition will recover. Specialized medical treatment may nevertheless be required, and gradual rehabilitation is essential before return to full activity.

Carpal tunnel syndrome

The terms tenosynovitis and carpal tunnel syndrome are often confused, although the conditions in fact are different in cause and effect. As already noted, the carpal tunnel includes the tendons supplying the fingers, as well as the median nerve and some small blood vessels. When inflammation and associated swelling takes place within the carpal tunnel, pressure is applied to the median nerve. This in turn gives rise to tingling and numbness in the distribution of the nerve, as well as pain, common at night, some loss of sensation, and weakness and clumsiness. For some reason which is not entirely clear, although perhaps from hormonal influence, the condition is more common in females, and tends to occur in pregnancy and after surgical removal of the ovaries.

Like tenosynovitis, the condition will respond to rest, although in some cases surgery will be required to release the carpal tunnel and allow subsidence of the swelling. The surgery is usually successful.

Epicondylitis

Epicondylitis is another common condition in manual workers. Although it is often called *tennis elbow* its occurrence is by no means confined to tennis players. It occurs because of repeated stress on the muscles which take origin at the epicondyles of the elbow and is found as a result of recurrent pronation and supination of the forearm, aggravated by excessive extension at the elbow joint. It too is relieved by rest and removal of the causative task, but it may also require treatment by a physician.

Cervico-brachial syndrome

Cervico-brachial syndrome occurs in the form of neck pain which radiates down one or both arms and may be accompanied with numbness in the hands. There is commonly limitation of neck movement, and aggravation of pain on neck movement. It tends to be aggravated where the general musculature of the region is poorly developed.

Multiple occurrence

More than one form of RSI can occur at the same time or sequentially in the same person. In particular, Zenz (1975) notes that carpal tunnel syndrome may be the result of a tenosynovitis of the tendons that pass beneath the carpal tunnel. In the original work of Phalen (1966), who was the first to recognize the occupational relationships, it was observed that out of 439 patients with carpal tunnel syndrome 96 had some form of associated musculo-skeletal disorder, including, but not confined to the following, not all of which are necessarily resulting from overuse:

Table 6.1. *Accompaniments of carpal tunnel syndrome (derived from Phalen, 1966)*

Condition	No. of Persons
Trigger finger	34
Periarthritis of the elbow	28
Tennis elbow	21
Tenosynovitis (and de Quervain's syndrome)	10
Thoracic outlet syndrome	5
Calcific tendonitis	3

The occurrence of multiple conditions was also reported by Brown and Dwyer (1984) in their study of 74 female workers who developed RSI over a two-year period. They noted several patterns of occurrence, namely:

(a) women with one initial problem continued working and gradually developed others;
(b) rest of the affected part was accompanied by development of other problems in the same limb;
(c) attempts to rest one arm were followed by the development of symptoms in the other arm;
(d) severe peripheral symptons were nearly always followed by development of strain in neck and shoulder musculature;
(e) inadequate rest from repetitive work following acute trauma or surgery to the arm led to the development of symptoms of RSI.

Occurrence in industry

In 1700 Ramazzini, the great Italian physician, who is considered to be the founder of occupational medicine, gave the first account of overuse injury, reporting its occurrence in clerks and scribes. Sporadic reports appeared thereafter in various forms of terminology including de Quervain's disease, bicipital tendinitis, and so on. It was not until the

1960s, however, that what amounted to an epidemic was reported from Japan. This was followed by further 'epidemics' in the United States, Europe, and Canada. The occurrence is widespread, and in Japan it showed a distribution affecting 16–28 per of keypunch operators, 13 per cent of typists, 10 per cent of calculator operators, 11–16 per cent of cash register operators, 12 per cent of packing machine operators, and 16 per cent of assembly line operators (McDermott, 1986).

In 1980, approximately 23 200 reported occupational injuries in the United States were associated with repeated trauma, while in New South Wales there was an increase of 220 per cent in the 1979–1980 period as compared with the 1970–1971 period. (Brown et al., 1984). Unfortunately, however, most definitive statistics are unavailable.

It will be apparent from its origins that RSI tends to occur in labour intensive industries where the work may not necessarily be categorized as heavy, but where there is commonly considerable use of the hands and arms, either in repetitive movements or some form of static holding. It can, of course, also occur in heavy industry. The worker may often be working in a seated position. Indeed, the seat may contribute to undesirable arm motions requiring raised elbows or work with the hands held forward of the body.

Not uncommonly RSI is found in electrical and electronic assembly work where operators sit for an 8-hour day at a work bench winding wires or using small hand tools. It can indeed be found in any assembly or fabrication work where there is much manipulation or hand tool usage, associated often with awkward posture, as well as lifting, bending, and stretching.

Poultry and meat flensing and packaging is a specific industry where RSI is common (Armstrong, et al., 1982). Much of this occurs from the use of knives and other small tools, often with the arms raised, and in a cold environment and may be aggravated by a need for heavy lifting of carcases.

Armstrong and his colleagues noted that the boning department accounted for the largest percentage of injuries/conditions, with 17.4 case per 200 000 hours compared with a plant average of 12.8. Kivi (1984) found the prevalence of epicondylitis and tenosynovitis among meat cutters to be 89 per cent and 45 per cent respectively. He also noted, that of all cases in all the occupations studied, the upper limb was involved in 93 per cent with the forearm in 63 per cent, tenosynovitis of hand tendons in 58 per cent, and epicondylitis in 24 per cent.

Manual sewing, particularly of leather, canvas, and other heavy material, requires intensive manual manipulation. It, too may have to be conducted from an awkward posture, with the arms upraised or the hands forward. Each of the latter requirements throws a strain on the arms, shoulders, and neck.

Of recent years, however, the condition has spread from the shop floor to the office, and now it is becoming common among word and data processors at video display units. In fact any kind of keyboard can be invoked as being incriminated, even a musical keyboard, or for that matter, any musical instrument (Fry, 1986), where the stress is applied to the fingers, wrists, forearms and ultimately to the neck. For the video display operator copying from documents an additional stress can be imposed on the neck as the operator moves his/her head from side to side checking document, screen, and keyboard.

A modified type of keyboard operation is found among workers on semi-automatic postal sorting machines, where the coder sits at a desk operating a keyboard to record the postal code from the envelopes presented to him/her. Although this is a one-handed operation it presents the same type of problems as a two-handed; indeed the problems could be even greater since most of the activity takes place with one arm (Arndt, 1981).

Many other tasks, of course, in variety of different occupations, can provide the conditions under which RSI can occur. The foregoing are merely a few of the more common.

Causative factors

The actual pathology is not at all clear, although swelling and inflammation are no doubt involved. Even the relevant influencing factors have not been very clearly defined, such as frequency of repetition, the total number of movements, and the duration of the total exposure on a daily, weekly, or longer basis. Clearly the more intense or prolonged the exposure the greater is the likelihood of developing problems. Yet there is no doubt some persons are intrinsically more susceptible than others. In all situations of environmental stress there is a wide variation in human susceptibility and human response, and RSI is no exception. Experience demonstrates that among a group of workers doing the same work, using the same equipment, and under the same conditions, some will develop RSI and others will not. It is not possible to define who will be affected in different circumstances.

Thus no relationship can be shown for example between such variables as height, weight, and the occurrence of RSI (Kuorinka and Koskenin, 1979), nor with hand size and greater or less use of force (Armstrong and Chaffin, 1979). Indeed Hymovitch and Lindholm (1966) point out that since there is no way of identifying persons who are particularly susceptible to this type of injury, early case finding is required to permit removal of the worker from the offending work activity.

There are, however, a number of factors which can determine, at least in part, the susceptibility to, and potential severity of, RSI. These can be classified as:

* factors intrinsic in the worker
* ergonomic factors
* environmental factors
* procedural (administrative) factors

Each of these will be examined in turn.

Factors intrinsic in the worker

Age

Although occasionally studies show clusters that might appear to be age specific, there is little evidence that age by itself is a factor in influencing the occurrence of RSI, with the possible exception of carpal tunnel syndrome, which tends to be more common in middle-aged and older women. It is possible that in industry, however, workers who have shown a susceptibility towards RSI may leave that kind of work at a relatively early age and seek work less physically demanding.

Sex

RSI, particularly in the form of carpal tunnel syndrome, tends to be 2–10 times more common in middle-aged females (Armstrong and Chaffin, 1979), particularly among those who have had removal of the ovaries. Why this should be is not clear. It might be that the occurrence is influenced by a change in the concentration of circulating hormones occurring after surgery, although the effects of a non-surgical menopause would not appear to be similar. The fact that women tend to be physically weaker and less well muscled than many men may also affect the occurrence.

Experience

RSI is more common mong inexperienced workers, either new to the job or returning after lay-offs, the more so if they have a poorly developed musculature, particularly of those muscles required for the task, or if their physical condition is poor.

Ethnicity

In the Australian experience, at least, (Mathews and Calabrese, 1982), the condition tends to occur more commonly among immigrants of non-European origin, particulary if they are migratory. This phenomenon, however, may well be a reflection of smaller build in many cases, and perhaps a lower nutritional status among migrant workers. In addition it may be that these workers tend to accept some of the less desirable jobs.

Leisure and home

One of the important factors in the minimization or prevention of RSI is rest from the responsible task. However, if the recreational or leisure activity of the worker requires the same type of action as that causing his/her problems, whether it be throwing a ball, polishing a floor, digging a garden, or restoring a house, then the condition caused by the work may be aggravated by the leisure.

Ergonomic factors

Ergonomic factors are those elements in the work station, the tools, equipment or machines, or the task itself that can influence the occurrence of RSI.

Work station

The design of the work station is significant in that an improper design tends to force improper work procedures on the worker. For example, if a work surface is too high it forces the worker to operate with the elbows raised, thereby increasing static work. The same effect occurs when the seat is too low. If the work surface is too low, on the other hand, he/she is forced to assume a stooped posture from the standing position, and so on. Thus adjustability, both in seat and work surface, is paramount. Ergonomic problems of work stations are considered in detail in Chapter 4.

Tools, equipment and machines

The use of tools that force the hand and arm into extremes of joint angle is one of the major factors contributing to RSI. Many tools, such as the standard screwdriver or pliers, are actually inappropriate to the ideal function of the hand-arm segment, forcing the operator to work with elbow upraised and hand deviated to the little-finger side (ulnar deviated). 'Bend the tool, not the hand' is a classic adage, and various tools have been designed to this end. The problems of hand tools are discussed in Chapter 10.

Machines can provide a different kind of problem. Many are not designed with human use in mind. Work height may be too high, or too low; operation may require repetitive raising of arms to operate switches so-placed in a misguided attempt to ensure safety; some operations indeed require repeated major body motions as the operator makes needed adjustments, while maintenance may demand unreasonable postures, and so on. Unfortunately, because the machine may be large, rigid, and expensive, a feasible solution to problems of the kind often cannot be found. Equipment design is dealt with in Chapter 9.

Not the least are the problems of the keyboard operator. These are considered in Chapter 11. The classic typewriting position forces the hands into extremes of movement, in a forward position, with the wrists and elbows unsupported. It is not at all surprising that tenosynovitis, along with elbow, shoulder and neck disorders are common among video display terminal operators who spend long hours in front of their equipment (McDermott, 1986). Much attention has been directed to this area, leading to the provision of adjustable work benches, seats, separate low-profile adjustable keyboards with wrist rests, adjustable displays, and so on, but many problems still exist, particularly where poorly designed equipment is casually located on an unsuitable desk or table and operated by a worker on an unsuitable seat.

Task

The same principles apply to the design of tasks. The way in which a task is done is determined to a large extent by the design of the work station and equipment. However, in setting up a task, consideration must be given to avoiding wherever possible the activities and motions which may induce the occurrence of RSI.

Therefore tasks should be designed to favour maintenance of optimum postures and limb positions, with the avoidance of repetitive motion, extremes of joint movement, and bending at the wrists. Static work should be minimized. Continued pressure on soft tissues by tools or handles should be avoided, along with the need for tissues to absorb impacts and vibration.

Environmental factors

While environmental factors are not directly causative of RSI they can influence its onset and severity. For example, there is an earlier onset and quicker progression towards a more severe condition in the presence of cold, and also of noise (Mathews and Calabrese, 1982). In the use of hand tools the conditon is further aggravated by the occurrence of vibration.

Administrative factors

Besides the constraints imposed by hardware, and the limitations of human capacity, there are a number of administrative factors that are influential in contributing to the occurrence of RSI.

The needs of work and task organization have already been referred to. Excessive pace demands, however, whether imposed by hard driving supervisors, or the push of financial incentive systems, may force the worker to work beyond his/her physical capacity. In other words the rate of doing the job is increased, which in turn leads to a faster repetition of movements. The effects of the increase in work rate can be aggravated by poor job training, and consequent development of poor work habits. Various administrative remedial actions also need to be considered and these will be examined in the next section.

Prevention and control

As with any investigation of a potential occupational health problem, it is first of all necessary to determine the scope of the problem that exists, or in other words, to obtain some understanding of its type and extent. There are five elements in the approach to prevention and control. These are:

* work hazard analysis
* health surveillance
* education
* inspection
* remedial action

Work hazard analysis

Work hazard analysis involves, firstly, a definition of those jobs which are relevant to the occurrence of RSI. This is most easily attained by way of discussions with managerial, supervisory, safety and health personnel, as well as by the examination of health and safety statistics, and personal observation.

The next step lies in the visual analysis of the various tasks, sub-tasks, and activities in each job, with the assistance of workers and supervisory personnel, to determine the type and extent of the hazard or potential hazard in each activity, from the point of view of causation of RSI. Tasks, sub-tasks and activities may be recorded on a form designed or obtained for the purpose, utilizing three columns, one for the tasks, one for the hazards, and one for the remedial action that should be taken. The third step lies in providing the remedial action. It may be useful to back up direct visual observation with video recording for subsequent re-examination.

Health surveillance

Health surveillance accomplishes more than merely providing shop floor health care, or reference to an external health care system. The presence of a trained and experienced occupational health nurse, perhaps backed up by an experienced occupational health physician, assists greatly in the development of an atmosphere of positive health and safety within a plant.

The duties of health care personnel in this connection include much more than the provision of first aid. They provide among other activities development of a reporting system for the injured; health and safety monitoring on the shop floor in conjunction with other health and safety personnel; provision of rehabilitation services, physiotherapy, and occupational therapy and the development of a system of health oriented records.

Education

Education is a vital part of health and safety awareness. In addition to discussions, films, and videos about the cause and prevention of RSI, and about the best methods of manual materials handling, educational programmes should be directed at outlining the relevant responsibilities of both management and labour for the control and prevention of RSI. Consideration should also be given to the role of government and legislation, and the education process should take place within the total context of safety and health awareness.

The educational programme can be conducted in-house by knowledgable instructors, along with video or other audio-visual aids. Alternatively, advantage can be taken of courses mounted by accident prevention associations, government and other bodies.

Inspection

Without routine monitoring of task activities and periodic formal inspections, the value of an RSI prevention programme soon recedes. The inspections should be conducted routinely by supervisors, along with periodic formal inspections by such personnel as health and safety committees, where they exist, as well as the person responsible for health and safety, the occupational health nurse, and so on, with the object of ensuring that the most suitable work procedures are being followed, that the proper personal and work equipment is being used and of course to identify new problems as they arise.

Remedial action

The nature of the remedial action depends on the problem, and in particular on its causative and contributory circumstances.

Where appropriate it might entail a change in administrative policy, a change in work procedure, or some engineering modification. Administratively, if the pace of work is excessive, whether because of demand or incentive, it might be valuable to reduce the pace of the work, along with an increase in the number and/or length of rest pauses, as well as to make an improvement in training with respect to both skill and physique. Consideration might be given to introducing job rotation, whereby a worker spends part of his/her shift on an RSI-inducing task and part on a non-RSI task, or even on a task where the muscle load is differently distributed.

Of importance also is the need for controlled rehabilitation on return to work, whereby the worker not only returns to his work on a gradual basis but at the same time undergoes physiotherapy to retrain his injured muscles. Indeed, physiotherapy, in the form of a controlled exercise training programme for all workers, injured, or not, has been successfully used in the prevention of RSI in injury-prone situations (personal experience).

Changes in procedure might include, for example, a change in the design of a fabrication process, such as assembly of an electronic device, so as to reduce the requirement for manipulating small hand tools, or it might include changes in the nature of the task so as to eliminate lifting the arms against gravity, and so on. Or it might be as simple as replacing hand operated tools with power tools.

Engineering changes can be more complex and expensive. At one extreme one can consider automation of a process, which of course eliminates the problem of concern, but may also eliminate many of the jobs. Even semi-automation, to remove, for example, the need for recurrent lifting, or excessive extreme arm movements, can be very helpful.

Much can be gained by designing the work station to meet ergonomic requirements, particularly those relating to working heights and arm reaches. Hand tools and keyboards can be reconsidered, and designed or selected to minimize joint and soft tissue stress. Powered lift or support devices can be added to reduce repeated lifting, and so on. Each situation has to be treated on its merits. What may be appropriate for one set of circumstances may be inappropriate for another. But wherever there is a problem some solution is applicable, and it is the responsibility of everyone concerned, manager, supervisor, safety officer, health personnel, health and safety committee, and the worker himself to define that problem and seek a solution.

Chapter 7
Human stress, circadian rhythms and shift work

The nature of stress

Much of the material from which this chapter is derived comes from the work of Fraser (1983a) published as a monogram by the International Labour Office, Geneva, outlining the interrelationships among stress, work, and job satisfaction. Because of its connotations and the diversity of its usage, stress has always been a difficult term to define. To some, stress describes the state of a physical body which has been subjected to pressures or forces close to or beyond its tolerance; to others, the term describes the phenomena which produce these pressures of forces. Stress can be seen as a physical entity associated with physical changes; or it is subjective and associated with psychological and emotional conditions. To some, stress and strain are synonymous; to others, they are descriptive of cause and effect.

Regardless of whether one refers to the condition in man as stress or strain, there is a tendency among many to consider it as *pathological* human response to psychological, social, occupational and/or environmental pressures. This, however, is not so. Hans Selye, the founder of stress physiology, has stated (Selye, 1974): 'Contrary to widespread belief, stress is not simply nervous tension nor the result of damage. Above all, stress is not necessarily something to be avoided. It is associated with the expression of all our innate drives. Stress ensues as long as a demand is made in any part of the body. Indeed, complete freedom from stress is death!' Consistent with this viewpoint, although perhaps inconsistent in terminology, the International Symposium on Society, Stress and Disease, sponsored by the World Health Organization and the University of Uppsala at Stockholm, 1972, adopted as a definition the statement that: 'Stress is the non-specific response of the organism to any demand made of it', thus placing themselves in the camp that considers that stress is the response and not the causative state.

While perhaps the point is academic, pursuit of which may lead to fruitless semantic discussion, the viewpoint that stress is the resulting and not the causative state is in direct opposition to the view of the physical scientists and engineers who originally defined the term. In fact, even Selye himself has publicly stated (Selye, 1973) that where he originally used the term stress he should have used strain. Perhaps, however, the term in its above-defined meaning is now too well established in psychophysiology and psychosomatic medicine to be changed. In physical terms, however stress exists when a force is applied to distort a body. The effect is manifest as elastic or non-elastic distortion and is measurable as strain. The relationship between stress and strain can be plotted in the form of a curve, as shown in Figure 7.1.

The foregoing curve demonstrates that, with the initial application of stress to an elastic object, some unmeasurable change will occur in the object, probably not rectilinear in its relationship to the causative stress. As the stress continues there is a measurable strain directly proportional to the stress. This is the region of elastic distortion, and if the stress applied to this level is removed then the strain will also disappear. In other words, at this level conditions are reversible. There is some point, however, at which the stress is such that there is no longer a direct proportional relationship between stress and strain. The stressed object changes its character and the resulting strain is not reversible. An elastic band, for example, will stretch beyond recovery.

There is a psycho-physiological analogue of this stress/strain relationship. The initial portion of the curve represents a non-measurable perhaps cellular or biochemical, response to the mildest of human stress. As the stress increases, either in intensity or duration, various physiological and psychological changes can be observed in relation to the stress, which

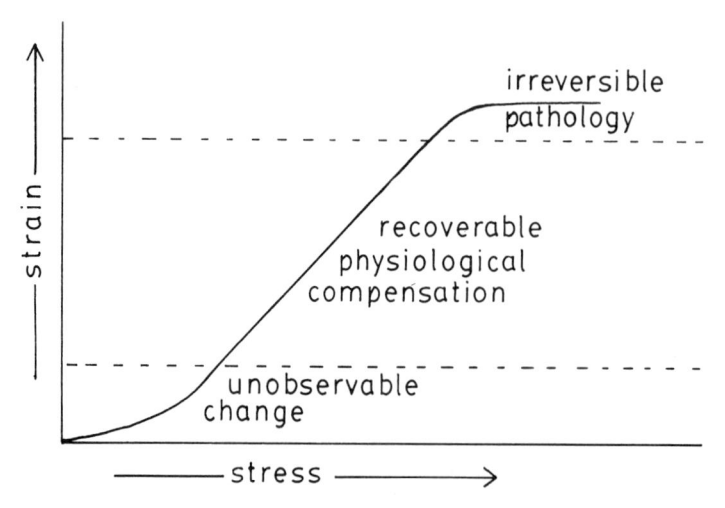

Figure 7.1. Schematic stress/strain relationship.

are reversible when the stress is removed. With still further increase in stress we enter the realm of pathology, beyond the level of adaptation, where the changes are not reversible, and some trauma, mental or physical, results.

By this analogy, stress describes the stimulus state, and strain the response, although the difference is more one of usage than meaning.

Homeostasis and feedback control

Regardless of the semantics involved, to examine the psycho-physiological mechanisms of human stress it is necessary first to examine the mechanisms responsible for control of the human organism in the relatively unstressed state.

As noted earlier, man is a complex of systems and sub-systems. Co-ordinated control is exerted over this complex to maintain it in a state of dynamic equilibrium to which is given the name *homeostasis* (Greek: 'steady state'). The major purpose of human physiological function is to maintain homeostasis, or more specifically, to maintain the internal environment of the body in a state of chemical and thermal stability in the face of constant change, or threat of change.

Some of this control is inherent in the individual cell, which is itself a complex system, the functions of which can be modified by specialized chemicals, or *hormones*, secreted by glands of the *endocrine system*. Overall control and co-ordination of the entire system is exerted via the central nervous system, through the network of the *autonomic nervous system*.

The principles of control are the same as for any other system of interactive variables. Preset standards are established by the autonomic nervous system at levels within which a variable may function. The resulting function is monitored by a sensory mechanism, again by way of the autonomic nervous system, to various co-ordinating centres deep in the brain. In these centres incoming information is compared with the preset standards. If significant variation is found, instructions are issued via the nervous system and glands to effect appropriate change in function. There is thus a closed-loop operation with an input, some means of processing, and an output, which in turn affects the input. The loop, and other interacting loops, are activated to reach and maintain equilibrium. Thus, there are acceptable limits within which the body function may be permitted to vary. Stress, however, provides the stimulus for change in those limits, or, in other words, for adaptation.

Neuro-endocrine control system

The ultimate co-ordination and control of involuntary bodily functions comes under the direction of the complex of nerve centres, nerve fibres, and glands known collectively as the *neuro-endocrine system*. This system controls and/or co-ordinates all metabolism, other than that inherent in the individual cell, and all bodily function including the co-ordination of some of the function that is conducted voluntarily, for example, ensuring an appropriate blood supply for muscles engaged in voluntary movement. It is therefore responsible for maintaining homeostasis in the face of internal and external demand, including the demands of stress.

The system in turn comprises two sub-systems, namely, the autonomic nervous system and the endocrine system.

Autonomic nervous system

The autonomic nervous system is a functional system of interlinked nerve cell clusters, or centres, in the base of the brain, with an input and output network of nerve fibres leading to and from these centres and bearing information about, or instructions concerning, such matters as heart rate, blood pressure, breathing rate, muscle tone, sweat production, urinary output, body temperature, and so on. There are, for example, respiratory centres, vascular (vasomotor) centres, thermal regulation centres, even entities like cough centres, along with their associated networks.

The autonomic nervous system can be further divided into two parts, namely the *sympathetic nervous system* and the *parasympathetic nervous system*. In general, but not always, the sympathetic nervous system serves to increase a given function, while the parasympathetic nervous system serves to decrease it. In practice they work together in opposition in a kind of 'push-pull' manner.

The sympathetic nervous system derives from centres in the brain and sends (or receives) nerve fibres to the spinal cord and thence into the chest. There they form two chains of nerves which proceed downwards, one on each side of the spinal column. These chains send out nerve fibres drawing from the different centres in the brain which unite in a multiplicity of relays called *ganglions*. Fibres from the ganglions (post-ganglionic fibres) pass to the various body organs or parts. On stimulation these fibres release a chemical called *norepinephrine* (sometimes known by the proprietary name of noradrenaline) at the end of the fibre. This norepinephrine then acts either to stimulate another fibre at a nerve junction, or *synapse*, for onward transmission, or acts directly on the organ itself to increase its activity. The effects normally last only a few seconds or less. If the norepinephrine were permitted to remain at the site the effects would go on until the material was exhausted. Normally this is not desirable, and consequently the norepinephrine is destroyed as soon as it is formed. This is acccomplished by the presence of an enzyme, or chemical catalyst, called *orthomethyltransferase*. If the initial stimulus is sufficiently strong, however, it also stimulates one of the glands of the endocrine system, namely the adrenal gland (see later), to produce epinephrine (adrenaline) the effects of which last for minutes.

The parasympathetic nervous system derives from its centres in the brain to form a nerve called the *tenth cranial nerve*, or *vagus nerve*, which passes out of the skull down the neck into the chest and abdomen where it gives off various branches. The ganglions are found in the various organs, and the post-ganglionic fibres are very short. On stimulation these fibres release the chemical *acetyl choline*, just as do muscle nerves. The acetyl choline is rapidly destroyed, like the norepinephrine, by still another enzyme, *acetyl choline esterase* which has already been noted.

Commonly, an organ or body system is dominantly controlled by either the sympathetic or the parasympathetic. For example, blood vessel tone is predominantly sympathetic. Specific actions are illustrated in the following table:

Table 7.1. *Sympathetic and parasympathetic responses (derived from Guyton, 1984)*

Organ	Symp. Stimulation	Parasymp. Stimulation
Pupil of eye	dilated	contracted
Sweat glands	sweating	none
Stomach	decreased mobility	increased mobility
Sphincters	increased tone	decreased tone
Heart rate	increased	decreased
Blood vessels		
Muscle	dilated	none
Skin	constricted	none
Hair muscles	contracted	none
Adrenal secretion	increased	none
Mental activity	increased	none

The *Alarm Response* occurs with sympathetic stimulation. In the alarm response, as a result of a perceived threat, the sympathetic nervous system is activated strongly. The pupils are dilated, the skin becomes sweaty and slippery from increased action of the sweat glands, and turns pale, particularly in the face, from constriction of the blood vessels in the skin. The hair stands on end and makes the body of an animal, for example, appear larger. The heart rate and force of contraction are increased to meet physical demand, and the blood supply is diverted from the skin and internal organs to the muscles. Alertness and vigilance are increased. Observing the overall nature of this response, Cannon, the great American physiologist described it as preparation for 'fight or flight'.

Control of the autonomic nervous system as a whole is exerted by a special area at the base of the brain called the *hypothalamus*, which is in close anatomical relationship to the various autonomic centres. It also exerts control over the endocrine system and is itself a component of still another sub-system called the *limbic system*. The limbic system is the end result of an early evolutionary stage of the brain and in a primitive way is concerned with the same activities as the higher centres of the brain. It is responsible for the subjective response to, and controls activities for, such concerns as survival, reward, punishment, primitive emotions and behaviour, pleasure, pain, and so on.

The higher cerebral centres in turn exert an influence, and some control, on the hypothalamus and the limbic system so that there is invoked a feedback system which, with extensive training and practice, allows the mind to exert some action on normally involuntary functions, or indeed may influence the normal management of involuntary functions to induce so-called 'psychosomatic' disease.

Endocrine system

The endocrine system comprises a heterogeneous group of anatomically unrelated glands scattered throughout the body. The word *endocrine* means without ducts, or distributing channels, contrasting with *exocrine* which means with ducts. The salivary glands in the mouth are examples of exocrine ducts. The products of endocrine glands are passed directly into the blood stream whereas the products of exocrine glands such as the salivary glands in the mouth, are passed by way of ducts.

The products of endocrine glands are called *hormones*. These hormones act selectively on appropriate body cells to produce a physiological effect, such as:

(a) control of rate of chemical reactions
(b) control of transportation of chemicals through otherwise impervious cell membranes
(c) participation in specific metabolic activities as constituent chemicals.

The various types of effects are determined by the hormone and the cell or cells on which it acts. There are six sets of endocrine glands, some of which occur in pairs. Not all are directly involved in the human response to stress. The endocrine glands comprise:

1. *Hypophysis* (pituitary): responsible for coordination of other gland activities and additional special functions.
2. *Adrenals*: responsible for mediating the stress response.
3. *Thyroid*: responsible for modulating the metabolic rate
4. *Parathyroids*: responsible for modifying the metabolism of calcium and bone.
5. *Pancreas*: exocrine and endocrine gland. The endocrine portion is responsible for controlling the transport across cell membranes of glucose and amino acids. The exocrine portion secretes digestive enzymes.
6. *Testicles and ovaries*: responsible for sexual maturation, development and function.

While the actions of all the endocrine glands affect every aspect of body function, the hypophysis and the adrenal are the two which are primarily involved in determining the human response to stress, along with the sympathetic nervous system. The hypophysis has a variety of functions, each controlled by its own hormone. It is in fact two separate glands joined together and located underneath the brain in close proximity to the hypothalamus and the autonomic centres. One portion of the hypophysis (the *adenohypophysis*) is responsible for coordinating the activities of all the other glands by way of secretion of specific hormones, one of which governs each of the other glands. It is in turn stimulated by nerve signals from the hypothalamus but is also influenced by feedback of hormones from the glands it coordinates. Thus there are complex feedback loops which involve external environmental stimulation and internal (body) stimulation to the brain, conducted through the limbic system to the hypothalanmus and thence to the autonomic nervous system output and the hypophysis, followed by hormonal and neural feedback.

In particular, both the hypophysis and the autonomic nervous system act on the adrenal glands. There are two adrenal glands, one located on the top of each kidney (hence ad-renal). Each gland has a shell, or *cortex*, and a core, or *medulla*, which act as independent glands producing different hormones. The medulla produces *epinephrin*, (sometimes known by the proprietary name of *adrenalin*), which acts to produce similar effects to, but more intense than, stimulation of the autonomic nervous system, that is, the effects that ready the organism for 'fight or flight'. The cortex produces several types of hormone collectively called the *corticosteroids* of which the most significant from the point of view of stress is *cortisol* (hydrocortisone). Cortisol acts, in effect, to protect the body against excessive physiological stress reactions. Whereas epinephrin acts to prepare the body for fight or flight, the corticosteroid hormones have been said to act as an asbestos blanket against fire.

The actual mechanism of action is complex. When a stress stimulus such as pain, cold or anxiety is perceived in the brain, a nerve signal is transmitted to the hypothalamus which in turn stimulates the hypophysis. The hypophysis releases a hormone known as *corticotropin* which is carried by the blood stream to the adrenal glands glands and causes the adrenal cortex to release cortisol and related hormones. These hormones are of two types, the *syntoxic steroids* and the *catatoxic steroids* (Selye, 1973). The synotoxic steroids act as 'tissue tranquillizers'. They raise tolerance to tissue damage, reduce the effects of inflammation, and minimize the severity of physiological responses. The catatoxic steroids on the other hand induce the production of enzymes to destroy harmful materials, both chemical and biological, and to increase resistance to the causative stress. Cortisol in particular has a specific action in inhibiting or reducing inflammation. Thus both cortisol and epinephrin are concerned in the body stress response, and in the adaptation process that takes place in the face of continued stress. A model of this process has been described by Hans Selye as the General Adaptation Syndrome.

General adaptation syndrome

Based on work begun in the 1930's, Hans Selye in 1950 published a comprehensive theory of the human response to stress which he termed the General Adaptation Syndrome, and which he continued to refine. To put it briefly, and in an over simplified manner, he states that there are three definable stages to the human stress response, namely the stage of Alarm, the stage of Resistance, and the stage of Exhaustion.

Stage of alarm

The stage of Alarm has already been considered. It occurs as a reaction to a perceived threat and is characterized by pallor, sweating, increased heart rate and redistribution of blood to the muscles. It is normally short in duration (a few seconds to a few days). It readies the body for whatever action is required in the face of the threat.

Stage of resistance

Should the stress persist beyond the immediate threat a physiological adaptation process begins. This has the objective of adapting the body to the new situation. It is termed by Selye the stage of Resistance. The resistance is a resistance to those factors which are considered to constitute the stress. During the stage of resistance most of the signs and symptoms associated with the alarm reaction disappear as the body develops its adaptation. The capacity to resist, however, is limited.

Stage of exhaustion

Should the stress be sufficiently prolonged, the stage of resistance will gradually be replaced by a stage of Exhaustion. Exhaustion in this connection does not mean fatigue, but refers to exhaustion of the body's resources and is characterized by some form of failure of the body's defence mechanisms. It is argued by Selye that this stage may be associated with the development of psychomatic diseases, for example, gastric ulcer, cardiovascular diseases, inflammation of the large bowel (colitis), skin conditions such as psoriasis, and so on. Cause and effect relationships of this type can be demonstrated in small animals, but since no *direct* cause and effect relationships have been demonstrated in humans, and since destructive testing of humans is frowned on, some workers in the field oppose Selye's view in this regard.

Selye (1974) goes on to suggest that the tendency to develop a specific psychomatic disease is determined by a pre-conditioning of some 'target organ' in the individual, by such factors as heredity, personal habits such as smoking, drinking of alcohol, or abuse of diet, previous exposure with tissue damage, or the specific actions of the causative stress such as a burn leading to a spreading skin ulcer.

It would appear further (Selye, 1960) that certain diseases occur not primarily because of any specific pathogen or trauma but because of a faulty adaptive response to the stressor effect of some otherwise relatively harmless stress. For example various emotional disturbances, headaches, insomnia, abdominal disorders, and so on, as well as recognized clinical entities such as rheumatoid arthritis, or certain allergic diseases, and sundry cardiovascular and renal diseases, are alleged to be initiated not directly by some external agency but as a result of faulty adaptive mechanisms.

Human stress, then, exists when an event or state occurs which disturbs homeostasis. The extent of the resulting response, or strain, is determined by the severity of the stress. The response is not specific to the stress and occurs regardless of its nature. It is mediated via the neuro-endocrine system and can be categorized in three stages of adaptation. Faulty adaptation, or failure of the resistive mechanism, is alleged to lead to disease. A given form of stress may in addition generate a specific response superimposed upon the non-specific, and may indeed overshadow the non-specific. Psychosocial stress induces a non-specific response. Some aspects of psychosocial stress, and in particular the problems associated with shift work and with work relationships, are considered in the next sections.

Circadian rhythms and shift work

Although shift work has always been common in many industries, notably mining and heavy manufacturing, there has been a marked increase in shift work throughout the industrial world in the last 25 years. The reasons lie partly in societal needs, and partly in economics. In terms of societal needs, there is an increasing demand for 24-hour services in, for example, transportation, provision of power and light, health care, security, and even consumer sales in grocery and other stores. From an economic viewpoint, where more and more plants rely on expensive, sometimes computer-mediated, machinery to achieve these purposes it is necessary to keep that machinery operating as much as possible to realize maximum profit.

As a result, an increasing number of persons is being called upon to undertake routine daily work during periods when the majority of persons are in one way or another at leisure; or worse still, workers are being called upon to switch fairly rapidly from working during a day shift to, for example, working on a night shift.

The problem, however, is not merely one of social inconvenience, it involves a disruption of basic physiological functions collectively known as *circadian rhythm* (Latin, *circa dies*, approximately a day), or *diurnal* (daily) cycles.

Circadian rhythm

Circadian rhythm is the term given to the basic patterns of physiological function which vary in their rates or other values throughout the day-night cycle. Normally the rhythms are such that functions are slowest, or least, during the night sleep period, and fastest, or greatest, during that portion of the day when alertness is most needed. There are other sub-patterns that can be determined in certain functions.

Probably all physiological and psychological function is affected to a greater or lesser extent but some phenomena are more obvious or more readily measured than others. The obvious easily measurable functions include heart rate, body temperature, blood pressure, metabolic rate, and urine production, which tend to be least during the night hours, reaching a low about 4 a.m., and greatest during the alert hours, reaching a high about 4 p.m.

In addition there is a significnt diurnal variation in the secretion by the kidney of various urinary electrolytes and body wastes, again being least during the night and most during the day. Other biochemical changes are found in the secretion of hormones. Notable among these is epinephrin, production of which is reduced during the night hours.

Behaviour, in the psychological sense, is also affected. For example, as is well known, sleep demand and sleep capacity is greatest during the night hours and is reduced during daylight. Similarly, alertness and mental capacity are reduced in a concomitant fashion while physical capacity is changed in the reverse manner. One interesting measure of a phenomenon akin to fatigue is changed quite markedly, namely, flicker fusion frequency. Flicker fusion frequency is the term given to the ability to fuse a flickering light into a continuous signal. With fatigue, and other conditions, the capacity to fuse is reduced, that is, the flicker requires to be faster before it can be fused. Fusion capacity is greater during the daylight hours.

Cycle periodicity

The periodicity of the cycle is not the same for all functions but tends to be in the range of 22–25 hours. The periodicity is determined by a kind of natural internal clock probably modulated by the pineal body of the brain which is an organ that appears to govern the general human response to light and dark. The 'clock' in turn is maintained in its time reference by what are termed *zeitgeibers*, or 'time-keepers'. Various time-keepers can be identified, some of which are physical and directly perceived by the senses, and some of which are psychologically inferred. The primary time-keeper is the physical night-day cycle itself. In the absence of the night-day cycle, either artificially or through residence in polar environments, the body has to rely on other sources such as clocks, or knowledge of clocktime, social contacts, work/leisure cycles, and so on.

Loss of all time-keepers over a period of several weeks gives rise to *desynchronization* of the various circadian rhythms. Normally these rhythms are kept synchronized by the time-keepers but when desynchronized each, over a period of time, adopts its own cycle. This results in disturbance of physiological homeostasis, and in turn leads to an impaired level of performance accompanied by subjective discomfort. The function is rapidly restored with recovery of the time-keepers.

Shifts in circadian rhythm with night work

Circadian rhythms can be grossly disturbed by night work. One of the major problems of night work is that it requires the worker to perform at a high level during a period when, at least initially, he is not prepared for it and is undergoing adaptation; secondly, shift work imposes the need for adaptation, and frequently imperfect adaptation, on a worker, sometimes for relatively short repeated periods. While complete reversal of circadian rhythm probably can occur in the industrial situation it is, to say the least, uncommon, and is normally incomplete.

When work is continued on a regular basis for a sufficiently long time, the rhythm can be altered to a certain extent so that the fall in temperature, for example, occurs during the day, in day sleep, rather than at night during working hours. The physiological functions affected, however, do not all change at the same time; various bodily and mental functions appear to change at different rates. A shift is normally apparent in about 6–10 days of continuous night work, but does not become complete even after three weeks of night shifts (Knauth and Rutenfranz, 1976). This complete change, however, does not occur during shift work. Instead, the temperature curve flattens out so that the peaks and low points are not clearly differentiated. Under these circumstances the worker is still being exposed to the stimulus of everyday 'zeitgeibers' from the society around him. With a complete change in time zones, as for example in moving across the Atlantic Ocean, a complete change can occur.

In the industrial world achieving even partial adaptation of circadian rhythm is in practice very difficult. Most shift workers have two or more days off in a week during which time they tend to follow the same rhythm as the society in which they are living. They readapt to the normal routine much more quickly than they adapt to the artificial routine, and consequently never become more than partially adapted to shift work.

Sleep

Sleep is one of the most significant factors in assisting the maintenance of circadian rhythms, and at the same time is one of the physiological phenomena most readily disturbed in shift work. Sleep has its own intrinsic rhythms, passing in the course of a night through several stages of varying depth and corresponding changes in physiological function. These stages include a mandatory requirement for what is known as Rapid Eye Movement (REM) sleep. This is a form of deep sleep paradoxically accompanied by muscle twitchings, increased heart rate, and rapid movements of the eye visible beneath the eyelids and associated with vivid dream states. According to some psychologists, although not accepted by all, it would appear to be a stage of sleep necessary for the welfare of the person. The capacity for REM sleep is diminished in night workers who are sleeping during the day.

Two large studies comparing day workers and shift workers have shown that the frequency of disturbed sleep is some 6-10 times higher for night workers than for day workers (Andersen, 1957, Ulich, 1964). Shift workers build up a sleep debt during the week and then sleep longer during their days off to catch up. It is not possible to 'store' sleep in advance. Lille (1967) analyzed the daytime sleep of 14 regular night shift workers and showed that the average length of daytime sleep was six hours on working days, and 8-12 hours on rest days, with the longer sleep on the second of the two rest days.

Extended working hours

Of recent years there has been a move among some major corporations away from the traditional 8-hour day towards a 12-hour day with more time off. There is no doubt that this regimen appeals to the worker and to organized labour, but, as pointed out in the discussion on fatigue, extension of manual working hours to this level on a regular basis 5-7 days per week, and particularly heavy manual work, tends to generate more fatigue and leads to lower levels of performance of manual work. Where the work is of a sedentary or monitoring nature, provided close vigilance is not required, there is less or no observable decrement, however, and where three to four days of shift work are followed by three to four work free days there appears to be little or no decrement. Indeed, in this connection, Lees, Workman and Laundry (1987) conducted an extensive comparison study on over 500 workers in a major manufacturing company which, during the course of the study, changed from an 8-hour system to a 12-hour system. Data were collected during the 8-hour shift period and again after stabilization following the initiating of the 12-hour shift period they found that:

(a) Male injury rates were significantly lower on the 12-hour shift than on the 8-hour for all first-aid cases and medical treatment cases.
(b) Male injury rates were not significantly different on the 12-hour shift system for restricted workday case injuries.

(c) Male injury rates were significantly higher than female injury rates on the 8-hour shift system for all first aid-cases and medical treatment case injuries.
(d) Male workers had significantly higher injury rates than females on 8-hour, but not on 12-hour shifts.
(e) Age and experience did not significantly affect injury incidence on either shift system for any severity of injury.
(f) Injury incidence was signficantly higher on the day shift than on either the afternoon or night shift for the 8-hour shift system.
(g) Injury incidence was significantly higher on the day shift than on the night shift for the 12-hour shift system.

Effect of shift work on health

In a study conducted by the National Institute of Occupational Health and Safety in the United States (Tasto *et al.*, 1978), some 1200 nurses and 1200 food processors were surveyed across the United States to determine the effect of shift work on health. The differences were most marked among those working rotating shifts, that is, working a few days on one shift and then moving on to another. These workers had significantly more visits for personal health care than the other workers and complained of a variety of vague but real problems, including digestive trouble, respiratory trouble, anxiety patterns and sleep disturbances. General effects on health tended to include weariness, irritablity, depression and loss of vitality, and a disinclination for work. Disturbance of the periodicity of the menstrual cycle may also be observed in women.

Effect of shift work on age

Resistance to the stress induced by night shift work, sometimes known as 'night stress', declines with age. Studies have shown that, for shift workers over the age of 45, psychic and somatic complaints increase and sleep length decreases with night work (Akerstedt, 1976). Indeed, the ability to sleep even for non-shift workers decreases with age, as does the ability to adapt easily to changes in time zone of greater than 2–3 hours.

Social aspects of shift work

Over and above the effect of shift work on physical and mental health, the social disruption that it induces is a significant feature. One of the major problems is the inevitable dislocation of family life that occurs. If both spouses are working on different shifts contact may be minimal. On the other hand if the wife is acting as a homemaker and mother she is not only left with most of the day-to-day responsibility but her husband is also expected to sleep during the bustle of daily living, perhaps with young children.

Of course not only is family life disturbed but also most contact with friends or associates outside of work is lost; nor is there opportunity for group participation in, for example, sports, institutions and religious activities. Even the after work drink in the local hostelry is not available. The worker, in fact, comes to consider himself as working in a kind of social isolation which for some people rapidly becomes intolerable.

Jamal (1987) researched workers across Canada for eight years in a survey of more than 2000 workers, and found that while daytime workers spend some 18 hours per week with their families, shift workers spend only 12.8 hours per week. At the same time, shift workers spend more time alone, 13.7 hours per week, as compared to 11 hours per week among day employees. In another study, by the Canadian Union of Public Employees in Winnipeg, Manitoba, a survey of 300 workers showed that 75 per cent were unhappy with their shift work schedules and would rather be working on day shift.

Effect of shift work on safety

The extent to which shift work can effect safety in the work place has been open to some argument. Until relatively recently no measurable effect was observed on the accident rate among different shifts, while other tangible evidence was not found to change significantly or in a manner from which one could reasonably infer a cause and effect relationship.

Within the last few years, however, extensive studies conducted by the National Institute for Occupational Health and Safety (Tasto et al., 1978) have indicated that those working the day shift have fewer accidents than those working second or third shifts, and also that those working rotating or swing shifts have more accidents than their counterparts on fixed shifts.

The net result of all the effects, however, is that two-thirds of shift workers suffer from some form of demonstrable ill-health and one quarter abandon shift work altogether.

Shift rotation

It is far from clear as to what should be the most suitable approach to shift rotation in industry. There is no doubt that some people are seriously affected by shift work; equally there is no doubt that some people enjoy shift work, particularly if it is accompanied by some financial or free-time bonus. As already noted, however, in the industrial environment full adaptation does not occur. Some authorities therefore tend to support a short shift rotation cycle on the basis that since adaptation is not complete even after several weeks there is no point in trying to achieve physiological adaptation at the expense of gross social disturbance. Hence, they recommend frequent shift rotation, for example 1–3 nights followed by 24 hours off. Still another schedule would involve work on two mornings two afternoons, two evenings, followed by two or three nights off. Workers in general tend to prefer rapidly rotating schedules over slowly rotating.

General recommendations

Certain general recommendations can be summarized. Firstly shift work should be minimized to the extent that it is feasible, and particularly to the extent that it involves nights. If short rotations are deemed necessary or are considered preferable, then night work should be restricted to a single night followed by 24 hours off, if possible. Wherever possible shift work should be organized to allow the worker to have a normal weekend. Regular schedules should be selected over schedules where starting times vary erratically from day to day. Finally, the medical history of a shift worker should be considered before placing him/her on shift. Particular care should be given to those with a history of nervous or gastrointestinal disturbance.

Chapter 8
Job satisfaction and work humanization

In his monograph on work and job satisfaction from which this chapter is derived, Fraser (1983) notes that there is a widespread belief amongst those concerned with influencing the social consciousness of the more advanced industrial societies that despite the amelioration of physical conditions, achieved over the last half century and more, the ambiance of work is becoming increasingly less tolerable in most levels of working society. The days of exploitation in an inhumane environment may be to a large extent gone, but the dangers and the indignities residing therein have been replaced, at least in part, by other more subtle and intangible threats.

The substance of this belief is predicated on recognition of the fact that while the requirements for the work force have changed with both industrialization and the development of technology, and while the character of the work force has also changed as its members have progressively become more educated, more skilled and more productive, the change in demand for work skills has not developed in parallel with the change in character of the workforce. It is argued that a considerable proportion of workers at all levels may be called upon to perform depersonalized or perhaps inherently stressful tasks in an alien, restrictive, and socially pressured environment, with resulting personal dissatisfaction or even sickness, social unrest and economic disruption.

If in fact this is so, and there is some justification for that belief, how has it come about? As noted, there would appear to be two elements: a change in the industrial requirements for skills and a change in the attitudes and aspirations of workers. As far as skill requirements are concerned, four factors may be observed. Firstly, with the development of technology and the assumption by machines and semi-automated processes of much of the work that was previously done by man, there is a greatly reduced need for the practice in industry of the creative arts and the manual crafts. These, by their very nature, demanded dedication and personal involvement, but they returned a dividend in the satisfaction of achievement. Secondly, even the requirements for physical strength and manual dexterity, of which a man might be proud, are disappearing from contemporary industry as machine power replaces human power. Thirdly, technology has created a new demand among relatively low-level workers for repetitive skills involving neuromuscular coordination, vigilance, minor decision making under externally paced conditions, along with the need for the emotional resources required to cope with the social pressures encountered. The combination of skills and resources so required is one found in few persons but demanded of many. Fourthly, the hierarchical system which has been developed for management of industry has demanded a new category of supervisory worker with an intellectual and executive capacity beyond the reach of a good proportion of the workforce.

These four factors, and no doubt others, have contributed towards a change in attitude about work among the workers, and particularly the younger workers, that has developed over the years. In addition, in developed countries in particular, the prosperity that has resulted from the application of technology and the resulting raised educational and cultural level of the workers, has changed their aspirations with respect to the work they seek, as well as to the conditions in which it is performed and the reward they expect to attain. With a reduced opportunity for personal involvement, there is at the same time, and no doubt in compensation, an increased demand for safe, healthy, and comfortable conditions, increased participation in organization and planning of work, and in decisions affecting their persons, and more human relationships with supervisors and management.

Paramount, however, and perhaps a motivating force for demand, is a feeling of increased insecurity induced by the existence of vague threatening factors, social, managerial, and environmental, that may have personal impact but are outside personal control.

Technology and industrialization have also brought about another change, significant in this regard, namely a change in the nature of the stress to which the worker is exposed. To primitive man, perhaps right up to the Industrial Revolution, stress was most probably a readily definable, clearly identifiable state, often urgent and life-threatening but, in a sense, tangible. Man responded to stress by the actions that his physiology and behavioural patterns demanded of him, and in so doing purged himself of the stress-induced anxieties. He may have lived at a survival level in recurrent fear, but it is doubtful if he suffered from psychosomatic disease induced by work. Contemporary man is still exposed to stress, and while at times it may be life-threatening and amenable only to 'fight or flight', more often it is relatively less intensive, not susceptible to personal corrective action, and leaves him strained and tense, ready for battle but unable to define the enemy.

Work and industrialization

In these circumstances, of course, the intangibility of the enemy derives from the change in nature of the work process. In his report to the International Labour Conference in 1975, Francis Blanchard, the Director-General of the International Labour Office, pointed out that the traditional views about work are being subjected more and more to question. He stated:

> The most widespread view in the industrialized societies is that remunerated work—normally performed within an employment relationship—is still the main means of personal fulfilment ... An entirely different attitude leads to such questions as: Is not work merely a constraint or even a necessary evil? It is not merely a means, from the personal point of view, of enabling us to earn a living and to do more interesting things in our free time.
>
> According to this view work should be merely an interruption in one's free time. It should be reduced to a minimum (from the point of view of the individual) and made as efficient as possible (from the point of view of society), while the question of job satisfaction should be considered not in relation to work itself but in relation to the other objectives in life.

These two viewpoints, of course, represent the extremes, and like any extremes they serve to define the boundaries. Neither viewpoint is completely tenable in our industrialized society, but while in the past our efforts have been oriented towards the former, it must be recognized that there is an increasing trend developing towards the latter.

Historically, the process of industrialization has been closely linked to organizational changes in agriculture, notably development of larger agricultural units which in turn facilitated new methods of cultivation, increased food production and more population growth. The combination of these effects formed a base for the subsequent process of industrialization.

Concomitant with the development of industrialization came the development of transportation and communication, opening new markets, and stimulating the need for mass manufacture. The worker who had previously been the focus of activity, was relegated to being a tender of machines, except for the fortunate few who retained executive authority. This viewpoint was crystallized by the work of F. W. Taylor (1856–1917) in the United States, who developed concepts of scientific management in which human function and performance were regarded as quantifiable and controllable variables.

He argued that for every task there is an ideal method which can be predetermined. Every function can be analyzed into its component tasks, each of which can be so designed as to be capable of performance by relatively unskilled labour. Several principles could thus be defined: Firstly, the independence of conception, planning, and execution; secondly, the division and standardization of tasks, equipment, and products; and thirdly, the intechangeability of operators.

From the point of view of productivity, there is no doubt that the approach and its implementation were effective, particularly at a time when the work force in the countries concerned was relatively unsophisticated and still primarily concerned with seeking the wherewithal for simple survival. This approach, however, carried with it the elements of its own destruction, in that the very productivity it has engendered has permitted development of sophistication and knowledge amongst the workers, such that they are no longer prepared to be mere mechanistic components of a manufacturing system, but seek instead recognition of their qualities and services as human beings, and intrinsic characteristics of satisfaction in the job itself.

Philosophy and theories of job satisfaction

Job satisfaction, or in its broader form, work satisfaction, is a difficult entity to define even in simplistic operational terms. For the individual worker, it exists when the perceived benefits of the work exceed the perceived costs by a margin deemed by the worker to be adequate under the circumstances. The costs and benefits, of course, are not necessarily measured in financial terms. Job satisfaction is not a static state, it is subject to influence and modification from forces within and outside of the immediate work environment. One school of thought (Goldthorpe et al., 1968), examines the problem in terms of its extrinsic or intrinsic orientation, that is whether the worker is primarily concerned with work as a means to provide fulfilment outside of the job, or finds fulfilment in the work itself, the former tending to be more of a working-class value, and the latter more of a middle-class one. Furthermore, job satisfaction is not the unitary or integrated state that the term would imply. There are multiple facets to the working state, some of which are more satisfying, or perhaps more acceptable, and others less so. Job satisfaction at best describes in comparative terms some integrated mean of that state at some point of time. There is no absolute on some infinite scale. At best one can state that at this particular time one is more satisfied with some aspect of one's job than at some other time.

Numerous authors have generated lists of characteristics considered to be desirable in the attainment of satisfaction at work, but the original organization from which many contemporary views has evolved owes much to the work of two American sociologists, working independently, namely, Maslow and Herzberg, despite the fact that their theories were largely developed from study of certain fairly limited sections of society.

Maslow's theoretical model postulates the existence in the worker of primary and secondary drives which serve to motivate him or her (Maslow, 1954). He argues that the primary drives are inherited, although the means for satisfying them can be learned. The primary drives stem from physiological needs and are oriented towards survival. They include basic appetites such as hunger, thirst, and sex. The secondary drives are not inherited but are learned and, to some extent at least, they may be culturally determined. They include such requirements as security, manifested in a need for protection and freedom from fear, as well as a requirement for organized structure, law and order. A need for love, affection and a sense of group identity, or belonging, is also defined. A third group comprises the need for self-esteem, represented by a desire for self-assurance, confidence and mastery along with feelings of achievement and the need for establishment of reputation and prestige. Maslow also defines a concept of self-actualization or the need to become more fully developed and to reify one's ideals.

The lower order needs, particularly the primary drives and the need for security and structure, are very largely met in today's industrial society. It is argued that what is needed now is satisfaction of the higher order needs, notably those pertaining to self-esteem and self-actualization, to use Maslow's terms. And, in fact, it is only in a society where the lower needs have been largely met that people can afford to seek satisfaction of those of higher order.

Herzberg (1966) states that the main factors involved in job satisfaction are advancement, recognition, responsibility, growth, and the job itself. These factors, termed 'satisfiers', will correlate, if optimized, with improved performance, reduced labour turnover, more tolerant attitudes to management, and general 'mental health'. Herzberg also recognizes 'dissatisfiers', which act in a negative direction. These include such things as working conditions and amenities, administrative policies, relationships with supervisors, technical competence of supervisors, pay, job security and relationships with peers. He argues that if the quality of the dissatisfiers is less than adequate, dissatisfaction will occur. Improvement in the degraded conditions will remove the dissatisfaction with beneficial effects on morale and perhaps on productivity. Raising the level above the adequate, however, will not of itself increase job satisfaction and performance, but it will provide a basis for the potential fulfilment of the 'higher needs' defined by Maslow. In this regard it should be noted that much of Herzberg's work was conducted among supervisors and middle-management employees, as so much motivational research has indeed been done. How much is applicable to the worker on the shop floor, or for that matter the worker in a different culture, is open to question.

Vroom, cited by Hunt (1971), added another dimension to job satisfaction theory. He argues that the choice of a job initially depends upon what he refers to as 'first level outcome', namely money or direct reward. Behind the first level outcome, and perceived by the worker with greater or less clarity, are second level outcomes which may be inherent in the job, such as prestige and power, or may be attainable by way of the money provided as a first level outcome. The effort that the worker is willing to expend, and the satisfaction that he derives in so doing, are directly related to the strength of the second level goals and the clarity of the perceived relationship between the primary and the secondary goals. Expectancy is a third factor. The higher the expectancy of achieving the secondary goal, as perceived by the worker, the greater is the perceived worth of the primary, and hence the satisfaction derived in attaining it.

Fox (1971) presents a more operationally oriented viewpoint. He defines three fields of concern in job satisfaction, namely content of the job (i.e. required skills), context of the job (i.e. the network of structure and reward within which the worker functions), and the needs of the worker. With respect to job content he refers to skills that require qualities of perception, motor coordination, intellect and education, and provide opportunity for creative expression and flexibility of response. The structural context includes the financial rewards, the location of the work, the nature of the work loads, and the adequacy of the equipment. More intangible factors are security of tenure, prospects for promotion, justice in promotion, and company attitudes; while the company structure itself, its planning policy and reputation are also significant. In the area of supervisory and peer relationships competent supervision, cooperation and communication throughout the hierarchy are important, while outside the actual task environment the provision of recreational resources is significant.

He defines the needs of the worker in personal and social terms akin to the higher order drives of Maslow or the satisfiers of Herzberg and recognizes that there must be orientation towards a personal goal, with however, an awareness of the system of priorities within which one may be permitted to achieve it. The individual must at the same time possess an appropriate level of physical and mental energy to achieve his ends, along with the capacity to conform, where required, and to tolerate stress. For many, there is also a need for social involvement.

One must recognize, in addition, that job satisfaction is seen to be different things by different workers, and that only where there is an expectancy of full employment can the worker afford to allow himself the luxury of being concerned with other more intangible requirements. When there is no expectancy of full employment the needs of survival become paramount — a phenomenon that has been exploited ever since man first went to work for his fellow man.

If there are definable characteristics associated with job satisfaction, it is pertinent to consider whether in fact there is a problem which can be attributed to lack of job satisfaction. There is a tendency to assume that if working, social, and other conditions are unsuitable, then dissatisfaction must exist and this in turn is an undesirable state, both for the worker and the employer. But is it in fact a problem? Few researchers have addressed themselves to this question. In a review Barbash (1974) noted that few people call themselves extremely satisfied with their jobs, but still fewer report extreme dissatisfaction. The majority claim to be reasonably satisfied. He observes from a United States Senate survey that 20 per cent will always dislike work regardless of how it is organized, and that a United Kingdom report places dissatisfaction in the area of 5 per cent, while two Japanese surveys show 23 per cent and 15 per cent respectively. But even taking the highest of these figures to represent the dissatisfied body, it would appear that the great majority of workers are indeed reasonably satisfied. But do these figures represent reality? Or for that matter do any of the surveys represent reality? One must recognize that to some extent at least a survey may incorporate some of the originator's prejudices. Most, if not all, attitude surveys rely on some form of questionnaire, in which the designer has pre-empted the questions and the areas of interest, normally of course with what he considers to be sound justification, but often oriented more to the negative than to the positive. The responder, however, then addresses himself to the questions that are asked, and not necessarily to the points that are specifically contentious to him, nor to what he finds good about his work. The questionnaire can indeed suggest to him some areas of problem that he might not have otherwise considered as being a problem. The resulting data are as a result inadequate, and insufficient as a base for drawing definitive conclusions. The observers, experts, and commentators on the working scene nevertheless consider there is a problem. And there probably is. A fairly large majority of workers, though perhaps inarticulate in their expression, are probably to a greater or lesser extent dissatisfied, although only a minority are prepared to voice their dissatisfaction and attempt to do something about it.

Specific factors

Herzberg conducted some 150 or more studies and found that, in order of importance, the features most desired by his subjects were security or steadiness of employment, opportunity for advancement, the worker's perception of the company and its management, salary, the intrinsic aspects of the job itself, the quality of the supervision, the social aspects of the job, the quality of communication, that is, the extent to which the workers are involved in communication, physical working conditions, and hours of work and the extent of the available benefits.

It will be recognized, of course, that these requirements are being reported by employees who are already for the most part, if not entirely, working under conditions which are reasonably tolerable, where wages are higher than starvation level, and supervisory relations, while not perhaps of the best, are certainly not on a master-slave basis. Various other studies, not detailed here, have confirmed the same general finding.

Job satisfaction and safety

Although the safety literature is replete with examinations of the motivations that may underlie safe or unsafe behaviour, not too many studies have been made to determine any relationship between job satisfaction and safety. Neuloh and colleagues (1957) found a higher incident of accidents among workers who had been moved from a job which they perceived to be good to a job which they considered to be poorer or less prestigious. Secondly, among skilled workers where the worker was exercising his trade, one worker in 20 became an accident victim, while if a skilled worker were not exercising his trade, one worker in six or seven was a victim. Where good relations existed with co-workers, one worker in 20 had an accident, but where relations were bad one worker in 10 had an accident. Other confirmatory evidence was provided by Kerr (1950) who showed that under the circumstances examined, the majority of accidents tended to occur where there was the least possibility of advancement.

Satisfaction and dissatisfaction in industry

As implied in the foregoing, it becomes apparent that the greatest satisfaction in industry tends to be found in jobs which maximize such conditions as possibilities for advancement, recognition of one's abilities and achievements, opportunities for affirming one's self-esteem, and opportunities for becoming involved in creative activity, although one must also remember that there are some, if not many, persons who do not wish that kind of responsibility and would prefer to be some unassuming cog in an anonymous wheel.

As has also been noted or implied, three basic causes of dissatisfaction can be defined, namely physical conditions, psychosocial conditions, and monetary reward. Physical conditions are examined in detail later. They include the definable physical agents in the workplace such as noise, vibration, heat, cold, reduced or increased atmospheric pressure, ionizing and non-ionizing radiation, illumination, as well as biological, chemical and particulate contamination. All of these agents, however, have to a greater or lesser extent been defined and are amenable to engineering solution or personal protection.

Adverse psychosocial conditions, on the other hand, are much less easily defined, and are not amenable to physical solutions. In summary, and expanding slightly, they include insecurity and unsteadiness of employment, poor perception of the image of the company and its management, the undesirable intrinsic aspects of the job itself, poor quality of supervisors, adverse social aspects of the job, poor quality of communication among management, supervisors, and employees, and inadequate fringe benefits, and although not amenable to simple physical solution they respond to a greater or less extent to enlightened administrative approaches.

The significance of salary and monetary reward is of interest. The stated views of workers seem to be somewhat inconsistent, and may even be industry specific. According to Walker and Guest (1953), for example, workers in the automobile industry at that time were almost exclusively concerned with money, whereas most other surveys, with a few exceptions, consider money as a significant but not necessarily primary factor in job choice and job satisfaction. However, in assessing the relative value of money as a motivating factor, it must be recognized that the reported data on attitudes to money are open to interpretation. Money, as part of a reward system, has emotional connotations. To some, regardless of the job, money is a goal; to others money may be secondary, while to all, their attitude may be coloured by the views of the culture in which they have developed.

Job satisfaction and life satisfaction

In the face of this consideration of the shop floor, it is imperative to realize that satisfaction or dissatisfaction at work is not inseparable from satisfaction and dissatisfaction with life. The dissatisfied individual is likely to be dissatisfied with both his work conditons and his life. Resolving work problems may improve the situation and allow him to view his life with a less jaundiced eye, but it will not turn him from a dissatisfied person to a satisfied person. Equally, stressful events in his life will be reflected in his attitude towards work. The sociological exercise that is known as Life Event Research (Cleary, 1974), which involves a long duration longitudinal study of the lives of individual persons, shows that stressful events in life, such as marriage, divorce, relocation, job change and death of a relative or friend, can render a worker less adaptable to work and more susceptible to physical stress.

Stress and satisfaction

It must be recognized that stress is always present to a greater or lesser degree and that paradoxically the total absence of apparent stress becomes in itself a stress. Thus, on the one hand, stress can be considered a load, increasing to an overload, arising from addition to the person-machine-environment system of qualities which are undesirable from the human point of view, such as intolerable working conditions, harsh supervision or unreasonable working hours. On the other hand removal of desirable attributes by, for example, the creation of a stulitifying environment, with reduced stimulation and intrinsically boring work, can act as a kind of negative loading which can be equally stressful. The stress experienced by an individual lies somewhere on the continuum between that arising from removal of desirable qualities and that arising from the addition of undesirable qualities. Thus there is some point where the stress level can be optimum.

One must also recognize that there comes a time when strain is equated not merely with reduction in satisfaction, but also in generation of dissatisfaction. Thus satisfaction has to be considered as one end of a satisfaction/dissatisfaction continuum, where one state merges into the other through a region of indifference.

By the same token, satisfaction is not an absolute. There is no upper bound of absolute satisfaction, while the lower bound merges indistinguishably into dissatisfaction, which itself has no absolute lower bound. Each is a relative term, relative to some previous state, or to the state of some other individual. Furthermore, the pursuit of satisfaction, like the pursuit of happiness, is seldom a consciously articulated human goal. One does not normally seek a state of satisfaction. One may seek various objectives, which one has to a greater or lesser extent defined, and in so doing one may find satisfaction, but normally one is more concerned with minimizing dissatisfaction than maximizing satisfaction. It is a quality, again like happiness, that tends to be seen more in retrospect than in prospect.

Not only, then is job satisfaction part of an unbounded continuum, it is also a personal state, as opposed to a group state, and its goals will vary from person to person, from circumstance to circumstance and from time to time in the same person. Furthermore, it is at least as much a function of the individual as of the job, with connotations of positive well-being which are barely consistent with reality and probably attainable at best by only a few. The majority of persons, the majority of the time, are neither particularly satisfied nor particularly dissatisfied. They occupy some shifting range in between, satisfied about some things, dissatisfied about others, dynamically adjusting to each in their individual homeostatic equilibria. Thus, data pertaining to the level of job satisfaction of groups have to be interpreted with caution. At best they are statistical indices which have often little or no application to the individual.

It is true that one can broadly define certain attributes of job satisfaction, such as those discussed earlier, and for that matter one can define even more easily certain attributes of dissatisfaction. It would be naive, however, to consider that if the defined dissatisfiers were minimized and the satisfiers were maximized than a state of job satisfaction would persist. This indeed is one of the goals of Work Humanization programmes, and a worthy goal it is, even if its only effect were to be a general raising of the quality of working life. But it should not be assumed that work humanization *per se* will lead to persisting job satisfaction. The goals of job satisfaction are ephemeral and recede as they are approached. The human capacity to adapt is such that were these goals ever achieved man would adapt to the new level of living, accept it as a norm, and seek still further levels of satisfaction. The nature of human physiology and psychology, as illustrated by the Hawthorne experiments and numerous other studies of many related and unrelated varieties in the field and laboratory, determines that, given an environment, or a machine-environment system, where homeostasis is not threatened, a person responds favourably to a change in state, not to achievement of a static state, provided that the combination of magnitude and rate of change is not too great. The change is preferably, but not even necessarily, towards a more favoured state. Where the combination of magnitude and rate of change is within acceptable limits, as perceived by the individual, then his arousal, alertness, interest, performance capability and indeed satisfaction are maintained at a high level. Where the environment threatens his homeostasis, where the magnitude and/or rate of change are too great, or not great enough (which also threatens his homeostasis) then there will be stress, expressed at least initially as dissatisfaction.

Bearing this factors in mind, and oversimplifying for the sake of clarity, it will be seen that the basic relationship between human stress and job satisfaction can be represented by a classic bell-shaped curve (Figure 8.1). The greatest satisfaction is found where the stress level is optimum. As desirable features are removed, or undesirable features added the level of satisfaction drops through a zone of indifference until it becomes dissatisfaction. It is emphasized that the curve is a schematic curve, and while no doubt it will always retain a bell shape it will not necessarily

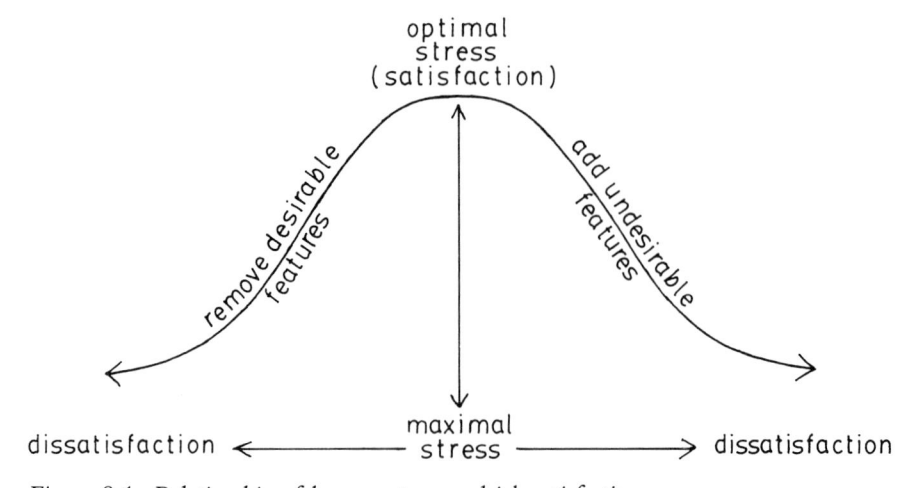

Figure 8.1. Relationship of human stress and job satisfaction.

retain the same proportions; indeed there is little doubt that the shape will be in a state of continuous change, reflecting the continuous adjustment of the system as it responds to disturbances in equilibrium.

Part III
Design for human use and function

Chapter 9
Equipment design for human use

Introduction

The design and/or selection of equipment for human use and function presents a significant series of problems in what is sometimes known as human factors engineering. It has unfortunately been all too common in the past to design and build equipment to accomplish a task while paying little attention to the capacities and limitations of the operator who has to use the equipment. It is the intent here to examine some of the principles that should be considered in design for human use. While some anthropometrically oriented specifications will be presented in passing it is not the intent to provide these in detail. The reader is directed for this purpose to such texts as *Fitting the Task to the Man*, by E. Grandjean, Taylor & Francis Ltd., or *Human Factors Engineering and Design* by E. J. McCormick and M. S. Sanders, McGraw-Hill Inc., or *Human Engineering Guide for Equipment Designers*, by W. E. Woodson and D. W. Conover, University of California Press.

To understand the needs of design for human use it is necessary to appreciate the concept of the person-machine interface.

Person-machine interface

The person-machine interface is the term given to that part of a person-machine-environment system where the operator imparts some type of energy to the machine (for example, by moving a lever or pushing a button to activate or control the machine) and derives information from the machine by way of some form of display. The concept is illustrated in Figure 9.1.

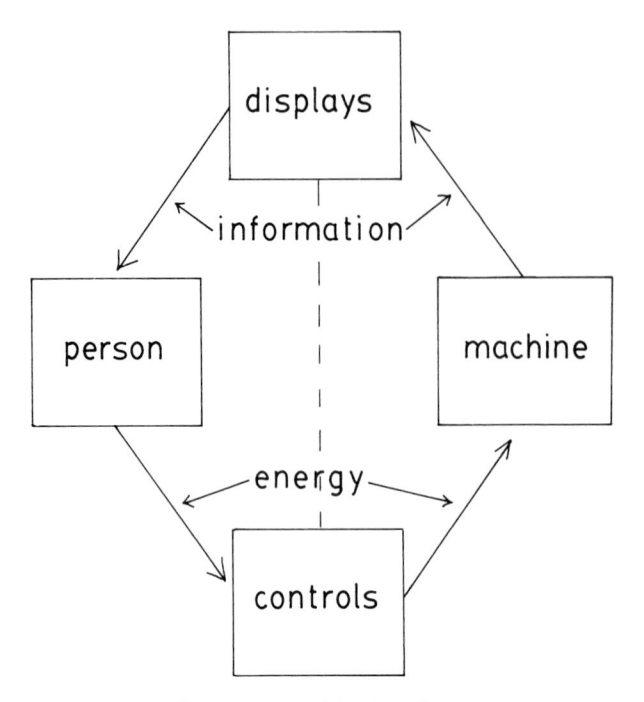

Figure 9.1. The person-machine interface.

Displays

A display then is a device, mechanism or channel by which information relevant to the task is transmitted to the operator. It should be noted, however, that despite its name a display is not necessarily a visual device, although it commonly is. It could, for example, be auditory, such as a buzzer or bell, tactile, such as a change in textured surface, or kinesthetic, such as the vibrating stick shaker that signals impending stall in an aircraft.

Functional criteria

There are certain functional criteria that must be met in a display. For example, depending on the needs of the system there may be a requirement for different priorities in speed of response, or in other words, how quickly should the display reflect any change in activity, how accurate should be the presentation, or for example, how closely should it reflect the change, and how small a change should it respond to. Consequently, certain design criteria need to be established before devising or selecting the appropriate display. These include:

NATURE OF THE VARIABLE
The particular variable that is to be displayed should be clearly established before any design is considered. For example it is no use to display a value when what is really wanted is a rate of change of that value.

TOTAL RANGE OF VARIABLES
The range of the value to be displayed will determine the design of the display. A range involving only a small quantity will require very different treatment from one involving a large quantity.

MAXIMUM REQUIRED ACCURACY
The maximum required accuracy determines the precision of the scale.

MAXIMUM SPEEED OF TRANSFER
The maximum speed of transfer determines how rapidly the required information must be presented to the display.

MAXIMUM EQUIPMENT ERROR
All equipment has a certain intrinsic error. The less the acceptable error the more complex and more expensive the display system may have to be.

DISTANCE BETWEEN DISPLAY AND USER
The distance between display and user largely determines the size and clarity of the visual components. On this basis one can begin to determine what kind of display is required to meet a given purpose.

Types of display

Displays can be real or artificial, static or dynamic.

REAL DISPLAYS
Although it is not commonly thought of as such, in a real display the environment itself provides the information directly. For example, by this definition, the windshield of a car is indeed a display and meets the criteria for a real display.

ARTIFICIAL DISPLAYS
Artificial displays provide a symbolic representational value of the information of interest, as for example the speedometer of a car. In the common design of a car speedometer the needle of the display points to a number which the driver interprets in terms of speed.

STATIC DISPLAYS
Static displays are not instruments and do not change with conditions. They give information by their existence and by the interpretation on

the part of an operator of the information contained therein. The adjustment setting for a lathe, for example, which is often attached as a plate or etched into the metal, is a static display.

Dynamic displays
A dynamic display is one which responds to the actual represented value, and changes as that value changes. An electrical voltmeter, for example, like a speedometer, is both an artificial display and a dynamic display.

Categories of display

Qualitative displays

These are used to distinguish between a small number of discrete conditions such as on and off, open and shut, normal and abnormal, and so on. There are three distinguishable types, which have in fact already been noted in passing. These are:

Auditory displays
Auditory displays include bells, buzzers, beeps, and the like. Their main advantage lies in their pervasiveness. One is warned without necessarily paying attention. They have the disadvantage, however, of being masked by competing noise. In this connection, it should be recognized that warbling sounds in short bursts are more attention-getting than continuous sounds of the same pitch.

Visual displays
Visual displays include lights, colours, and even the position, for example, of a lever which has been moved. They can be made distinctive in a variety of ways including varying sizes and shapes according to demand. Colour can be useful provided that one remembers that some 8 per cent of the population may be to a greater or less extent colour blind. Varying intensity of lighting can be used to emphasize different characteristics, but this approach is only of real usefulness when the light is changing. The great disadvantage of visual displays of course is the fact that they have to be looked at. It might be remembered, however, that flashing lights tend to attract the peripheral vision and direct the attention of the operator towards the source of the flashing.

Tactile displays
Tactile displays in this connection include kinesthetic. While texture has been used or at least incorporated into other forms of tactile display they are mostly represented by some form of 'positioning' which is perceived by the operator's hand or foot. For example, one can tell by 'feel' whether

the handbrake of a car is on or off, or whether the gear lever is in first gear or reverse, and so on.

SIGNAL AND WARNING LIGHTS

Probably one of the most critical uses of qualitative displays lies in the provision of signal and warning lights. These are used to attract attention, to denote alarm, or simply to indicate the status of a system or component.

For a signal to be detectable it must meet certain criteria. The absolute threshold of visual detectability depends on the size, that is, the angle subtended at the retina, the luminance, or the amount of light emitted, and the duration of exposure. The threshold of luminance in fact varies inversely with the exposure time (McCormick and Sanders (1982)).

The addition of colour to the signal can change its detectability, but where there is a high absolute brightness of the signal, with a high brightness contrast against a dark background, then there is nothing to be gained by adding colour. Where there is a low brightness contrast the fastest response to colour, other things being equal, is found with red, green, yellow and white, in that order (Reynolds, White, and Hilgendorf, 1972).

Flashing lights are more readily detectable than steady state lights, hence flashing lights should be reserved for emergency purposes.

As the flash frequency is increased, eventually the observer perceives the flashing light as being continuous. The frequency at which this occurs is referred as the *flicker fusion frequency*, which is generally considered to be about 30 flashes per second. The recommended flash rate for emergency flashers is 3 to 10 per second with a duration of not less than 0.05 seconds per flash (Woodson and Conover, 1964).

The background against which a signal is presented can modify its detectability. The worst case is found, as might be expected, where there is a flashing signal against a flashing background; and, of course, the best case occurs where the signal is flashed against a steady background. Intermediate to these is a steady signal on a steady ground.

It is also preferable to have one warning light rather than multiple lights. Multiple lights can be very confusing, the more so under emergency conditions. Where several warnings are required at the same time it may be desirable to have a single warning light with an annunciator panel to indicate to what the warning refers. The intensity of a light should be twice that of the immediate background and its location should be within 30 degrees of the operator's normal line of sight (McCormick and Sanders, 1976).

Pseudoquantitative displays

As the name would imply, pseudoquantitative displays provide non-numerical comparative information. They are useful in determining the status of a function which has a limited number of predetermined ranges, for example high, medium, or low; or for selecting and maintaining some desirable range, or for observing trends in the activity of a function.

Commonly a dial with a pointer is used to present the information, the dial in this case being uncalibrated and without numbers. The degree of change can be represented on the dial by position, for example by having segments of the dial marked high, medium, or low, or perhaps by colour coding with the background of the dial passing from green for safety, through yellow, to red for danger.

In some cases, particularly in the monitoring of multiple machines, it may be necessary to have numerous dials. The most effective monitoring in this situation is undertaken by organizing banks of dials, for example in 3 × 4, or 4 × 4 matrices, all referring to the same type of function. Perception of change is then based on monitoring the totality of the pattern presented by the multiple pointers. If normal is defined by having a pointer at, say, 12 noon or 9 o'clock, then any deviant pointer will be readily observed (Dashevsky, 1964). The observation can be enhanced by extending graphic lines from the tip of one pointer to the base of another. In addition, sub-patterns of dials can also be defined. Figure 9.2 illustrates these concepts.

Quantitative displays

Quantitative displays present numerical information in a visual medium. There are two basic types, namely *analogue* which present the information

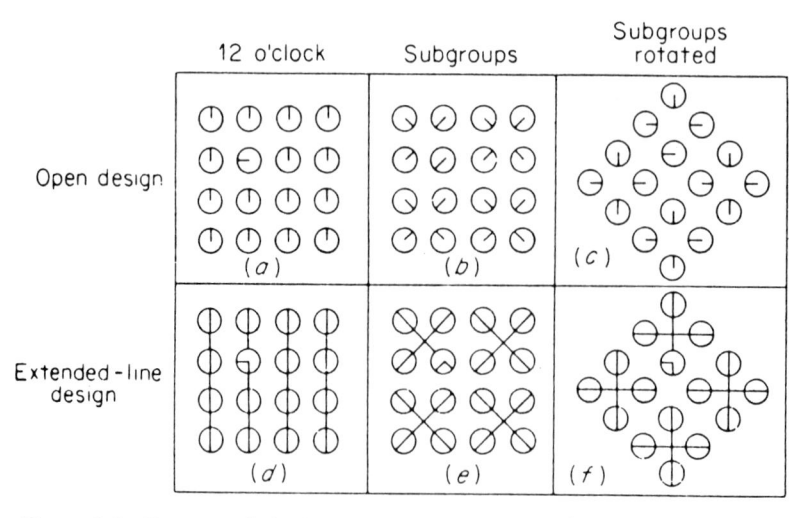

Figure 9.2. Patterns of check-reading dials (after Dashevsky, copyright 1964 by the American Psychological Association. Reprinted by permission).

by means of a pointer and a numerical scale, and *digital*, which present the information in the form of changing numbers.

The analogue display most commonly uses a pointer moving against a fixed scale, although it may also be used with a moving scale operating behind a fixed pointer. It is commonly easier to assimilate information from an analogue display than from a digital although it is less precise. In fact assimilation of precise information is faster from a digital than an analogue (Zeff, 1965). An analogue display also shows trends, or rate of change, more readily. Heglin (1973) lists certain characteristics which should be borne in mind in the selection of analogue displays noting that in general people prefer a moving pointer to a fixed pointer.

Representational displays

A representational display presents a picture or a model of the system to be controlled, as for example a graphic and dynamic two-dimensional model of a rail marshalling yard, or the process control system of a refinery. The design in these cases is specific to the needs and may involve various qualitative and quantitative sub-displays integral with the total pattern.

Scale and dial design criteria

In selecting a scale it is necessary firstly to determine the range of scale to be covered. Within that range it is also necessary to determine the reading precision required, which in turn determines the number and positioning of the scale markers. Thirdly it is necessary to determine the format and terminology of the readout, for example, velocity, pressure, voltage, and so on, and whether the format should be in direct readout or in computed readout. For instance, it might be more useful to have a scale in percent of some predetermined value rather than in the actual value.

In determining the design of the scale itself, there are three areas for special consideration. These are the number of marked divisions, the size of the estimated subdivisions, and the organization and structure of the scale.

Number of marked divisions

The number of marked divisions turns out to be a compromise between the need for speed of reading and the need for accuracy. Where there are too many divisions the reading is slow but more precise; where there are too few reading may be more rapid but tends to be less precise. The compromise is the smallest number compatible with the requirement, allowing the operator where necessary to subdivide by eye.

Size of subdivisions

In general, the size should be large enough for easy discrimination by an operator at normal reading distance, which lies between 25 and 75 cm (10–30 in).

There should in fact be enough separation to make reading easy, along with clear cues, such as length or thickness for ready differentiation between major and minor graduations. From a number of sources, Bailey (1982) has defined that under normal illumination, with high contrast, adequate lighting on the face of the dial, and an expected reading distance of 30–70 cm (12–28 in) the following criteria can be established:

minimum width of major mark: 0.32 mm (0.125 in)
spacing:
  white marks on black dials: not less than twice stroke width
  black marks on white dials: not less than one stroke width
minimum distance between major marks: 12.8 mm (0.5 in)
height of major marks: 5.6 mm (0.2 in)
height of intermediate marks: 4.1 mm (0.16 in)
height of minor marks: 2.3 mm (0.09 in)

Under low illumination the following would apply:

height of major marker: 5.6 mm (0.22 in)
width of major marker: 0.89 mm (0.035 in)
height of intermediate marker: 4.1 mm (0.16 in)
height of minor marker: 2.5 mm (0.10 in)
minimum separation between centres: 1.8 mm (0.07 in)
width of intermediate marker: 0.76 mm (0.03 in)
width of minor markers: 0.64 mm (0.025 in)

Interval values and numerical progression

The interval values or numerical progression of scales require consideration (Grether and Baker, 1972). Normal values should be in ranges of 1, 2, and 5, or decimal multiples of these (for example, 10, 20, and 50). Values of 2 have been found to be less desirable than values of 5. In fact scales marked in 1, 10, or 100, in general, have been found to be superior to other values. There should be no more than 9 marks between intervals and the design should also be such that where possible there is no requirement for interpolation between markers.

Alphanumerics

In addition to the graduation markers some kind of alpha-numerical legend is normally necessary. Alphanumerics should be designed for reading at 70 mm (28 in) unless there are circumstances to the contrary, with values as follows (Grether and Baker, 1972):

	Height (mm, in) (Low Luminance)
Numbers on counters, moving scales	5.1–7.6 (0.2–0.3)
Numbers on fixed scales, controls and switches	0.04–0.08 (0.15–0.3)
Identification labels, instructions	1.3–5.1 (0.05–0.20)
Width (all numbers except 4, 1)	3/5 of height
number 4	1 stroke width wider
number 1	1 stroke width wide

	Height (mm, in) (High Luminance)
Numbers on counters, moving scales	0.12–0.20 (0.03–0.05)
Numbers on fixed scales, controls, switches	3.0–5.1 (0.10–0.20)
Identification labels, instructions	1.3–5.1 (0.05–0.20)

Organization and structure

The organization and structure of a display is based on the expected reading norms under the expected viewing conditions, bearing in mind the expected visual acuity, the ambient lighting and the reading posture. The following items require special consideration.

Counters versus dials

For precision reading a counter is better than a dial, although it is commonly easier to assimilate information from a dial, or analogue display, than from a digital. Assimilation of precise information, however, is faster from a digital than from an analogue.

Moving pointer versus moving scale

In general, people prefer a moving pointer to a fixed pointer. Because of limited space, however, it might be necessary to select a moving scale. For example a vertically mounted dial might require a diameter of 7–10 cm, whereas if the scale were turned at right angles to the face and allowed to move against a fixed pointer in a window the required vertical space would be much less. It is, however, undesirable to mix the two types of display in the same format. If the display includes a control function, that is, if the pointer can be set at a particular point on the scale, or if it is used for tracking, it is better to move the pointer on the scale rather than the scale on the pointer. Relatively small variations in motion are observed more readily with a moving pointer than with a moving scale. For immediate access to stationary or slowly moving numbers the digital display is superior to the analogue.

Scale layout

Certain conventions have arisen in the layout of scales, largely because of habit and usage. In particular, an increase in values should normally be associated with movement of a pointer in a clockwise direction on a circular dial, from bottom to top in a vertical dial, and from left to right in a horizontal dial. It should be remembered that these conventions are learned, rather than natural, and may not apply in all societies. Another aspect of the same phenomenon is discussed in consideration of control display stereotypes.

The end-point of a scale is also subject to conventional treatment. Normally, in a single revolution scale there should be a break between the two ends of the scale, with zero at the bottom. On the other hand in a multirevolution scale it is common to have zero at 12 o'clock, while, as already mentioned, with multiple dials set up for check reading the normal position is commonly found at 9 o'clock. It is also desirable for the numbers on scales not to be obscured by the pointer. For this reason they are frequently placed outside the circle inscribed by the pointer.

On some scales it is desirable to have zone markings indicating, for example, some operating range, or some unsafe range. These may be colour and/or shape coded as required.

The pointer, of course, is critical to the design; simplicity should be the keynote. Grether and Baker (1972) have indicated some of the criteria. The pointer should extend from the centre of the scale up to but not overlapping the minor scale markings, such that the latter should readily be distinguishable. While it may extend from the centre it is commonly desirable to have a tail extending beyond the centre, half the length of the head. In order to minimize viewing parallax, which might lead to reading error, the pointer should be as close to the face of the dial as feasible. A two-colour (black and white) pointer assists in discrimination. The portion from centre to tip should be the same colour as the markings (commonly black), and the portion from centre to tail the same as the dial face (commonly white). The angle made by the tip of the pointer should be 20 degrees.

Coding of visual displays

The presentation of information on visual displays can be coded in various ways. The use of alphanumerics, that is, letters and numbers, is of course obvious. Other methods include colour, geometric shape, area and/or size, such as different areas of squares or rectangles, and, of course, the angular orientation of a pointer. Number of items and, for example, frequency of flashing, are a form of coding as is the use of different kinds of lines, thick, thin, dotted, or broken.

A coding scheme may be single in that it involves one piece of information, such as the appearance of a red light, or it may be redundant in that the same piece of information is reinforced, for example, by presenting a light in a special location. The term 'compound' is used when a piece of information is presented using two or more different design concepts, such as a warning light and an annunciator panel.

There is no clear experimental or experiential indication of the utility of different types of coding. The choice tends to be somewhat arbitrary on the part of the designer, dependent on the nature of the task to be performed and the context in which the display is going to be used. Table 9.1, compiled from various sources by McCormick and Sanders (1982), summarizes significant characteristics of various methods of visual coding.

Table 9.1. *Usefulness of various visual coding systems (derived from compilation by McCormick and Sanders 1982, from various sources)*

Alphanumeric	Good, especially for identification; uses little space if there is good contrast. Certain items may be confused.
Colour of surfaces	Particularly good for searching and counting tasks. Affected by some lights. Problem with colour defective persons.
Colour of lights	Good for qualitative reading.
Geometric shapes	Generally useful coding system, particularly in symbolic representation; good for CRT's. Shapes used together need to be readily discriminated; some sets of shapes more difficult to discriminate than others.
Angle of inclination	Generally satisfactory for special purposes such as indicating direction, angle, or position on round instruments like clocks. CRT's, etc.
Size of shapes (e.g. squares)	Takes considerable space; use only when specifically appropriate.
Visual number	Use only when specifically appropriate, such as to represent numbers of items. Takes considerable space; may be confused with other symbols.
Brightness of lights	Use only when specifically appropriate. Weaker signals may be masked.
Flash rate of lights	Limited applicability if receiver needs to differentiate different flash rates. Flashing lights have possible use in combination with controlled time intervals (as in lighthouse signals and naval communications) or to attract attention to specific areas.

Electronic display technology

In recent years more and more attention has been paid to the development of electronic display technology. There is a wide variety of different types, some of which are examined below.

Cathode ray tube

The cathode ray tube (CRT), such as is found in an oscilloscope or television screen, is the most common and popular electronic device for large displays.

It has the advantages of high resolution, simple addressing, and high writing speed. It can of course present material in full colour, dichromatic, or grey, and is capable of providing large-scale storage, with appropriate ancillary equipment peripherals. It can come in a variety of sizes.

It is, however, bulky, expensive, relatively fragile, and generates much heat. It has the disadvantage of a curved screen which can distort the presented signal. Of major significance is the fact that it requires high operating voltages which may in turn give rise to secondary emissions, as well as presenting intrinsically hazardous maintenance problems.

Many attempts have been made to develop a flat-panel CRT. While the technology exists the hardware is still very experimental and expensive.

Light emitting diode

The light emitting diode (LED) is commonly found in calculators, watches, sundry types of instrumentation, and even as discrete miniature lamps. It is, as the name would imply, a diode which when activated gives off light. It is cool to operate and, for its size, gives off good luminance with a low power requirement. The cost is low and the reliability high. It is readily incorporated with integrated circuits. Although commonly used for small area signals large arrays can be coupled together to present messages and more complex readouts.

Liquid crystal display

A liquid crystal display (LCD) is formed from a drop of fluid sealed in a transparent container and including a crystalline substance. When an electric field is applied the crystals will align themselves to vary the transmissibility of externally generated light. According to the layout of the crystals various alphanumerics can be generated. Initially there is a high contrast ratio which degenerates with time. The LCD is useful for single or multiple alphanumeric layout with or without multiplexing. The characters can be made any size at low cost with low voltage and power requirements.

Electrophoretic display

The electrophoretic display is a light modulating display. It results from the process of electrophoresis in which charged particles suspended in a liquid are moved by application of an electric field. The particles may be pigmented, or of different optical density from the liquid. Movement to the front or surface of the liquid renders them visible while returning to the rear takes them out of view. The technique has found application in on-off displays, matrix displays, and seven-segment alphanumerics, particularly where colour is desirable.

Plasma display

Plasma displays are developed as a matrix of neon bulbs between transparent plates. They require power of high voltage and are complex to produce. They do, however, provide a viable alternative to the CRT, particularly for large applications. They can be used in single rows for alphanumeric readout, or in multiple rows or large matrix panels, for graphics. They can also be used as dot matrices, or in seven-segment forms, for the generation of numbers.

Electroluminescence panels

Electroluminescence panels employ capacitance storage to excite an electroluminescent field using very little power. They are commonly produced in sheets the thickness of plastic which can be produced in appropriate shape, in size from centimetres or less to metres. Dependent on the state of field excitation they can produce a full range of grey scale. The contrast, however, is low and they are difficult to read in bright light.

Selection of electronic display technology

In order to select an appropriate display it is first of all necessary to define the information to be displayed, along with the environment in which it is going to be displayed, while bearing in mind the voltage and power constraints.

Specifically, one needs to determine what kind of symbols are going to be used and whether the information presentation will be dynamic or static, and if dynamic at what rate of change. The format of display can be alphanumeric, which is the most common; it can be vectorgraphic or even pictorial, and in some instances it may be a combination.

The display, however, cannot be examined alone without consideration of the totality of the workspace layout. It must be integrated into the design of the workplace and compatible with the environment in which it is to be used. It is of course ridiculous to have a sophisticated electronic display on a machine that is going to be used for some simple mundane purpose. Not the least of the considerations is ambient lighting. Lighting must be assured or it may be necessary to provide integral lighting within the display.

Controls

Within a person-machine system controls provide direction from the operator to the machine in response to the information passed from the machine to the operator via displays. The direction is achieved by a transfer of energy which may involve movement of a lever, a knob, a wheel, and so on.

There are human limitations which constrain the design of a control system. These include *stature* or height, which governs the positioning of a control in relation the ground or floor; *weight*, which can be significant in determining the application of large forces; *human anatomy*, or more specifically *anthropometry*, which determines such matters as the nature of the function or operation of levers, buttons, foot pedals, and so forth, as well as the direction of control motion and the positioning and separation of controls; *sex* and *age* are also of anthropometric interest since they also can modify specifications for control design. Lastly, the complexities of ethnic origin and cultural habitat can, for example, determine physique and influence habit patterns.

Control function

Five types of control function can be distinguished.

ACTIVATION
This is the act of initiating or terminating a control function, that is, whether it is placed in the 'on' or 'off' position. It also can act as a display by its position, and provide information on the status of the function.

DISCRETE SETTING
This provides pre-set adjustment points known as detents, or 'click' positions, which hold the control until a slightly greater force is applied to overcome the resistance. Thus the position of the control can provide information both on status and on quantitation.

QUANTITATIVE SETTING
In this mode the control can be set at any position marked on the scale of a continuum. It also provides information on status and quantitation.

CONTINUOUS CONTROL
A continuous control allows for tracking of a time varying function as expressed on a display. It provides quantitative, qualitative, and representational information.

DATA ENTRY
Data entry is also a form of control although of a different genre. It is implemented by keyboard and provides alphanumeric and symbolic information.

Control types

A wide variety of control types can be defined. Some of the more common are discussed below.

PUSH BUTTON
A push button is commonly operated by the hand or finger. It can be coded by label, shape, size, and less usefully by colour and location. It is easily identified and can be operated quickly. A foot push button is less common. It is normally identified by location. Foot controls are much less readily operated than hand controls. In either case a push button normally has only two settings, on and off.

TOGGLE SWITCH
A toggle switch is normally operated by the fingers or hand. It can be flipped forwards and backwards, or sideways. It can have three positions, namely central, forwards, and backwards, although commonly it has only two of these. It may be identified by label or location. As will be noted in later discussion of control display stereotypes, the switch is best installed to operate vertically. Horizontal installation should only be used if it has to be consistent with the function being controlled.

ROTARY SELECTOR SWITCH
A rotary selector switch is turned by the hand or fingers. It should normally have no more than 24 settings, that is with 15 degrees between each setting. Each position should be identified by a detent or 'click' sensation. Normally the pointer is an integral part of the turning knob, pointing to a scale around the selector. Increase in value occurs in a clockwise direction.

CONTROL KNOB
A control knob is normally used to move the pointer on a display. It may have a one-turn or multi-turn relationship to the display. Values should increase with clockwise rotation.

Thumbwheel

A thumbwheel is a special form of knob which is operated by the thumb while the fingers and the rest of the hand may be engaged in some other control activity. It is found, for example, on the control column of an aircraft. The same general principles apply to its function.

Pedal

A pedal is a foot-operated control, normally acting in a downward or forward direction depending on whether the operator is seated or standing. It is normally undesirable, although not uncommon, to have a pedal operated by a worker from a standing position. It is generally understood that foot controls are slower and less accurate than hand controls.

Other controls, such as the lever, the wheel, and the crank can also be found. Each has specific applications to the function that has to be controlled.

Control identification

To avoid error it is necessary to ensure that controls are easily identifiable and easily distinguishable one from the other, sometimes in conditions where visual identification is not available. Some of the techniques have already been noted in discussion of control types. The general principles are presented below.

Shape and texture (size)

A large number of clearly identifiable shapes has been defined which can be distinguished by touch alone. These include round, square, rectangular, trianglar, vertically elongated, and so on, with various indented, textured, or sculptured forms. The shapes are useful not only for identification, but also for standardizing groups of functions; for example, all round knobs might serve the same type of function. It might be noted, however, that texture, and even to some extent shape, is ineffective as a distinguishing feature when the operator is wearing gloves.

Location

Placement of a control in some standardized position serves as a means of identification. For example in a car, according to standards fixed many years ago, the brake pedal lies between the clutch pedal and the gas pedal or accelerator. The latter, incidentally, also according to convention, has a distinguishable shape. Location can also be used for standardizing functions. There is however a limited number of useful locations in the immediately available space, and furthermore location by itself may be inadequate for identification.

Colour
Colour can be effectively used, bearing in mind that some 8 per cent of males are colour blind. Utilizing hue (colour), saturation (intensity), and brightness (clarity) some 24 differences can be clearly identified, but in practice these tend to be limited to about nine of the more commonly recognized (Jones, 1962). For safety, however, it is probably wiser to use a smaller set of colours. Conover and Kraft (1958) identified sets of eight, seven, six, and five colours, respectively for this purpose. While colour provides good visual identification for the colour-normal subject and is very effective in standardizing a group of related functions, it has the disadvantage that it has to be viewed under adequate lighting to be effective.

Operational usage
This term refers to the method by which the control is used. A lever has a limited and distinctive set of functions, commonly in fact one unidirectional function, which sets it apart from other controls. In the absence of vision, however, this property may have the disadvantage that its function has to be tested for the operator to know what it is.

Labelling
Labelling refers to the naming, either alphanumeric or symbolic, of a control. This approach has the advantage of precision, and applicability to any form or number of controls. It also requires some learning, and indeed the ability to read the language printed. However, to be effective the label has to be viewed and understood under adequate lighting, and because of the physical dimensions of the label, extra space may be required on what may be a surface restricted in area.

Requirements for control function

There are four basic considerations that must be taken into account in analyzing the requirements for control function. These can indeed be incompatible and require an ultimate compromise in selection. In conducting the task analysis, then, one must consider the need for speed, accuracy, force, and range.

Speed
Where speed of operation is deemed to be the primary need the control task should be assigned to the arm, hand, or even finger, where the greatest dexterity can be found with the least effort. For easiest operation the control itself should be located at or just below elbow height to be operated with the elbow at or about 90 degrees of angle. It should be noted that maximum speed is of course not compatible with application of maximum force.

Accuracy

The need for accuracy is commonly found in association with the requirement to track a moving target, which of course could be a pointer. To achieve accuracy there is a need for a high quality of information feedback. This in turn brings up the need to understand what is referred to as the control/display (C/D) ratio.

Control/display ratio

The control/display ratio, or control/display response, is the ratio of the extent of movement of the control device to the extent of movement of the display indicator (for example, the pointer). Where there is a low C/D there is a high sensitivity and where there is a high C/D there is a low sensitivity (Jenkins and Connor, 1949). Furthermore, for control operation there are two other elements to be considered, namely the travel time, that is the time to move the control to the approximate position, and the adjust time, which is the time to place it in the position required. Where the C/D is low the travel time is low but the adjust time is high; where the C/D is high the travel time is high but the adjust time is low. Thus when designing for accuracy the C/D ratio becomes vital in determining the relationships between adjust time and travel time. The selection will ultimately depend upon the results of a detailed task analysis.

Force

The requirement for an operator to use excessive force should be minimal and restricted to emergency or occasional use where other power is unavailable.

Applicable forces have been examined by Hertzberg (1972). For maximum torque the most favoured posture is standing with the body supported to allow application of the shoulder muscles. For maximum foot push, which is the strongest force applicable by the body, the best position is seated with the body supported and the knees angled 20 degrees below the horizontal. The maximum hand push or pull is greatest in the vertical plane, that is, up or down. Because of the additional need to maintain isometric contraction in addition to useful work the least effective hand push or pull is found in the horizontal plane. Thus, for example, in the vertical plane one can push 130 per cent of body weight or pull 100 per cent of body weight, whereas in the horizontal plane these numbers are reduced to 15 per cent and 10 per cent respectively.

Range

The range refers to the scope of action, or extent of motion demanded of a control. For example where a wide range of motion is required a multi-turn hand crank might be appropriate whereas a handwheel or even a knob would need the requirement for a small range.

Other factors

Certain other factors also need consideration. For example, dependent on the position of the operator, such as seated or standing, a control may be placed in the vertical plane, in the horizontal plane, or at some intermediate angle. In general also, where no other considerations intervene, a clockwise motion of a rotary control is preferable to an anti-clockwise.

Various requirements for control selection, derived from Grandjean (1980) are presented in the following table:

Table 9.2. Requirements for control selection (after Grandjean, 1980).

Type	Speed	Accuracy	Force
Range			
Crank			
large	poor	unsuit.	good	good
small	good	poor	unsuit.	good
Handwheel	poor	good	fair/p	good
Knob	unsuit.	fair	unsuit.	fair
Lever			
horizontal	good	poor	poor	poor
vertical (to and			short: poor	poor
from the body	good	fair	long: good	poor
Pedal	good	poor	good	v. poor
Pushbutton	good	unsuit.	unsuit.	unsuit.
Rotary switch	good	good	unsuit.	unsuit.
Joystick	good	good	poor	unsit.

Control and display stereotypes

Controls are commonly associated with displays. Normally, there are certain expectations that movement of a control will produce a display or machine response in a predicted direction. In designing controls and displays then it is essential that the expected control/display relationship is maintained. For example, increase in speed, volume, voltage, and so on, would normally be produced, and presented on a display, by downward motion of a pedal, or by motion upward, to the right, or away from the body, for a lever, or by clockwise rotation or motion to the right, of a pointer.

The classic exception is the light switch. In the Western Hemisphere the up position of the light switch indicates on, and the down position off, whereas in Britain the reverse is the case. In selecting a vertical toggle switch, then, care has to be taken to consider where and by whom the switch is going to be used. Toggle switches in the horizontal position have no fixed stereotype, although there is a tendency to consider that movement to the right indicates on.

Task analysis for control layout

The layout of displays and controls on a console depends to a large extent on whether one is dealing with an operation in which the sequence of operation is fixed, repetitive, and consistent, a so-called fixed sequence operation, or whether there is no obvious fixed sequence but only irregularly occurring functions. In the former case the controls and displays can be laid out to match the sequence; in the latter it becomes necessary to group controls and displays according to the nature of the function or functions being controlled. Thus a careful task analysis is needed.

One of the requirements of the task analysis is to prioritize the elements of the task and define the controls and display items required for each element. These can then be categorized in terms of the items considered most important, the items used regularly most frequently, the items used together in any sub-sequence, and the items which are related by function. One can then place in the centre of the console, or the interface of the machine, those items which are considered most important or are most frequently used, in a manner to make them easily accessible and well differentiated. Peripheral to these, and again organized by priority or functional relationship, are the advisory displays and secondary controls.

Logical groupings for the purpose might include a group of sub-sequence operations, or functionally related operations. These should be aligned in columns and rows, discriminated by separation, perhaps with a graphic line around the group, or by some form of coding such as a common colour, or common dial shape, along with, for example, alignment of pointers to some suitable zero point such that any deviating pointer can be readily distinguished. Controls should be placed in relation to displays, commonly below the display in such a manner that the display is not obscured when the control is operated. Labelling should be easily seen, and placed, whenever possible, above the respective control or display.

Chapter 10
Hand tools

Guidelines for tool design

In any area with so broad a background and so long a history as that of the design of hand tools it is difficult to make any dramatic changes simply by the application of the expertise of human engineering. Designs that have stood the test of millennia tend to be good. However, while the broad principles of tool design have been handed down from generation to generation, not all designs have survived without occasional loss of the subtleties that distinguish the good from the bad. Nor is it necessarily true that no improvements can be made in even the simplest of tools; and furthermore, with the development of power tool technology much can yet be done to ensure that the interface between the tool and the user is optimal.

General requirements

Essentially a tool comprises a head and a handle, with sometimes a shaft, or in the case of a power tool, a body. It may be difficult to determine where the junctions of head, shaft, and handle actually occur. In a double-ended wrench or spanner, one head and a portion of the shaft act as the handle for the other. In the case of a hammer the handle and shaft are continuous, while in the case of the screwdriver the same applies to the head and shaft.

A tool requires some motive power which, in the simple hand tool, is supplied by the musculature of the user, while in the caes of the power tool it is provided by a motor mounted in the body or handle, or from some external source. In every case, however, there is some form of handle and it is at the handle that the greater portion of the human interface of any tool is found. Thus the handle is of prime significance as the means whereby human input is applied to the system as a motive force, or guiding and stabilizing force, or some combination. In a power tool the controls and the means of mounting the head are also part of the interface.

In other more complex person–machine systems the information is also the site for the provision of information on the state of the system. That information is usually provided in the form of visual displays. In a person–tool system, however, most of the feedback, other than tactile and kinaesthetic (which are indeed important sources), occurs by direct visual monitoring of the result. Formal visual displays are almost non-existent except in such tools as the torque-wrench, devices for measurement, and a few special purpose displays. Some of these requirements will be discussed later. In the meantime consideration will be given to the requirements for handles.

Grasp

To understand handles it is first necessary to understand grasp. Several approaches have been made towards defining the characteristics of grasp. Some of these have been purpose oriented, and some have been function oriented. A simple and yet embracing viewpoint has, however, been outlined by Napier (1956) who defines the prehensile movements of the human hand in terms of a power grip and a precision grip. Each grip has different functional characteristics but virtuallly all manual activities, excluding the hook grip which will be discussed later, can be classified in terms of their requirement for a precision or power grip, either separately or in combination.

In a power grip the object is held in a clamp formed by the partly flexed fingers and the palm, with counter pressure being applied by the thumb lying more or less in the plane of the palm. Such a grip is found for example in holding a heavy hammer. In a precision grip the object is pinched between the flexor aspects of the fingers and the opposing thumb, as for example in tapping with a light hammer. In fact the position of the thumb and forefinger in relation to the handle determines the relative amount of power versus precision. For precise tapping movements with a light hammer the thumb will be aligned along the handle and the forefinger separated from the others such that the hammer is held in a triangle comprising thumb, forefinger, and middle finger. For a heavy force with a large hammer the fingers are curled around the handle with the thumb giving additional support in a firm power grip. The posture of the thumb and fingers in a precision grip ensures that the sensory surfaces of these digits are used to the best advantage in providing the greatest opportunity for delicate adjustments of grip in response to sensory feedback.

Napier outlines other factors which influence the posture of the hand during function. Of these, the shape influences the grip only in so far as the eventual use of the object will determine how it is going to be held. Thus, other things being equal, it may be more convenient to hold a cylinder in a power grip and a ball shape in a precision grip, but if the cylinder-shaped object is going to be used for a precision purpose, for example a chipping hammer, it will be held in a precision grip, while if the ball-shaped object is going to be used for forceful activity, for example the front handle of a power sander, it will be held in a power grip.

From the point of view of design and operation it is difficult to combine a grasp function with a control function. This indeed can be done, as witness the trigger or lever of many power tools. Greenberg and Chaffin (1976) however, cite one study of a sander with a near-dome shaped handle on its top. The handle of course is used for directing the sander. Downward pressure of the dome, however, also activates the sander. The fingers, therefore, not only are required for grasping but also have to exert counter pressure upwards to allow the palm to compress the dome downwards. This action, which requires prolonged and intensive activity of the small muscles of the hand, is of course intrinsically more fatiguing than a simple grasping action. Such complex requirements should be avoided.

It must also be recognized that while two varieties of grip may be defined for the sake of analysis, the two types of prehensile activity, namely precision and power, are not of course mutually exclusive. One or other may dominate in a given action, but one may yield to the other during the course of an activity, and consequently the design of a tool may have to meet the needs of both. Thus in driving a screw the initial activity is one of precision but as the screw becomes set the requirement for power becomes dominant and the grip changes.

The relationship of the hand to the forearm shows differences between the two grips. In the precision grip the wrist is dorsiflexed and stabilized while the tool lies in the axis of what would be the extended forefinger. In the power grip the hand is deviated towards the ulnar side and the wrist is held in the neutral position between full extension and full flexion.

As noted earlier, another form of grip, namely the hook grip, is found where there is no requirement for precision but where something heavy, such as a power tool, needs to be carried. In the hook grip the object is suspended from the flexed fingers, with or without support from the thumb. Since this grip can be maintained for more prolonged periods than a power grip, heavy tools should be designed in such a manner that they can be so carried.

Handedness

Consideration of the requirements of grasp leads to consideration of the problem of handedness. For a single-handed activity the vast majority of people have a hand preference, some 92 per cent favouring the right hand. A few are completely ambidextrous and all can learn to function adequately with either hand, although few will have the strength and dexterity of their favoured hand in the less favoured hand, even with training.

While the number of persons who are clearly left-handed is relatively small, their requirements should be borne in mind wherever possible. The fitting of handles to tools should make the tool applicable to both left and right-handed persons, for example, in the position of controls in a power tool, unless it is clearly inefficient to do so. It might be noted that the driving of screws and fasteners utilizes the powerful supinating movement in a right-handed person and the less powerful pronating movement in a left-handed person. This limitation has to be accepted since the provision of left-handed threads is not a feasible solution for the purpose.

Hand strength

Hand strength has been the subject of several surveys, although largely confined to somewhat selected populations. The studies quoted by Damon, *et al.* (1966) refer to United States military and civilian workers and have a mean value for hand strength ranging from 41.9 to 59.8 kg (94 to 134 lb) for males and, in another population group including British and United States workers, 24.5 to 33.0 kg (55 to 74 lb) for females. Other studies, quoted by Greenberg and Chaffin (1976), suggest an average grip strength of 43.3 kg (110 lbs) for men.

Shape of handle

The shape should conform to the natural holding position of the hand. In the resting stage, the right hand of a right-handed person holding a tool in such a manner as to meet the requirements of both precision and power will be held more than half-supinated, with the wrist abducted about 15 degrees and slightly dorsiflexed, the little finger in almost full flexion, the others less so, the first finger less than half flexed, and the thumb adducted and slightly flexed. The combination of adduction and dorsiflexion at the wrist with varying flexion of the fingers and thumb generates an angle of about 78 degrees between the long axis of the arm and a line passing through the centre point of the loop created by the thumb and first finger, that is, the transverse axis of the fist (Figure 10.1). While the wrist is being used, of course, that angle will not be continuously maintained. For example, in hammering, the wrist will move from full adduction to full abduction; on the other hand during the operation of a power drill the angle will very largely be maintained continuously.

In general, the shape of a handle corresponds basically to that of a cylinder, or a truncated cone, or occasionally a sector of a sphere, although the basic shape may undergo flattening, or other curves may be

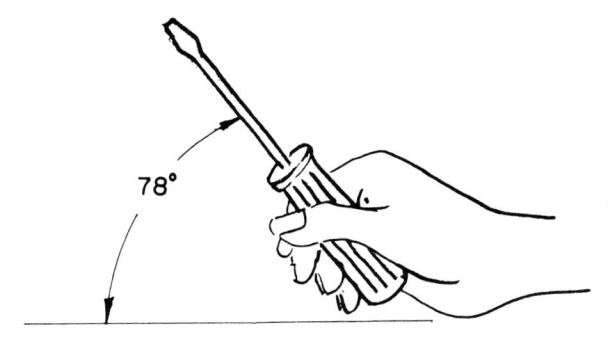

Figure 10.1. Angle of grasp (after Fraser, 1980, derived from Ergonomic Principles in the Design of Hand Tools, p. 27, copyright 1980, International Labour Organization, Geneva).

superimposed upon it. Because of its attachment to the body of a tool, a handle may also take the form of a stirrup, a T-shape, or an L-shape, but the portion that is held by the hand will commonly be in the form of a cylinder or cone.

While a cylindrical form is the basic shape for most handles, a true cylinder is indeed not the desirable shape, except where a handle is intended virtually solely as a hook grip for carrying. Instead, the cylinder should be modified into the form of a curved and truncated cone, such as is found in hammers, screwdrivers, chisels, files, and so on, or in more complex modifications as in the handles of saws and power tools. The truncated curved cone derives from the varying degree of flexion in the fingers during the resting grasp. The space enclosed by the grasp is of course not in fact cylindrical but is complex and multicurved.

It would be a relatively simple matter, of course, to make a casting of that shape and build a handle to match it. This, however, would be a highly undesirable approach since the resulting handle would be appropriate only to the hand on which it was modelled, and only under the circumstances of use under which the model was made. The shape also varies in its dimensions from hand to hand and during use of the hand. Instead of using a contour-matching shape, it is necessary to develop a shape which, with minimal obstruction, will meet the requirements of both hand and function. In fact, any form of specific shaping to a hand is undesirable, such as ridges and valleys for fingers, fluting, indentations, and so on, since with varying shape and size of hands, and varying mode of function the resulting shapes do not in fact fit the hands of a significant number of users. Shapes then should be generalized and basic, sectors of spheres, flattened cylinders, long contoured curves, flat planes, put together in such a manner as to conform in general to the contours of the space of the grasping hand, but not specifically. Particular examples of this shaping will be discussed later in connection with examination of individual tools.

Consideration must also be given to other shapes. The sphere, or portion of a sphere, is in fact not commonly found except in some forms of stabilizing handle. It has the advantage that it can be readily grasped from many angles. It is desirable, however, to use a sector of a sphere as the dome at the head of tools where some of the drive comes via the palm of the hand, such as screwdrivers, chisels, and planes. It provides a good contour fit to the palm with no sharp projections. In this regard it should be noted that no edges should be permitted in the handles of tools. All potential edges should be smoothly curved off. Many contemporary tools, particularly some screwdrivers, still have relatively sharp edges to the fluting that has been designed into the plastic handles with the object of improving the grip. Continued pressure from the edges of tool handles give rise to discomfort, inefficiency, and eventually damage to the hand of the user.

Still another shape, which can be regarded as a modification of the cylinder, is the hexagonal section. It is of particular value for small calibre implements which would otherwise be cylindrical in section. It is easier to maintain a stable grip on a hexagonal section of small calibre than on a cylindrical section. Square or cuboidal sections can be rotated less freely, and have sharper edges, but are also of value. A hexagonal section for the handle of a small screwdriver can be very effective.

Where the grasp has to undergo dynamic change, such as in the use of pliers and shears, rather than remain static, such as with a hammer, there are still other considerations with respect to shape, but the same principles apply. One is still concerned with a cylindrical or conical form, but in this case one is concerned with sections of the cone or cylinder which continuously change in diameter along the handles. The specific requirements for the handles of such tools will be examined later.

Thickness (width) of handles

With respect to thickness, again it is desirable for the handle to conform to anthropometric requirements. Surprisingly, however, very little work has been done to determine the specifications of human grasp, although fairly adequate information exists on the recommended dimensions of knobs, selector switches, and so on. One study on human grasp involving construction workers (Bobbert, 1960), indicates an average maximum inside grasp between thumb and index finger, encompassing at least 70 degrees of a cylinder, to be 7.4 cm (2.9 in).

Design requirements for knob controls, which have some of the requirements of handles, include the following (note that all dimensions are approximate):

Table 10.1. Recommended diameters of knobs (derived from Chapanis and Kincade, 1972)

Grasp	Fingertip	Palm of hand
Diameter		
Minimum	10 mm (0.4 in)	40 mm (1.5 in)
Maximum	100 mm (4.0 in)	75 mm (3 in)

It is clear that these numbers cannot be used directly to determine handle sizes but they provide some reference background. Recommended dimensions for levers are perhaps even more applicable. These are shown as follows:

Table 10.2. Recommended dimensions for levers (derived from Chapanis and Kincade, 1972)

Parameter	Minimum	Maximum
Finger grasp	13 mm (0.5 in)	75 mm (3.0 in)
Hand grasp	40 mm (1.5 in)	75 mm (3.0 in)
Length of grasp area	75 mm (3.0 in)	no limit

From a review of several studies (particularly Ayoub and LoPresti, 1971; Konz, 1974) it is apparent that a grip diameter of 40 mm (1½ in) is most appropriate for a power grip. The grip force at 50 mm (2 in) is 95 per cent of that at 40 mm, and 70 per cent of that at 65 mm (2½ in). For precision grip, diameters of less than 6 mm (¼ in) tend to cut into the hand and do not give sufficient control.

Although 75 mm (3 in) is given as the recommended maximum, a handle of that size would be unsatisfactory. In practice, most handles should range between 25 and 40 mm (approximately 1½ in). Indeed the capacity to apply torque becomes reduced when the diameter of the handle exceeds approximately 50 mm (2 in), (Pheasant and O'Neill, 1975). This latter figure should be considered the operational maximum. For a hook grip a diameter of 20 mm (¾ in) is recommended (Woodson and Conover, 1966). For females all the foregoing recommended limits should be reduced by 10 per cent, and indeed additional as yet unspecified limitations might be required by reason of ethnic heritage.

The actual width for the individual tool will of course vary with its function and size. Thus, where the tool is large and a power grip is required, for example in a heavy hammer or the handle of a power drill, the width will be found at the upper limit of the range. Where the tool is small and demands a precision grip, for example a jeweller's screwdriver or a dental drill, the width of the handle will lie at the lower end, or even below, for special purposes. Where a power grip is required, however, even where the tool head is small, as in a small driver for wood screws, it is necessary to provide a grip at the higher end of the calibre range. Where rotation of the tool handle is unwanted, as in the case of a hammer, it is desirable to have a bilateral flattening, rather than retain a circular or near-circular section.

In general, the thicker the handle the less is the load on the hand muscles. However, because of the shape of the space enclosed by the grip, the calibre of a handle should not normally be the same throughout its length. It will, of course, normally be wider at the thumb end and narrower at the little finger. Representative dimensions for the classic 'pistol grip', which is applicable to many forms of tool handle, are shown in Figure 10.2.

Length of handle

While in some cases the handle of a tool merges indistinguishably into the shaft, as for example in a hammer, in others the length is determined by the working position of the hand. It is thus fixed by the critical anthropometric dimensions and the nature of the grip used. Particulars will be considered later when examining specific tools, but ideally the length must meet the maximum expected dimension at the level of the 97.5th percentile or higher. Thus, for example, a power tool handle must accommodate the maximum closed grasp at the 97.5th percentile, that is approximately 100 mm (4 in), bearing in mind the possible need for gloves, while the heavy screwdriver, used partly in a precision grip and partly in a power grip, must accommodate the length from palm to flexed knuckle of the forefinger, again approximately 100 mm (4 in). Short handles are unsuitable for tools requiring a power grip. A handle with a length so short that it cannot be grasped between thumb and

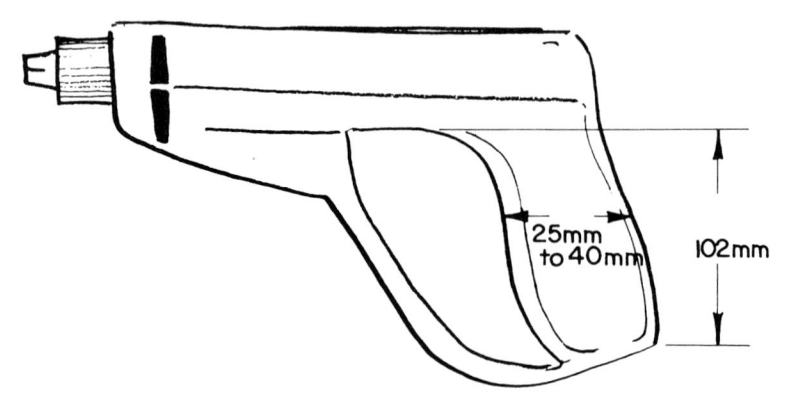

Figure 10.2. Representative dimensions for pistol-grip handle (after Fraser, 1980, derived from Ergonomic Principles in the Design of Hand Tools, p. 44, copyright 1980, International Labour Organization, Geneva).

forefinger, that is approximately 19 mm (¾ in), is unsuitable for any tool. Drillis (1963) notes that the length of handle for a tool such as a file, or for that matter a screwdriver, should be one 'thumb-ell' in folk terminology, or the distance between the ulnar edge of the hand and the tip of the outstretched thumb.

Weight of tool

The weight of the handle should be considered in relationship to the weight and balance of the tool. In the case of percussion tools it is desirable to reduce the weight of the handle to the minimum, and have as much weight as possible in the head. In other tools the balance should be evenly distributed where possible. In tools with small heads, and bulky handles, such as small screwdrivers, this may not be feasible, but the handle ideally should then be made progressively lighter as its bulk increases relative to the size of the head and shaft.

In their study on hand tools and small presses Greenberg and Chaffin (1976) note that the weight of a tool determines its ability to be moved and hence that a heavy weight reduces the proficiency of the worker. While advocating an 11 kg (25 lb) limit on weight they also recommend that requirements for lifting the tool be limited to five times or less per minute. While 11 kg can be considered an upper limit, it would be well to add that for most efficient purposes the weight of a tool should not exceed 4.5 to 5.5 kg (10 to 12 lb). Lifting handles or designated grasp points should be provided in heavy tools so that two persons can cooperate in moving the device. A lifting bolt or hole should be provided for the better use of a hoist or tool balancer to assist in lifting and lowering the tool. The grasp points or handles should allow the fingers to wrap around the surface for at least 270 degrees.

Angulation of handles

The line of transmitted force in using a hand tool passes along the fingers, then through a centrally located carpal bone in the wrist to the radius bone, and up the arm. Although the middle finger is the central finger, and also normally the longest finger, the axis around which the hand operates is not in fact that of the middle finger but that of the index finger. Thus the axis of function of a tool grasped by the hand, whether it is a hammer, a screwdriver, or a power drill, is along the line of the pointing index finger, a fact that has been sometimes overlooked in the design of tools with angled handles. Therefore any angulation of handles that is necessary, for example in power tools or single-handed shears, should be undertaken with this anatominal relationship in mind. Thus the handles should not only reflect the axis of the grasp (that is, about 78 degrees from the horizontal), but the handle or handles should be so oriented that the eventual axis of function of the tool is an extension of the index finger. When two hands are operating in parallel the axis of function is between and parallel to the axes of the two index fingers, as when using two hands, one above the other, to operate a saw. When one hand is supplementing the activities of the other, as in supporting a power drill, the axis of function is that of the dominant hand. This fact will be considered when examining the placement of handles in power tools.

Texture and materials

It is not by accident that for millennia wood was the material of choice for tool handles. In addition to being readily available and easily worked, it has qualities that make it desirable as the link between hand and metal. Its inherent elasticity provides the degree of resistance to pressure and shock absorbency that permits comfortable application of force; its thermal conductivity permits a rate of heat exchange between the tool and the skin such that subjectively it feels neutral in warmth; its frictional resistance allows the application of torque with minimal discomfort at the skin, and also even when the skin is wet with sweat or other liquid; it is light in relation to its bulk, and it is visually and tactually pleasing. On the other hand, although moderately hard wearing, it can be damaged fairly readily and it will easily become stained and impregnated with grease and oil. Even more significantly, it will break and become detached from its mounting. Inevitably the wooden handles of many tools have given way to handles of polystyrene or some other form of impact resistant and stain resistant plastic, which has something of the same qualities as wood but is more durable, more colourful, and more economical. In one quality, however, namely the aesthetic of texture, plastic cannot replace wood. Texture of course is not merely an aesthetic quality; it is also functional. A tool handle requires a readily identifiable texture to provide an input to the sensory nervous system to assist in maintaining the grip. It is desirable in fact to ensure that some distinctive surface texture is incorporated into the otherwise smooth plastic handles for this purpose. As already noted, flutings, ridges, and indentations, which were intended to provide texture and increase frictional resistance, may in fact cause pressure injury on the fingers. Some dull roughening, palpable to the skin of the hand, but neither sharp nor injurious, can serve the purpose better.

Deep recesses of greater than 3 mm (⅛ in) are not recommended because of the variation in morphology of the finger throughout the population. In particular, a person with large fingers may create compression forces on the lateral surfaces of the fingers, which are abundant in superficial nerves, arteries and veins; or a person with small fingers may be forced to attempt compression of two fingers into one recess with similar results. In general finger recesses should only be provided when the primary force is pulling across the palm, as in a tool used to insert or pull apart objects, and then only when small forces of no greater that about 6.5 kg (15 lb) are expected. If larger forces are required a pistol grip tool should be provided. The use of a flange and thumb stop on the handle is also recommended for tools used to insert or press parts together.

If recesses are placed on both handles of, for example, a pair of pliers, there is a possibility of damage to the palm of the hand as well as to the fingers; it is therefore recommended that the force bearing area, where high or repetitive forces are expected, should be designed to span the breadth of the palm, which for 95 per cent of male workers would require a length of no less than 10 cm (4 in), and it should have a curvature of no greater than about 1.3 cm (½ in) over its entire length.

Several other recommendations in this regard can be quoted from the work of Greenberg and Chaffin (1976). Bright highly polished surfaces should be avoided. Smooth surfaces should only be provided when small forces are needed frequently to actuate the tool. Non-reflective ripple coatings should be used in most cases. Cast or machine surfaces should, if possible, be coated with matt paint or other similar material. Should this not be practicable, such surfaces should be sandblasted or otherwise surface treated so that the sharp surface peaks are rounded, thereby reducing the abrasive characteristics of the surface.

All exterior edges of a tool which are not part of the functional operation and which meet at an angle of 135 degrees or less should be rounded with a radius of at least 0.8 mm (1/32 in) approximately. Similarly, all corners formed by the intersection of three or more surfaces of which two form an angle of 135 degrees or less should be rounded to at least 1.6 mm (1/16 in). If rounding is not feasible a layer of plastic or rubber may be overlaid and securely attached. The same principle applies to distinct surface protrusions which should be removed, relocated or countersunk; failing that, the part should be covered with a rounded pliable material. In some situations, for example, a projecting bolt, it might be necessary to cover it with a guard.

A related problem may occur with openings. Exterior openings may be found in large power tools. These openings can catch clothing, tear skin, or even injure joints if, for example, a finger should be caught. Although this is perhaps unlikely, such openings should be eliminated wherever possible, or covered where otherwise necessary. Where the openings are inherent in the action of the tool, such as in the jaws of pliers, or clamps, consideration should be given, where it is compatible with the intended action of the tool, to so design the jaws that even when fully closed they could not compress a finger. Even if this is not feasible, at least it should be ensured that the handles of other moving parts of the tool do not come sufficiently close together that they could trap the skin of the operating hand, as may happen with improperly designed pliers or clamps. Furthermore, operational procedures should be designed to minimize the likelihood of entrapment of body members.

Metal handles may be used in some tools in place of wood or plastic, but to meet the requirements of shock absorbency, thermal conductivity, frictional resistance and texture, they have to be covered with a rubber, leather, or synthetic sheath of a thickness appropriate to the material used.

Size of tools

Various studies have shown that the size of an object, taken in conjunction with its weight, have a multiplicative effect over and above that of either size or weight alone. The size of a hand tool, however, is only of significance when it is unusually large, a situation that is relatively uncommon. A significantly stressful factor in this regard is the horizontally measured displacement of the centre of gravity of the load from the torso (Greenberg and Chaffin, 1976). The size of the load may be a major factor in determining this effect. Where the centre of gravity is displaced from the front of the torso by greater than 25.4 cm (10 in) the ability to lift and lower an object is greatly decreased for both men and women. Consequently, objects that have to be handled, including heavy tools, should be designed to locate the centre of gravity as close to the person's torso as possible. Intructions or markings should indicate either where to grasp the tool to achieve this result, or where the centre of gravity is located, if it is not close to the perceived geometric centre.

Sex

Ducharme (1975) surveyed the opinions of female United State's military personnel on the suitability of a variety of tools and equipment in a variety of craft skills, such as electrical maintenance, vehicle maintenance, aircraft maintenance, metal working, structural work, electronic maintenance, and so on. Each craft had at least one tool or piece of equipment considered to be unsuitable by more than 10 per cent of the female work force. The average age of the workers was 21½ years, the average height 165 cm (65 in), the average weight 50.2 kg (127 lb) and the average hand length 17.5 cm (6.9 in). A selection of some of the offending tools and pieces of equipment from different trades is shown below, with the percentage of workers complaining, and the cause for complaint (Table 10.3):

The foregoing represent only one small selection from a small study, undoubtedly there are many more. For example, soldering guns found unsuitable for the above noted population have a somewhat similar operational action to that of power drills but are lighter. Presumably power drills would also be found to be unsatisfactory. Similarly shears and rivet cutters which were found unsatisfactory by up to 22 per cent of the female work force examined are similar in action to a variety of

Table 10.3. *Tools Unsuitable for Women (after Ducharme, 1975)*

Item	%	Reason
Soldering iron	15–17	too heavy — handle too large
Soldering gun	15	too heavy — can't reach trigger
Crimping tool	13–25	handles too far apart
Wire stripper	11–19	handles too far apart
Metal shears	22	too large — need two hands
Rivet cutter	17	too hard to squeeze
Box hammer (1.1 kg)	11	too heavy
Carpenter tool chest	11	too heavy
Plane	16	too big
Goggles	11–100	poor fit

cross-lever tools, as are also pliers (29 per cent) and crimping tools (13–25 per cent). Even the light hammer caused problems (11 per cent).

Generic tools

Over the 10 000 years or more of their specialized development tools have assumed many forms, but certain clearly discernible functional groups can be identified. The primary difference today is in the distinction between manually driven and power driven tools. In the former, of course, the motive power for operation is derived from the operator alone, while in the latter, although the operator may hold and direct the tool, the motive power is from some external source. The development of each of these fundamental groups is examined below.

Manually driven tools

1. Percussive tools

Percussive tools are defined as those which require a propulsive force to deliver a blow. The basic representatives are the axe and the hammer, the axe having a cutting edge. Other differences are determined by the weight, shape and material of the head, and the length and angular relationships of the handle.

In analyzing a percussive tool one must distinguish between the centre of mass and the centre of percussion (Figure 10.3).

The centre of mass is found at the balance point of the hammer-handle system, normally along the handle from the head, while the centre of percussion is the ideal point at which striking should occur to deliver an optimum blow. The line of action passes through the centre of percussion, which is normally close to the centre of mass of the head, at right angles to the plane of the hammer contact face. Using geometric

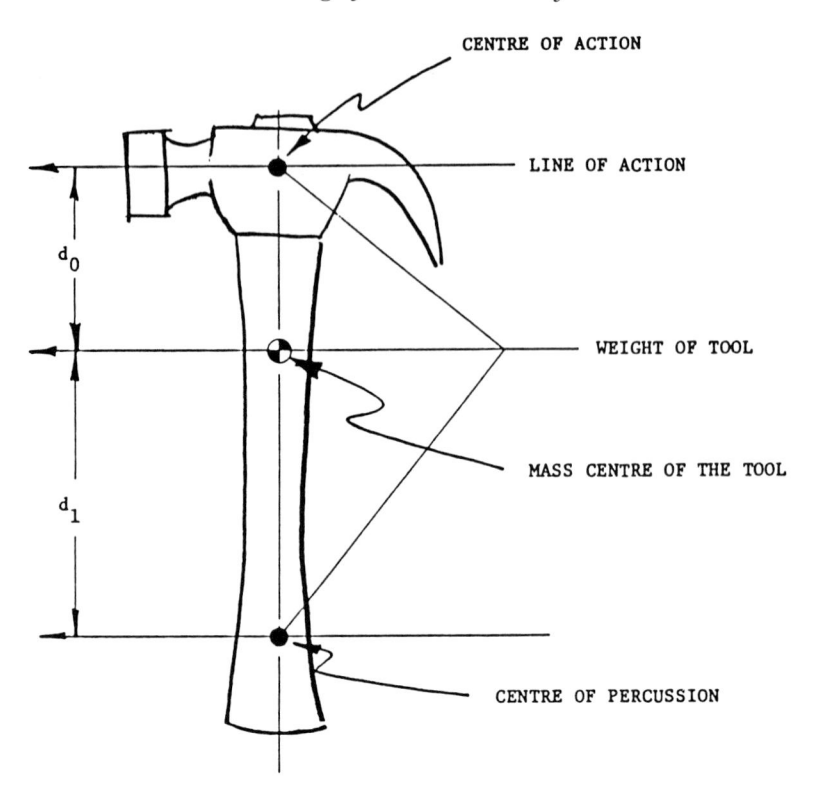

Figure 10.3. Functional centres of hammer (after Fraser, 1980, derived from Ergonomic Principles in the Design of Hand Tools, p. 49, copyright 1980, International Labour Organization, Geneva).

methods. Drillis and his colleagues (1963) showed that the efficiency of the system is a function of the distance from the mass centre to the line of action and the radius of gyration with respect to the centre of action.

Thus for the efficiency of the system to be maximum the centre of mass must lie in the head of the tool on the line of action, an impossibility of course with a shafted tool. The stone hand-held hammer may well have been mechanically efficient, even if the delivered blow was weak.

The hammer, although occurring in many varieties, is one of the simplest tools ever developed, comprising a shaped head and a shaft or handle. Normally it is used in a power grip, but the ordinary carpenter's hammer is not uncommonly held in a precision grip which merges into a power grip as the character of the work changes. Light-weight carpentry and panel-beating hammers, or chipping hammers, and so on, are also commonly held in a precision grip. Thus a hammer handle must meet the need of a wide variety of activities. In fact a straight cylindrical bilaterally flattened wooden handle of calibre within the range of 25–50 mm (1–1½ in), appropriate to the weight of the head, is indeed very effective. The length of the shaft is also a function of the activity. It has been shown that the mean weight for a chopping action should be no greater than 2 per cent of the operator's weight, 6.5–7.5 kg (3–3½ lb), and the mean length of the handle, 35 per cent of the operator's height (Drillis, 1963). Appropriate dimensions for several types of hammer are shown in Figure 10.4.

There are many varieties of hammers. The contemporary model originated in the Age of Metals and has changed little. A sledge-hammer, weighing for example up to 3.3 kg (7½ lb) with a handle approximately 60 cm (24 in), may be used double-handed for heavy driving or working of wrought iron; a fitter's hammer with a 0.9 kg (2 lb) head and a handle of 25–30 cm (10 to 12 in) is a single-handed tool which combines strength with speed; the ballpeen or hemispherical back of the head is used for rivetting. Geologist's and boilermaker's hammers have longer narrower heads for use in restricted space and for concentrating the blow on a smaller area. The back of a carpenter's hammer may have either a narrow straight edge for driving nails with small heads or a claw for extracting

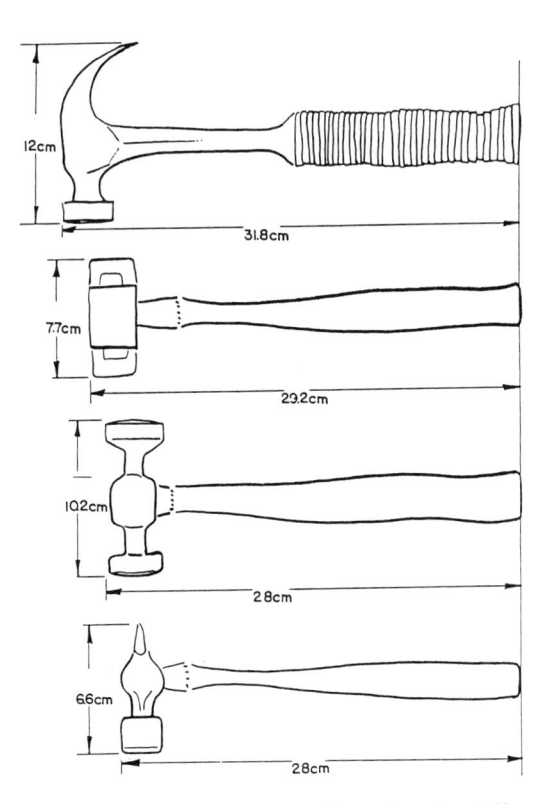

Figure 10.4. Representative dimensions for different types of hammer (after Fraser, 1980, derived from Ergonomic Principles in the Design of Hand Tools, p. 51, copyright 1980, International Labour Organization, Geneva).

nails. The claw hammer, in fact is known from Roman times. Heads made of soft metal, rubber, rawhide, or synthetic materials may be used to avoid damage to the material being struck. Some indeed may be hollow and weighted with lead, while lead itself is used as the head of a plumber's dresser, and wood as the head of a carpenter's mallet or a wooden maul.

Axes and adzes are also striking tools, but with a cutting edge. The essential difference is in the relationship of the head to the handle. In the adze the plane of the head is at right-angles to the handle. The weight and shape of the axe head is adjusted to the operation it has to perform, varying from 0.5 to 2.2 kg (1 to 5 lb) or more.

2. Scraping tools

Saws: Primitive stone tools were of course scraping tools as well as percussive tools, but the saw as an implement did not become specialized until the seventh century B.C., with the beginnings of the Metal Age. It was originally used with a pull cut only. The push cut utilized by most saws today originated with the Romans. Pruning saws, fret saws, and coping saws, with thin narrow blades, may have pull cuts as also do powered reciprocating and sabre saws. The concept of the M-shaped teeth with variable set used in contemporary hand-saws was developed in the Middle Ages, but the modern saw blade originated from rolling mill stock in the eighteenth century.

The action of heavy sawing essentially involves a fixed grasp in the power position, with repetitive flexion and extension at the elbow, while the action of light sawing, such as with a fret saw, requires a precision grip with some manipulation at the wrist. Very heavy crosscut sawing with a two-man saw may indeed require the use of two hands, one superimposed upon the other, but the grasp is the same as for one hand. Thus basically there are two types of handles required for saws, one for a power grip and one for a precision grip.

For the heavier work the broad compound curve of the pistol grip provides a comfortable efficient handle, where the major limiting feature is the width of the gloved or ungloved hand. To conform to the flat planes of the saw the sides of the handle can be flattened without loss of effectiveness, provided that the bilateral width of the handle does not become less than 25 mm (1 in). Narrower widths increase the pressure loading on the thenar eminence and palm and give rise to limiting discomfort. While the compound curve of the pistol grip is desirable it is not mandatory, providing the angle of the handle between vertical and horizontal conforms to the approximate 78 degrees which represents the angle of the resting grasp. All edges however must then be rounded. A high quality design which uses an almost rectangular section with curved edges has been marketed.

Representative dimensions for saw handles are shown in figure 10.5.

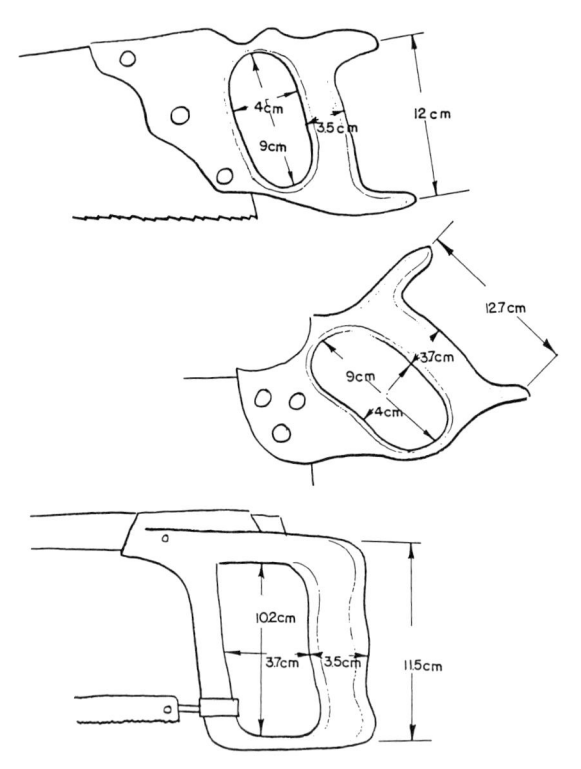

Figure 10.5. Representative dimensions for saw handles (after Fraser, 1980, derived from Ergonomic Principles in the Design of Hand Tools, p. 63, Copyright 1980. International Labour Organization, Geneva).

For light hacksaws, fret saws, and the like, where precision is the keynote, a handle comparable to that of a screwdriver (see later) meets most purposes. Since no rotation of the saw is required, however, the handle can be shaped to conform more easily to the precision grasp, while allowing for a power grasp. Thus a pistol grip angled almost horizontally would indeed meet the requirements even better than the oval handle commonly found.

Files: Files also show their basic origin in the antiquity of rough Stone Age tools. Bronze files, characterized by their teeth oriented in one direction, appeared as long ago as 1500 B.C., but could not of course maintain their abrasive cut. They became popular in the Iron Age and were common by A.D. 1100. Today there are many varieties, distinguished by shape, size and by the presence or absence of a handle, but essentially there has been no change since the Middle Ages.

Chisels: The chisel originated in the stone hand-held axe. As the Age of Metals began, copper chisels appeared, their edges hardened by hammering. By the time that bronze replaced copper a wooden haft had been added to the chisel. With the use of iron different varieties appeared, distinguished by size and shape, until by the eighteenth-century some 70 different types could be counted, including curved gouges. While most of these are no longer in use, the basic chisel has remained almost unchanged.

Planes: No definite line of descent has been identified for the plane, although one might suspect a derivation from the adze. The first clearly identified users were the ancient Romans whose plane was very similar to that of today. The major difference in fact has been the addition of the top iron, or double iron, in eighteenth century England. This device is an inverted plane iron wedged over the cutting iron, which limits the thickness of the shaving and assists its curl.

3. Drilling and boring tools

Drilling and boring tools, such as the awl, gimlet, borer, and drill itself, are derived from the primitive use of abrasive sand on the end of a stick. For many thousands of years the motive power was supplied by rotating the stick between the hands. This activity was eventually replaced by the use of a bow string wrapped around the stick. The stick was then rotated by a sawing action of the bow. The pump drill, which operated by the vertical movement of a handle on a screw, was developed by the Romans, who even added a flywheel to the system to maintain the motion. This method continued to the present day where it is still used (without the flywheel) in the operation of some rapid action screwdrivers. The modern rotary-action spiral drill with handle and ratchet is a product of nineteenth-century mechanical development. Power drills are considered later.

4. Screwdrivers and wrenches (spanners)

In 300 B.C., Archimedes utilized a screw system for raising water; this may be the first recorded use of the screw principle. By the first century B.C., however, very large wood screws turned by hand spikes were in use in wine and oil presses. The same type are still used today.

Metal screws, hand cut, began to appear in the fifteenth century, followed by bolts and threaded nuts in the sixteenth century which were fastened by a T-handle socket wrench. Wood screws, started by an awl or drill, also came into use at that time.

With the increasing use of wood screws came the need for a screwdriver, which initially was a slot-bladed bit used with a carpenter's brace, which itself was an early form of drill. The handled screwdriver, however, did not appear until the nineteenth century, although it became common after 1850 in a variety of shapes and sizes with the mass production of tapered gimlet-pointed wood screws.

Box and socket wrenches also appeared by the early nineteenth century, along with the adjustable sliding-jaw wrench which was originally held in 1830 by a wedge tapped into the appropriate position. Screw wrench patents appeared in 1835 and the familiar monkey wrench in 1858.

The handles of screwdrivers, and tools held in a somewhat similar manner, such as files, scrapers and hand chisels, have special requirements. Each at one time or another is used with a precision grip or a power grip; each relies on the functions of the fingers and the palm of the hand for stabilization and the transmission of force. The screwdriver in particular must also submit to torque. In each case the handle must also be capable of being approached equally effectively from different angles.

The most effective shape has evolved to be that of a modified cylinder, dome shaped at the end to receive the palm and slightly flared where it meets the shaft to provide support to the ends of the fingers. Torque is transmitted by the palm which is passively maintained in contact with the handle by way of pressure applied from the arm and frictional resistance at the skin. The fingers, although transmitting force, occupy more of a stabilizing role. Thus the dome of the head becomes very important in handle design. If uncomfortable, and particularly if there are sharp edges or ridges where the dome meets the rest of the handle, then either the hand becomes injured and calloused, or the transmission of force is transferred towards the fingers and thumb which is more fatiguing and less effective.

The body of the handle is perhaps less significant; different tool makers use different shapes and designs to distinguish their product. Many of them, unfortunately use ridges and flutings which can be undesirable. Pheasant and O'Neill (1975) conducted a study in which they compared the effectiveness, as handles for the exertion of torque, of various sizes of smooth and rough steel cylinders, and a range of commercially available screwdrivers of varying size, shape, and surface quality. None of the screwdrivers was significantly better for the exertion of torque than roughened cylinders of the same mean diameter, nor were any worse than a comparable smooth cylinder.

In these experiments the amount of torque that could be exerted increased as the handle increased in size from 1–5 cm (3/8–2 in). Muscle strength was found to deteriorate when using polished steel handles of greater than 5 cm (2 in). Knurled cylinders were found to be significantly better than smooth up to as much as 7 cm (2 3/4 in) when they became marginally better. The simple knurled cylinder had an optimum diameter of 5 cm (2 in).

5. Holding tools

Spanners, wrenches, and screwdrivers also act as holding tools, but a variety of specialized holding tools has also been developed.

Tongs or pliers date from the first working of metals. Later forms differ chiefly in the shape of the jaws, which may be narrow and rounded for twisting wire, elongated crosswise and grooved for gripping, or with special hinges to ensure a parallel relationship of the two faces at all times. The most complex devices are found in the very wide variety of faces, sizes, linkages, and handles of surgical and dental forceps, each of which is designed for some special purpose.

A number of related but different tools use the same principle of function as pliers. These include wire strippers, grip pliers, pincers, nippers, single-handed cutters and shears, and even scissors. Basically these tools have a head in the form of jaws which have many configurations, a joint which may be simple or complex and forms the body, and two handles. Although occasionally the handles are straight, generally they show bilateral outward or even compound curvature to conform roughly to the position of grasp. Depending on function, the grasp may be either of the precision or power type. General specifications for representative examples are shown in Figure 10.6.

In their common form, with relatively small calibre metal handles, each showing the same degree of curvature, and without any angulation

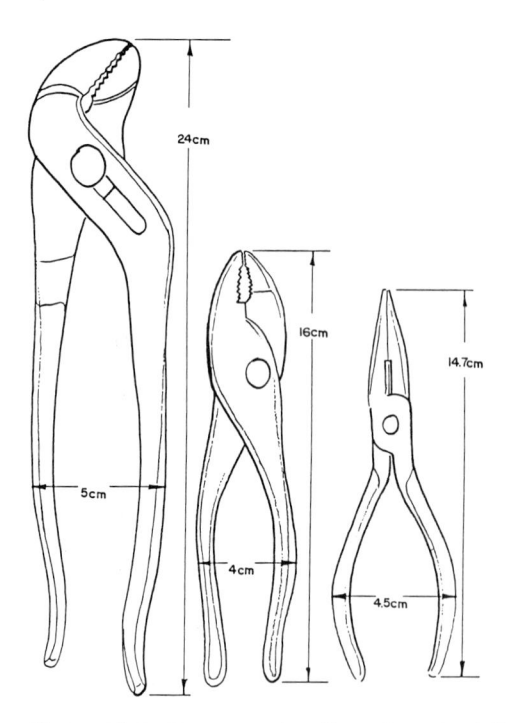

Figure 10.6. Representative dimensions for different types of pliers (after Fraser, 1980, derived from Ergonomic Principles in the Design of Hand Tools, p. 54, copyright 1980, International Labour Organization, Geneva).

of the head, pliers are a relatively simple type of tool which can be used casually with reasonable efficiency and little discomfort for short periods. The relationship and angulation of the handles to the head and each other, however, forces the wrist in normal usage into nearly full pronation and extreme adduction, a posture which cannot be held repeatedly or for prolonged periods without undue fatigue, or even the occurrence of tissue pathology such as tenosynovitis.

For some operations it is also desirable for the pliers to remain in the open position against spring tension. Various types of spring can be used for this purpose. Where extreme accuracy is required, however, spring tension may reduce kinaesthetic and tactual feedback, and it should be avoided. In addition the requirement to repeatedly close the pliers against the tension can readily lead to fatigue and repetitive strain injury among the operators.

Ideally pliers and similar tools should be designed to meet the requirements of the position of grasp at rest, whether in precision or power form. This requires angulation of the handles to the head, as well as compound and different curves in the handles with the lower curve conforming to the inner surface of the flexed fingers, and the upper curve conforming to that of the inner surface of the thumb and the thenar eminence. In addition the handles should be made thicker to provide a more substantial grip, and flatter to increase the area of skin contact and distribute the force more widely.

6. Cutting tools

Cutting tools are derived from the primitive stone axe. The most common form is the knife in which a hardened steel edge is pressed with a sliding action against the material to be cut. The variety is very great, from the general purpose pocket knife to the agricultural scythe and sickle, or from the grossness of the sabre to the precision of the scalpel. Each is distinguished by characteristics of size, shape, handle, balance and weight, although in all a sliding cutting action is used, the proportion of sliding to direct forward movement being very high for the cleanest cut.

Power-driven tools

Any kind of tool can be operated by some form of external power. The decision to use manual or external power is generally determined by the type of work and the productivity demanded. For casual light work manual power is common; for heavier work with a higher productivity requirement power tools should be considered. Electric power tools, for example will pay for themselves in an industrial setting if used for eight hours or more per week (HMSO, 1969). The sources of power for common power tools are electricity and the internal combustion engine. Compressed air and explosive charges are also used for special purpose tools which will not be considered here other than to mention them in a broad general outline.

Electric power tools

Electric power tools are generally compact with one or two hand grips. Most have a finger-operated trigger switch incorporated into the main handle for starting and stopping the motor. Action is sustained by holding the switch in position, or it may be maintained by operating a secondary catch which must be released to stop the action. Depending on requirements, the power may be from main supply or battery.

The tool may be single purpose, such as a circular saw, or multipurpose, with a chuck at the drive end to accommodate twist drills or spindles for other types of functions. The tool is generally driven by a universal motor with a wire-wound rotor supplied through a commutator and carbon brushes, and geared to drive the chuck. Alternative speeds may be available. The motor and gears are contained in either a metal or insulated casing. Typical examples with some comments are presented below:

POWER SAWS:
Electric power circular saws are ten times or more faster than hand saws. The motor, depending on type, may operate on AC or DC current. The motors operate at a high no-load speed, depending on size and design. They are connected to the saw arbour through gearing to reduce the motor speed of 2500 to 7000 rpm to a speed more practical for sawing. Cutting takes place on the upward portion of the cycle, by blades of 15 to 30 cm (6 or 12 in) in diameter, allowing cuts of 45 to 115 cm (1¾ to 4½ in). Most are equipped with an adjustable foot or base which permits tilting of the blade through a controlled angle and also limits the depth of the cut.

Fixed upper guards and movable guards are fitted over the blade, the lower guard commonly held in place by a spring which compresses as the depth of cut increases and allows the lower guard to slide into the upper guard.

Saw blades are available for wood and nonferrous metal cutting, friction blades for thin sheet steel, and abrasive discs, or diamond grit blades, or tungsten tips, for concrete, stone, brick, tile, iron and steel.

Powered jig saws for complex cuts in wood or for dovetailing, as well as reciprocating and sabre saws for straight cuts, curves, and scrolls are similar in general principle.

Power drills

Power drills use electric motors similar to those of power saws with different gearing and different heads. Whereas the circular saw is commonly at the side of the motor, the drill is normally in direct line. The head comprises a clutch into which various shapes and sizes of bit can be fitted. Depending on size and weight the tool is designed for one or two-handed action with one or two handles. In addition, heavy drills may be supported by an external sling and may be convertible to a bench drill.

The larger drills weigh up to 6 kg (13 lb), and have two speeds, consuming 600 watts of power. They may penetrate up to 90 mm (3½ in) in steel (HMSO, 1969).

Power screwdrivers

Screwdrivers are similar to drills. The operating rpm are slower and the tool includes a sensitive adjustable clutch which operates when the driver is pressed into the screw head and slips when the torque reaches a predetermined tension.

Percussion drills

In a percussion drill, which has the same general form of power source, the rotary motion is converted into a powerful fast percussion, up to 160 strokes per second, for penetration of concrete and stone with appropriate impact bits.

Electric hammer

An electric hammer is another percussion tool in which the rotary motion is converted into impacts at a rate of 25 per second. It is generally heavier than a percussion drill, weighing up to 4 kg (9 lb).

Sanders and grinders

In a disc sander the rotary motion from the motor drives an abrasive coated disc of approximately 11.5 to 30 cm (4½ to 12 in) in diameter. It is particularly suitable for carpentry and wood scraping. In a belt sander the drive is applied to a rollilng continuous abrasive belt mounted below the motor. The weight of the tool, 7 kg (16 lb) or less, assists in maintaining the friction. It is valuable for fast sanding of flat surfaces. An angle grinder is similar in principle to a disc sander but is used for weld dressing and cutting off a welded part.

Special tools

A variety of special electric power tools also exists, including a brickwork chaser used for cutting grooves up to 35 mm (1⅜ in) in depth and for compacting concrete by delivering vibrations in the order of 200 per second (HMSO, 1969).

Compressed air tools (Pfeffer, 1971)

Compressed air or pneumatic tools are inherently simple and offer the advantage over electric tools in that they are free of electrical hazard. Because of the requirements for a compressed air source and connective tubing they are clumsy, but they offer advantages of lightness and ease of control. They are also useful under potentially explosive or wet conditions where electrical power should be used either not at all or with great care. Size for size they are generally capable of heavier work than electrical. Both percussive and rotary pneumatic tools can be defined.

PERCUSSIVE TOOLS

In a percussive tool compressed air actuates a piston which may strike freely against the driven toolpiece, for example, riveting gun, caulking, shipping or drilling hammer, or which may be rigidly connected to the toolpiece, for example rammers, percussive drilling and cutting machines. In some cases the pistol itself may act as the toolpiece, for example in a scale chipper. A system of ports or valves controls the input and output of air.

ROTARY TOOLS

In rotary tools the compressed air is applied to reciprocating pistons actuating crankshafts, or to rotors; the resulting motion is used to drive rotary cutting or abrasive toolpieces such as drills, reamers, screw cutters, grinding wheels, and so on. Various types of low speed rotors are found, while for very high speed precision work requiring little power, such as a dentists's drill, a turbine mechanism is employed.

Depending on size and weight, pneumatic tools are designed for one or two-handed use with appropriate hand grips. Like electric tools, most have trigger control which can hold a sustained action.

Internal combustion tools

The only portable hand tool to use an internal combustion drive is the chain saw. Although some chain saws may be driven by electricity, most are used in situations where electricity or other sources of power are unavailable. The chain saw commonly operates from a light petrol (gasoline) engine integral to the tool, which also carries stores of fuel-oil mixture, and oil for lubricating the chain. The reciprocating motion of the piston is converted to a rotary action which drives a toothed chain. The chain running along the rim of a metal blade, projects forward with the blade from the side of the motor for a distance of up to 60 cm (2 feet) or more.

The engine is commonly started, after switching on the magnetic ignition, by pulling on a rope attached to a flywheel. The movement of the saw chain is effected by a clutch mechanism operated by a trigger incorporated into the rear handle of the machine. The rear handle lies in the vertical plane. There is also a front handle lying in the horizontal sideways plane. The device is heavy and clumsy and normally operated two-handed. Action of the saw can be stopped by releasing the trigger, while the motor can be stopped by switching off the ignition. Various types of teeth are available for the chain. Some of the problems associated with chain saws are examined in connection with the effects of segmental vibration.

Explosive drive tools

Explosive charges are used in devices such as bolt guns to force a nail or stud into brick, concrete, or steel. They are also used for cutting various sizes and types of cable, splicing or attaching terminal fittings to cable, removing rivets, tightening rivets, joining pipe, and so on. The principle of operation is essentially the same in all cases. The explosive powder in a cartridge is ignited by means of a percussion-type primer and the resulting force is directed to perform the desired function.

The tools are designed so that the resulting gases are confined and directed. Accurate control is accomplished in various ways. Some tools use a selection of powder charges over a varying range. Others control penetration by means of calibrated powder plugs or a positioning rod to place the sliding device a predetermined distance from the cartridge to reduce combustion volume. All models must be held in close contact with the work and forcibly pressed against it before they are operated. In one model an indicator button informs the operator when the tool is in firing position. In another design it is necessary to depress the tool and rotate a firing ring, thereby using both hands. Still others require the release of safety devices before the tool can be fired.

In each case the cartridge is placed into the breach between the firing pin and the barrel; the bolt or stud is then introduced through the muzzle or breach. The cartridge is ultimately detonated by a firing pin activated either by a lever or twist-type trigger or by a blow by a hammer.

Power tool handles

The majority of power tools use either a pistol grip, sometimes with a supporting side-grip in the case of drills and similar tools, or a front dome in the case of sanders. Some small power tools, such as a light powered screwdriver are essentially the same shape as a manual screwdriver with a somewhat larger handle. All should conform to the general principles previously outlined.

Chapter 11
The office work station

Basically an office work station is no different from a work station on the industrial shop floor. In either case an operator is required to work within an artificial environment, using fixed or mobile equipment to perform a function. Clearly the environments are different; commonly the office environment is less noisy, less dirty, and less hazardous than the industrial. Obviously the equipment is also very different. It is of course because of these differences, and the apparently congenial working conditions of the office, that until recently much less attention has been given to defining and solving the problems intrinsic in office work.

Over the past 5 to 10 years, however, an increasing number of studies has been conducted to examine the quality of the office environment and the potential hazards to health and safety therein. These have been undertaken in a wide variety of fields. It is not feasible to list all the papers that have been presented but some of the more significant are noted here. A number of papers, including an authoritative paper from the World Health Organization (WHO, 1979) and another from a United States government office (United States General Accounting Office, 1980) have stressed the significance of air quality, while problems associated with the presence of asbestos (Anon, 1976), and urea formaldehyde (Harris, 1981) have also been noted. Questions pertaining to heat and cold have been examined (Altman, 1976), and even dermatitis as an allergic response to paper has been discussed (Wikstrom, 1969). Lighting has received attention, both in general (Shoskes, 1976; Maas, 1974; Ott, 1976) and with specific reference to word processing. Much work of course has been presented recently on word processors. This has included work on illumination and glare (Ferguson *et al.*, 1974; Hultgren *et al.*, 1974; Crouch and Buttolph, 1973), stress and fatigue (Smith 1980; Komoike *et al.*, 1971), as well as work station design (IBM Corp., 1979; Kroemer and Robinette, 1969; Stewart *et al.*, 1974), and general health hazards (Hunting *et al.*, 1980; Murray *et al.*, 1981). In addition there is an authoritative text on the topic (Cakir *et al.*, 1980), and a collection of working papers (Grandjean and Vigliani, 1980).

Through much of the discussion of office problems, however, there runs a common thread, namely that many of the problems are ergonomic in origin. Various ergonomic studies have been made of the office environment, including those by Fraser (1983b), Stammerjohn (1981), Nemeck and Grandjean (1973), and Harris Associates (1980), and much attention has been paid, for example to keyboard and work station design, as exemplified by the papers of Klemmer (1971), Kroemer (1971), Duncan and Ferguson (1974), and Ferguson and Duncan (1974).

It will be apparent that the advent of the computer-mediated word processor, and the increasing number of workers utilizing its resources, has brought to light a host of previously unrecognized problems which, although seldom if ever dangerous to life, generate varying degrees of discomfort, reduced proficiency, and disability.

Why the increase in problems is associated with word processing is not at all clear. It is probable, however, that the operator with a word processor works at a higher rate, with a greater degree of concentration, a more demanding visual input, and a higher output than his/her counterpart with a standard typewriter or calculator. Because of the impact of the word processor this chapter is primarily oriented to problems associated with its use. The material utilized in the chapter is derived in part from the above-noted studies. Information pertaining to air quality and other matters in the field of occupational hygiene will be discussed in later sections.

Computer-mediated word processing is of relatively recent origin, coming into prominence in the decade of the 1980's. It of course involves the use of a type keyboard and a visual feedback device known as a video display. The total complex is known as a video display unit (VDU) or a video display terminal (VDT). The latter name is more common in North America, although it is also widely used elsewhere.

Role of radiation

Before exploring the problems of office ergonomics, however, it is well to examine, and place in perspective, the role of radiation in VDT usage. This matter has produced much discussion amongst investigators and a great deal of concern amongst operators. The concern is based on fear of various radiation-induced effects, such as cancer, cataract, sterility, spontaneous abortion or the birth of deformed children, derived in turn from occasional clusterings of such conditions among VDT workers.

There is no doubt that a VDT generates and disseminates radiation. All electrical equipment, and particularly high voltage equipment, does so, in the form of particles or photons. This is not the place to examine the nature of radiation in depth, but is useful to have a cursory review.

Video display screen

Although new approaches are being made to the development of video display screens, such as plasma screens and various forms of electroluminosity, display screens at this time are in the form of cathode ray tubes (CRTs). A CRT is an evacuated glass tube similar to a TV tube. The core of the operation lies in an electron 'gun' which fires a high voltage (12–15 kilovolt) beam of electrons at the inside of the front glass. The glass is coated with one of a number of different 'phosphors' which glow for a short period when hit by a stream of electrons. The beam is focused and moved by electrostatic or electromagnetic devices in such a manner as to place a series of miniature dots on the screen, which in turn 'paint' a picture. According to the nature of the phosphors the picture will be in black and white or colour, and each picture will be retained on the screen for a very short but varying length of time. Since the picture quickly fades it has to be repeatedly refreshed by re-application of electrons from the electron gun.

In addition to the refreshing process then, control circuitry is required to sweep the beam across the screen surface in a series of lines to paint the picture. Additional circuitry is required for data handling and display control, in the form in particular, of very high frequency digital clocks. All these mechanisms, and others, emit radiation, which of course is a form of energy and is characterized by its frequency. The frequency is measured in the unit of hertz, which defines the number of times a given cycle is repeated per second. One hertz equals one cycle per second. Different electromagnetic devices emit radiation at different frequencies, hence a spectrum of radiation energy can be defined.

Ionization and related concepts

All matter is made from elements; the smallest portion of any element is the atom. All atoms with the exception of the hydrogen atom have the same basic structure with a positively charged nucleus surrounded by electrons. The nucleus is composed of protons and neutrons with almost the whole mass of the atom concentrated in the nucleus. Overall electrical neutrality is maintained by the circulating electrons, whose total negative charge exactly neutralizes the charge of the nucleus. Thus an atom in its normal state has the same number of protons as electrons so that the atom is electrically neutral. A neutron has no charge, but its relative mass is the same as that of the proton.

Nuclei are either stable, in which case they show no chemical or physical reactions, or else unstable, in which case they display the property of radioactivity. The unstable nucleus of an atom has all the properties of the stable nucleus of the same type of atom. The instability of the nucleus is due to a change in the number of neutrons. Such unstable atoms are called radioisotopes. These will change spontaneously into more stable isotopes by the process of radioactive decay. In doing so they give off some of the excess energy in the form of ionizing radiation.

Radioactive decay proceeds at a specific unalterable rate for each type of radioisotope. The time required for a radioisotope to decay to half of its initial strength is called its half-life. The half-lifes of known radioactive substances range from a fraction of a second to thousands of years. Obviously the longer the half-life the greater is the potential for biological damage.

Biological damage may occur by reason of interaction between biological systems and ionizing radiation. Ionization is the process by which a fast moving quantity of energy is transferred to some of the atoms of the material through which it is travelling, leaving them as electrically charged ions. An ion is an atom or molecule which carries an electric charge because of the removal or addition of electrons from its shell. Infrared, visible and ultraviolet radiations are classified as non-ionizing because their photon energy is too small to remove orbital electrons from atoms or molecules.

Ionizing radiation is radiation which has sufficient energy to remove or add electrons to an otherwise stable atom. Ionizing radiation can occur from natural or artificially induced instability, or spontaneous disintegration, of an atomic nucleus with the subsequent emission of nuclear particles and photons with a high level of energy.

Natural radiation can result from emissions of cosmic or solar rays, either non-ionizing, as for example in the case of light, or ionizing. Ionizing radiation can also derive from ionizing materials in the soil or in building materials developed from the soil containing such chemicals as unstable isotopes of uranium, thorium or even potassium.

Artificially induced ionizing radiation can also occur. In the extreme conditions it occurs as nuclear radiation from atomic explosions. Much more common, however, although much less intense, is the radiation deriving from clinical use in diagnosis and therapy, as well as in industrial radiography. As already noted, however, radiation, both ionizing and non-ionizing occurs in conjunction with all electrical activity. The following list is merely representative:

* TV, radio, VDT, radar, microwave transmission
* digital calculators, photocopiers, microwave ovens
* suntanning and lighting equipment
* radiofrequency and infrared heating and sealing
* automobile engines
* electric motors

The spectrum of radioactive emissions includes four different types, the first of these is alpha particles, which are in fact the nuclei of helium and contain two positive electric charges. Although they are indeed radioactive they have little ability to penetrate objects and do not enter the body except by such means as ingestion. Beta radiation comprises fast moving electrons which have the ability for limited penetration of the skin barrier. From the point of view of human exposure, gamma radiation is much more significant. Gamma radiation comprises photons which are emitted from the disintegrating nucleus. They have great capacity for penetration and great ionizing potential. Gamma radiation occurs spontaneously. Radiation similar to gamma but generated artificially is x-radiation. It is emitted when a high energy electron beam is directed to hit a controlled metal target in an x-ray machine, or for that matter in a cathode ray tube.

VDT radiation

A VDT does not generate significant x-radiation. It is true that a cathode ray tube, which is the basis of a VDT, can produce x-radiation by way of the beam emitted from the high voltage electron gun. This occurs, however, at voltages higher than 25 kilovolts. The standard VDT operates at 18–20 kilovolts. Other forms of radiation, however, are produced including infrared and ultraviolet which occur as a result of the interaction of the beam with the phosphors on the inside of the glass of the CRT. While both infrared and ultraviolet can give rise to irritation of, and damage to, the skin and eyes when sufficiently severe, as for example in extreme sun exposure or in welding, there is no evidence of any such effect on VDT operators.

Pulsed energy in the radiofrequency band occurs from operation of the digital clock mechanism that controls the picture generation, and from the flyback transformer that is used in scanning control. Again, however, there is no evidence that radiofrequency energy can cause adverse effects in human tissue. Microwave energy has observable effects on skin and body tissues, although there is considerable discussion as to the exact nature of these effects. Fortunately, however, there are no sources of microwave radiation in video display terminals.

There are three other sources of energy of interest in VDTs. These are extremely low frequency radiation (ELF), electric and magnetic field fluctuation, and the effects of static fields.

The effects of exposure to ELF are still under discussion, but there is no clear evidence of any adverse results, nor is there any evidence of adverse results from exposure to electromagnetic fields. It has been suggested in the Swedish literature, however, that the negative electric charge on the front of the skin may attract positive ions from the face of the operator and give rise to a punctate skin rash occasionally found on the face.

Radiation exposure guidelines

Guidelines for radiation exposure have been established over many decades, both nationally and internationally. The two most significant international bodies are the International Commission of Radiological Protection which is an association of scientific experts independent of political and commercial interests. It provides guidance within the field of radiation as a whole, revising its recommendations in the light of developing knowledge. The second is the International Atomic Energy Agency which has among other functions that of providing health protection standards for minimization of danger to life and property. These standards include basic safety standards, and specialized regulations and codes of practice relevant to particular fields of operation. As a result there has been general conformity on radiation health and safety standards throughout the world, although with one major exception namely with respect to microwave energy, where the acceptable exposure in the United States is ten times higher than the equivalent exposure in the U.S.S.R.

The energy imparted by radiation per gram of tissue is called the 'absorbed dose'. Its unit is the rad (radiation absorbed dose) which is defined as 100 ergs (units of energy) per gram. The rate at which the dose is delivered, that is, the dose per unit time, is called the 'dose rate'. The roentgen (R) is the unit in which radiation exposure is measured. The rem (roentgen equivalent mammal) is a unit of relative biologic dose, or dose equivalent. Current dose limits as recommended by the International Commission on Radiological Protection in its 1965 report are shown below.

Table 11.1. Current dose limits (International Commission on Radiation Protection)

Organ or Tissue	Maximum Permissible Doses Adults in Radiation Work	Members of Public
Gonads, bone marrow	5 rem in 1 yr	0.5 rem in 1 yr
Skin, bone, thyroid	30 rem in 1 yr	3 rem in 1 yr
Hands, forearms, feet and ankes	75 rem in 1 yr	3 rem in 1 yr
Other single organs	15 rem in 1 yr	1.5 rem in 1 yr

In addition to agreement on permissible levels of ionizing radiation, there is general agreement on permissible exposure levels for operators on video display terminals and on the emission levels from these terminals. Typical findings from one such study are shown in Table 11.2 below (Muc, 1981):

The actual values of radiation levels found are not of particular importance at this time. What is of greater significance is recognition of

Table 11.2. *Typical VDT radiation (after Muc, 1981)*

Spectrum	Emission Level	Guideline Level
X-ray	<0.05 mR/hr	0.50 mR/hr
Microwave	<0.05 mW/cm2	5 mW/cm2
Radiofrequency	<0.05 mW/cm2	1 mW/cm2
Electric field (RF)	<14 V/m	60 V/m
Electric field (ELF)	<1 V/m	1 kV/m
Static	<1 kV/m	10 kV/m

the fact that the emission levels are very much smaller than the permissible levels.

Management of VDT radiation

While there is still considerable discussion, some of it acrimonious, most authorities recommend that no particular control measures are required. While noting the scientific evidence, some institutions and corporations permit workers in late stages of pregnancy to stop VDT work because of the emotional stress that may be generated by compelling them to continue. This approach, however, is motivated more by politics than by science, since if there is any effect on the unborn foetus any such effect would be more greatly manifest in the early months of pregnancy than the later.

In this connection, one of the most definitive statements ever made by any government body was given by the Health and Welfare Department of the Government of Canada as follows: 'There is no reason for any person, male or female, young or old, pregnant or not, to be concerned about radiation health effects from VDT's' (Anon., 1982).

Role of ergonomics

While in certain aspects, at least, the role of radiation as a hazard in the office may remain in some doubt there is little doubt about the role of ergonomics, which is well substantiated.

As with all problems of ergonomic origin the ultimate effects on the operator are those of increased fatigue, or earlier onset of fatigue. The resulting fatigue may be muscular, with pain, stiffness, and physical discomfort, or even musculo-skeletal injury in the form of various types of overuse or repetitive strain injury; it may be mental or emotional, with weariness, staleness, loss of concentration, irritability, and even dizziness; it may be visual, with eyestrain, eye irritation, headache, abnormal after image, and even disturbed acuity; or more commonly it is some combination of all three. Whatever the effect, the end result is loss of proficiency, and its concomitant, loss of productivity.

Ergonomics in the office

Productivity in the office is no less a function of good ergonomic design than is productivity on the shop floor. The Steelcase studies in particular (Harris Associates, 1980) have shown that the ergonomically oriented complaints of the office worker are real, associated with reduced productivity, and not primarily related to rate of pay, other studies have confirmed the relationship. In a later study, for example, (Humantech Inc., 1986), an anlysis was made of employee problems and productivity over a five year period among workers using VDTs for varying periods of time during the day. It was shown that workers using VDTs for as little as one hour a day had twice as many complaints of discomfort in the head and shoulder region as those not using VDTs. Members of the VDT group showed higher rates of absenteeism, less job satisfaction, and a higher turnover rate among new employees.

In the same study, a special sub-group of 123 workers was evaluated in conditions where their work stations had been modified to meet more appropriate ergonomic requirements. A variety of factors was monitored including absenteeism, per cent of time spent using VDT equipment, number of errors per document, mean time to complete tasks, postural discomfort, and perceived well being.

Analysis of the findings in members of this group showed that during the period under examination absenteeism dropped from 25 per cent to 11 per cent and in particular Monday morning absenteeism dropped from 7 per cent to 1 per cent. Absenteeism, of course, is not necessarily related to the use of unergonomically designed equipment, but its occurrence, and in this case its decrease, can be considered as a general index. During the study period also the use of VDT equipment increased from 60 per cent to 86 per cent, while time on the job increased by 40 per cent. At the same time, according to the subjective questionnaires, postural discomfort was significantly reduced.

The office system

In this regard, one may recognize that the real purpose of the office is not merely to process or manage information as an end in itself. The purpose is to permit pertinent control and proper decision making. The office, in fact, is part of a system the function of which is to assist the decision making process. The office worker, of course, is an integral part of that system. Equally, unless the system is oriented to meet the needs of the worker its functional efficiency will be reduced. The concepts and methodology of ergonomics can be used to delineate these relationships and in particular to provide knowledge for such activities as the definition of standards, the development or specification of designs that are compatible with human capacities and limitations, the evaluation of existing conditions, and the modification of existing conditions to meet the defined ideal. Four components of this system can be defined, namely the task, the worker, the work station, and the environment.

The task

The task amongst other things determines the requirements for equipment and workplace organization. Fundamentally, an office task demands an interaction between an operator and his/her equipment. If it is a VDT task it involves continued interaction between an operator and a computer. Thus we have a classic person–machine system where the input is provided by an operator and the output by a computer. The interface comprises an input device, namely a keyboard, and an output device, namely a display, in the form of a cathode ray tube or other display surface.

In this system three different types of task can be defined, each with some degree of overlap. These tasks can be referred to as word processing, data processing, and enquiry with dialogue.

Word processing

Word processing involves typing of words, correction of errors, preparation of tables, figures, and so on, along with manipulation or relocation of words, phrases, sentences, and even whole paragraphs. In the course of completing the task there may be a requirement to copy from documents, copy from dictation, or to initiate an entry on one's own. The particular characteristics of this task require a keyboard for two-handed operation with facilities for reading, writing, and manipulation of documents.

Data processing

Data processing is concerned just with the manipulation of numbers and consequently requires a numeric keyboard only. The work commonly requires the operator to copy from documents, manipulate data already in the system, and add new material to the system. It can be entirely a one-handed operation. It also requires facilities for document storage, manipulation, and usage. The task not infrequently may accompany word processing and, of course, the keyboards may be combined.

Enquiry and dialogue

The third variety of task, namely enquiry and dialogue, involves querying the computer base with subsequent interaction between the operator and the computer, perhaps at the behest of, or on behalf of a third party. Typically it is the task conducted at an airline ticket or hotel reception desk. It combines the requirements of a normal VDT operation along with facilities for face-to-face communication with the third party. Commonly, although it is not desirable, operators may work from a standing position. A sit/stand operation might be more effective.

The worker

Generation, manipulation, and duplication of words and numbers has always been a function of clerical and secretarial staff. The advent of the word processor, however, has not only brought greater proficiency to this work, but also greater demands on the part of the operator.

These demands in turn have given rise to problems among the operators, manifest as two interrelated states, namely stress and fatigue. Without defining either of these terms at this time, it may be intuitively understood that the interrelationship is close. Stress, be it physical or emotional, leads to fatigue, and fatigue in turn constitutes a stress.

As has already been discussed in Chapter 3, fatigue can be considered from an operational point of view as a physiological and psychological end-result of work, regardless of how caused, that ultimately leads to reduced proficiency, and may indeed lead to impaired health.

The work fatigue of VDT operations may be considered to take three forms. Musculo-skeletal fatigue, such as is induced by prolonged maintenance of an undesirable posture, can ultimately involve stiffness, discomfort, pain and a temporarily reduced capacity for further work. Mental or emotional fatigue, or the fatigue associated with skilled or intellectual work, can cause subjective weariness, reduced capacity to concentrate, irritability, and ultimately reduced proficiency. Visual fatigue associated with the concomitant vigilance of VDT operations can give rise to eye strain and eye irritation, and may result in temporarily altered function.

The characteristics of the worker, however, are significant in determining the type of severity of the stress imposed. Two types of worker may be defined in this connection, namely, the executive and the clerical. Each of these uses the equipment with a different motivation and a different objective. Executive usage, in this connection, however, is not confined strictly to usage by a line manager. It includes usage by such persons as engineers, accountants, or systems specialists of one sort or another. From the point of view of VDT operation, however, the important characteristic, of the managerial user is that he/she uses the system as and when he requires, to achieve an end significant to himself of herself. The use is irregular, discontinuous, and controlled by the user. As a result any dissatisfaction or perturbance that occurs is determined more by the job and the nature of the objective than by actual use of the VDT.

The clerical user on the other hand uses the system as an end in itself. The use *is* the job, continuous, perhaps 6–8 hours per day, and outside the control of the operator. The user is in effect driven by the system and has little or no control over the objectives. Indeed, it is not uncommon that the worker may have to seek approval via the system before leaving the equipment for any purpose, personal or otherwise. Dissatisfaction then tends to be directed to the perceived cause, namely the VDT, aggravated by a variety of disconnected factors.

It was shown, for example, in another portion of the studies conducted by Harris and colleagues (1980), that among executive and clerical workers using VDT's some 76 per cent of the executive workers found the work interesting, as opposed to 24 per cent of the clerical workers. 48 per cent of the clerical workers were bored with the work as compared with 14 per cent of the professionals and 41 per cent of the clerical workers were dissatisfied with the workplace as compared with 17 per cent of the professionals.

Similarly among a host of other complaints, 91 per cent of VDT workers complained of eyestrain, versus 60 per cent of control office workers; 82 per cent complained of painful or stiff neck and shoulders against 55 per cent of controls, and 75 per cent complained of back pain in comparison with 56 per cent of controls (Smith *et al.*, 1980).

These observations not only demonstrate some of the problems that can arise, but also the extent of the differences between the executive and clerical workplace.

Thus, in the light of these differences, the greatest emphasis in the provision of guidelines for VDT work stations should be placed on the needs of the full-time VDT worker. Indeed, while it may be aesthetically and even politically desirable to provide a sophisticated work station for occasional use by a manager it is probably not justifiable in terms of cost-effectiveness.

In this regard, consideration might be given to the recommendations of Bell Canada, who suggest that special treatment of VDT work stations should be provided under the following circumstances:

Table 11.3. *Need for special work stations (after Bell Canada, 1986)*

Proportion of time at Work Station	Provision of Special Work Station
24% or less	Not normally provided
25–49%	Considered
50% or more	Normally provided

Work station

The factors considered above, namely the task and the operational characteristics of the worker assist in defining the work station. From an ergonomic viewpoint, there are three elements to be considered in the layout of an office work station, namely the furniture, the equipment, and the organization of work surfaces.

Office furniture

Office furniture is a major component of the work station, and, indeed office furniture manufacturing is a major industry.

As far as furniture design is concerned the primary ergonomic criterion is that the furniture should permit adoption of a proper working posture that can be maintained in reasonable comfort for prolonged periods. Ideally, the VDT worker should be seated with the legs approximately at right angles to the thighs and the forearms horizontal or sloping slightly downward. Normally this is achieved by establishing the height of the work bench or desk and adjusting the height of the seat accordingly.

To account for both seated leg length and knee clearance it therefore becomes necessary to provide a height of some 720–750 mm to the centre line of the keyboard, of which 650–690 mm should be free leg room. Since the desk itself is rarely less than 20–30 mm in thickness, and since the keyboard is rarely less than 30 mm at its thickest dimension, it also becomes necessary to ensure that the desk is adjustable at two or more levels, one level on which to mount a keyboard, one for documentation and terminal, and perhaps still another for the display screen. The total depth of the work surface should be 750 mm.

The requirement for adjustability will be repeated in other contexts. Probably the single most important factor in ergonomic design of office furniture, and for that matter work stations in general, is adjustability to meet the requirements of different human dimensions. It is necessary also, however, to ensure that the adjustability is simple. It is useless having adjustable furniture that requires awkward manipulations with tools to effect the adjustment, or, worse still, that requires the services of a maintenance man. Such equipment does not get adjusted. Simple adjustment, preferably one-handed, push-button, or lever action, of which a variety of commercial types are available, must be the aim.

Of at least as great significance as the desk is the chair. A number of ergonomically designed chairs is beginning to appear on the market. Indeed marketing specialists have seized on the concept of ergonomic design as a major selling feature in these seats, to the exclusion of other aspects of ergonomics. It is inadequate, however, to take even a properly designed chair and merely place it in front of a standard desk on which is mounted a VDT and its appurtenances and expect thereby that all pertinent ergonomic problems are solved. Instead, the entire work place must be considered as a whole. There are nevertheless desirable characteristics that can be defined for an office chair.

Normally the height of the seat above floor level should be in the range of 450–520 mm, but, as with the desk, it should be adjustable to allow the operator to work with the forearms in an approximately horizontal position. This requirement may indeed conflict with the requirement for knee clearance in persons, for example, with high percentile leg dimensions, and occasionally a compromise may have to be effected. The seat should have a rolled front so that the front edge does not cut into the soft tissues of the thigh and reduce the blood and nerve supply to the lower part of the leg. Similarly, it should not be specifically contoured to fit the form of the buttocks since if the fit were improper the ridges of the contour might again compress soft tissues. Instead of a buttock contour the seat should have slight 'dishing' to allow it to accommodate a wide variety of shapes. The seat should not have a slippery finish since such a finish causes increased effort to maintain a proper seated position. Instead it should be lightly cushioned and upholstered, with a rough textured and flexible surface. The covering should not be compressible in use more than 30 mm. It should allow body heat and perspiration to dissipate, and have not localized pressure points such as seams, buttons or pleats.

Recommended dimensions include a seat width of 400 mm, seat depth of 380–420 mm, a slight upward fixed angle of 2–5 degrees from horizontal to the top of the cushion, a seat back height 400 mm above the height of the seat cushion, with a back angle of 5–10 degrees, a seat back width of 300–450 mm, detachable seat arms 230–280 mm above the seat cushion and 150 mm in length. The arm rests should be padded and not cut into the arm or create pinch points.

Any operating controls should be easy to grip with rough surfaces to prevent slippage. The force required to operate them should neither be so high as to create difficulty nor so low as to permit unintentional operation. All controls should automatically return to the locked position when released.

The base of the seat should be stable to prevent accidental tipping. In this regard five feet are preferable to four. The base diameter should be 400–450 mm.

Since the ideal posture requires that the thighs be horizontal with the knees bent at 90 degrees and the feet either flat on the floor or angled upwards, it may become necessary to provide a footstool. Ideally this footstool is separate from the desk and seat, adjustable in height and angle, and should have a friction surface on the underside to prevent it slipping away from the user. The height should be ajustable to 50 mm at the upper end, with an inclination angle of about 10–15 degrees from the horizontal. Where footrests are not feasible, it is possible, but much less desirable, to provide a foot ring around the chair or a foot bar under the desk.

Equipment

The equipment comprises the keyboard, display, and other ancillary devices such as the printer, document holder, telephone and so on.

Keyboard

Keyboards have been the subject of concern since the invention of the typewriter. It has long been recognized that the format and design of the standard QWERTY keyboard (so named because the term comprises the first five letters of the top row of keys) is inappropriate for prolonged efficient use since it places too much load on the weak fourth and fifth fingers of the left hand. The design, however, is now so well established that it is difficult if not impossible to change it.

The fact of the design, however, and the relationship of the keyboard to the operator's wrists and hands, does give rise to the significant and increasingly more common problem of repetitive strain injury, and specifically tenosynovitis, among keyboard users. Although always a problem with keyboard operators it has become more of a problem with increasing usage of VDT's, as the demands on the operator increase and as the working posture of the operator's hands and arms is forced into more awkward positions by poorly laid out equipment. The problems of RSI are discussed in Chapter 6.

To minimize the likelihood of tenosynovitis the keyboard should be separate from the display, adjustable in the horizontal plane, particularly for one-handed operation, provided with a rest for the wrist, and made with a low profile. Ideally the keyboard should be as thin as possible to allow the forearms to operate in the horizontal position. In practice the thinnest keyboards tend to have a maximum thickness of 30 mm at the home row of keys (that is, the middle row of letters). The angle of the slope should range from 10–35 degrees at the rear, to horizontal at the front. The back row of keys should be no more than 400 mm from the front of the desk, preferably with a free area of 60 mm between the front of the keyboard and the front of the desk.

Key dimensions, by design, and by evolutionary convention, should be square, indented with rolled edges to stabilize the fingers, 12–15 mm on a side, with intercentre spacing of 18–20 mm. They should have a matte surface to minimize glare. Colour is largely a matter of choice with a preference for black or beige-grey.

Other significant features include a key pressure of 0.25–1.5 newtons, and a key travel of 0.8–4.8 mm. Membrane type keyboards, (that is, where the key has no travel) are considered undesirable for VDT work. Indeed, desirable features also include acoustic, tactile and kinaesthetic feedback (that is, perception of motion or acceleration) as the keys are depressed.

Display

Of ergonomic importance in the display is the range of view, which should be 32–40 degrees below the horizontal, with an optimum line of sight of 20 degrees below the horizontal. Of equal importance is the viewing distance, which is essentially determined by the compromise between the need to minimize accommodation, (and hence eye fatigue), and the need to maximize clarity of the visual image on the display. It is also important to ensure that documents and other frequent sources of visual input are maintained at the same viewing distance to minimize the need for repeated change in focus or accommodation. The ideal viewing distance is considered to be 450–500 mm.

Optical characteristics

The visibility of characters on a printed page is determined by the extent to which light is reflected by these characters as compared to the background; the greater the difference the more legible is the character. Hence, for print, legibility can be improved by increasing the illumination.

With video display terminals, however, the brightness of the character is determined by the characteristics of the VDT. Increasing the illumination will decrease the contrast. Contrast represents the difference between character luminance and background luminance. In practice there is a slight difference between the contrast values achieved by different screens, but good terminals, under normal office lighting conditions, can achieve values similar to those of the printed text. Both luminance and contrast are significant in influencing the ultimate legibility of the characters on the screen.

One must distinguish between legibility and readability. Legibility is the ability to recognize and distinguish between individual characters; readability on the other hand is a measure of the ability to comprehend the characters in meaningful units. Both legibility and readability, however, are dependent on the maintenance of an appropriate image quality which in turn is a function of the optical and physical characteristics of the VDT, as well as the characteristics of the generated characters, symbols, and the processes of their manipulation.

Screen size

The viewing area of a VDT screen is normally specified in terms of its diagonal dimension, commonly 12 or 15 inches. In the former case the screen should have a viewing area of 6¾ × 9 inches, and in the latter 8¼ × 11 inches. The relationship between height and width is referred to as the Aspect Ratio and is commonly 3:4. Occasionally, special purpose VDT's may be mounted with an aspect ratio of 4:3. In determining appropriate screen size the objective is to effect a suitable compromise amongst providing a large viewing area, minimizing distortion from curvature at the edges, and achieving a particular aspect ratio.

Resolution and character generation

The size of the screen, as well as the size, brightness, and contrast of the symbols, is a determinant of resolution, or sharpness of image. Each character is defined by a set of dots selected from a rectangular matrix of dots by the scanning control of the VDT. The number of lines within which a symbol is generated affects the quality of the character shape. In general the number of lines should not be less than 10. The resolution is defined by the number of horizontal and vertical dots. A matrix of 5×7 dots, extracted within the 10 lines (raster lines), is normally adequate for capital letters and numerals. For text display, with upper and lower case, a 7×9 dot matrix is better. Upper and lower case provides for a more readable text and renders words more recognizable as a unit. Individual letters such as 'b' or 'p' may require 'ascenders' and 'descenders'. Ascenders, and upper case, should project above the lower case characters; descenders should project below the baseline.

The minimum character size is determined by the ability of the eye to detect a point of light. This point can be shown to subtend an angle of 1 minute of arc. In practice the minimum height for a character is determined to be 16 minutes of arc, or 1/200 of the viewing distance, which in turn defines a height of 2.5 mm at a distance of 500 mm. Better still would be 20 minutes of arc, or 3.1 mm at 500 mm.

Since the fixated eye can recognize objects within a viewing angle of 1–2 degrees of arc then to optimize the number of characters that can be seen in one fixation some compromise is needed in character width and spacing. The width should comprise some 70–80 per cent of height. The stroke width to height ratio should be 1:6 to 1:8. The intercharacter spacing is also significant; it should be 1/5 of the character width. Vertical spacing between the lines has an important effect on making the lines readable. It has been determined that the best compromise lies between 20 per cent to 15 per cent of the character height.

Colour

The presence of colour may enhance a graphic image and may be useful in presenting or highlighting particular aspects of a display. It serves very little purpose in data or word processing. Provision of colour is technically complex, expensive, and presents a display with less intrinsic brightness than an equivalent black and white VDT.

The basic colour of the screen background is determined by the type of phosphor with which the back of the screen is coated. Colours ranging through white, yellow, green, blue, orange, and red can be generated, although with widely differing persistences. Provided that luminance, contrast and other qualities are satisfactory there seems to be no physiological or psychological reason, other than personal preference, for selecting one colour over another.

Stability

Stability of the image, and in particular freedom from flicker, is an important quality for a screen display. The major determinant of flicker is the refresh rate. The duration of exposure of a character on the screen is a function of the persistence of the phosphor on the back of the screen which is activated by the beam of electrons derived from the electron gun at the neck of the cathode ray tube. To keep the image fresh it must be continually regenerated or refreshed. The refresh rate is the frequency with which each point on the surface of the screen is re-established by the electron beam as it sweeps through the raster. Normally the refresh rate is determined by the alternating current frequency of the source, which may be 50 or 60 Hertz depending on the country of origin.

The ability to perceive flicker is also determined in part by the ability of the eye and brain to fuse a flickering image, or the Flicker Fusion Frequency, which is normally about 30 Hertz. A large and bright area, however, will appear to flicker before a smaller and dimmer area.

There are several ways of ensuring a flicker free image. Phosphors vary in their persistence. With a shorter phosphor the persistence is not usually greater than 2 milliseconds; medium phosphors can persist for about 2 seconds, while long phosphors may persist several minutes. While television screens would tend to have relatively short phosphors to avoid the 'smearing' that would tend to occur with a moving image, the VDT, which has a more static image, can use relatively longer persisting phosphors.

Work surface organization

Many of the ergonomic problems occasioned by reaching and stretching can be minimized by organization of the work surface. The work surface can be organized according to the requirements of the task in relation to both reach distance and viewing distance. Thus one may define a Close Area, a Convenient Area, and a Maximum Reach Area.

Close area

The close area can be defined as the work area that can be encompassed with the arms held close to the body. The work equipment in the Close

Area comprises units that are in frequent use, such as the keyboard, the display, and immediate documents.

Convenient area

The convenient area is the area that can be accessed by arm stretch with minimal body movement. The work equipment comprises units which are in occasional use, such as the telephone and secondary documents.

Maximum reach area

The maximum reach area is the area which can be accessed by arm stretch with major body movement. The work equipment comprises units in infrequent use, such as reference documents, the printer and so on.

Environment

In addition to special ergonomic considerations the general office environment is of importance to the health and well-being of the worker. Factors to be considered include noise, air quality, thermal conditions, illumination and glare. The first three of these items are examined in separate chapters, and will only be considered briefly here.

Illumination

General problems of illumination are considered in Chapter 16. They are only briefly examined here with particular respect to video display terminals. It is argued by some that general illumination of VDT workplaces should be kept low (e.g. 100 lux) to minimize distracting glare. The Illuminating Engineering Society on the other hand recommend general office levels of 1000 lux, which are much too high for VDT usage. The consensus is that VDT work places should have an illumination level between 300 and 500 lux with care being taken to shield against both direct and reflectance glare. Local lighting should be used to illuminate the work areas where required.

Glare and reflectance

Glare occurs when the range of luminance is too great in the line of sight. Glare may be direct, that is, caused by actual light sources in the line of sight such as windows or luminaires, or reflected from illuminated surfaces such as walls, desktops, equipment casing, and so on. Glare gives rise to visual discomfort, impairment of visual input, and can, if sufficiently great, or prolonged, produce actual impairment or disability. This latter is not found with VDT work.

Direct reflectance from an untreated screen can amount to some 3–5 per cent of the incident light. Consequently if the room lighting amounts to 5000 to 6000 cd per square meter the reflectance could be 200–300 candela (cd) per square meter, while that of paper or desktop would be only 150–250 cd per square meter.

An unstructured source, that is, diffuse light from no specific shaped source, produces a more diffuse reflectance than a structured source. Hence in the management of glare one should, for example, avoid the direct reflectance of luminaires, as well as the reflection of patterned drapes and venetian blinds with light shining through them.

Direct shielding of overhead luminaires can be accomplished by installing translucent covers. Many are commercially available. The most effective (and most expensive) are prismatic shields, followed by grid or louvre shields, and lastly smoked glass shields. It might be noted that grid shields do not reduce the luminance directly below. Thus reflectance from work surfaces may still be a problem.

Ideally, the VDT should be located at right angles to the source or sources of glare; the luminaires should be parallel to the windows and to the line of sight, and the intensity of glare sources should, whenever possible, be reduced. Failing these measures, the application of screening filters should be considered as follows. Note that any filter, regardless of type, causes some reduced character brightness and resolution.

Colour filters

A colour filter changes the apparent colour of the screen but does little either to reduce the reflectance or improve the readability. The actual background colour of the screen has no effect on its reflectance. It is argued that since green and amber lie towards the centre of the visual spectrum, and tend to stimulate all retinal cones, then green print on an amber background, or vice versa, should be more readily perceived than other combinations. While the evidence for this viewpoint is not conclusive there is certainly no reason to consider that the use of these colours might diminish reflectance.

Polarization filters

With a filter of polarized glass or plastic, the incident light is dispersed. The filter is effective on both direct and diffuse reflectance. The polarized filters themselves, however, tend to be reflective. This reflectivity can be reduced by bonding polarized plexiglass to a vapour-coated anti-reflective glass, a relatively expensive technique which is very useful for small displays (e.g. meters), but not so good for VDT's.

Micromesh filters

Micromesh filters comprise a fine nylon mesh placed in a frame directly over the surface of the screen. A similar type of effect can be achieved using micro-louvres, akin to miniature venetian blinds. The mesh filter gives the screen a black appearance. It is effective but presents problems. Because of the mesh the display is obscured at oblique angles. Thus, while it can be viewed by the operator, it cannot be viewed by someone looking over his/her shoulder. There is some scattering of light and a reduction of character brightness amounting to 30–70 per cent. The mesh tends to collect dust and may be distorted by bending. Either of these phenomena can cause still further reduction in character sharpness. There is also some loss of contrast. The technique is applied best to a VDT operated by one operator working in an area with at least 400 lux of ambient lighting.

Surface etching

Surface etching is a technique of careful roughening of the screen glass. The roughening may be applied directly to the screen, as is found in many contemporary displays, or it may be applied to a panel bonded to the screen glass. The technique has the capacity to reduce the reflectances by half, but with an accompanying loss of character brightness.

Spray-on coating

Several chemical sprays have been developed for application to the screen. If properly applied they can be effective, but generally for a short time only, since they tend to be susceptible to the collection of dust and to smearing by fingers. In some circumstances, indeed, a coloured halo effect may be created. There is also some scattering of light and loss of sharpness.

Vapour deposited coating

Another form of coating requiring a special technique is a vapour deposited coating, which allows for retention of the resolution of characters. The deposit is expensive to apply, but, while the results are subject to smudging, the smudging is less so than with the spray-on approach.

Thin film layer

A special form of coating is the thin film layer, which is deposited by a unique technique that applies a film, with a controlled thickness of one-quarter the wavelength of light. For this reason, since lambda is the Greek symbol for wavelength it is sometimes known as the quarter-lambda layer.

The technique has the capacity to reduce the luminance of the reflections by a factor of 10 without significant reduction in the luminance of the characters. There are of course penalties, chiefly the high cost, although it is also sensitive to fingers and dust.

Tube shields

Tube-shaped shields, which fit over the screen and into which the operator can look have been used for preventing reflectance. They are, of course, clumsy and awkward, tending to fix the position of the operator's head and force him/her into an unergonomic posture.

Filter selection

While the most effective and least costly protection against reflectance and glare is to minimize them at the source and adjust the location of the VDT accordingly, the various filters can be ranked in order of effectiveness. Ignoring the cost factor, the most effective filter is the quarter-lambda thin film layer. The next most useful technique is etching, followed by a polarization filter bonded to anti-glare vapour coated glass. The micromesh filter is probably the next most effective, while a simple polarization filter is least. Taking into account cost as well, regardless of operator preference, the most cost effective filter is the micromesh filter.

Air quality

Office air quality is a relatively newly recognized problem, occasioned in part by an increasing move towards more closely sealed buildings to conserve heat, and in part by a forceful movement to reduce smoking in the workplace. It is nevertheless a real problem, manifested by a build-up in the office air of gases such as carbon monoxide, carbon dioxide, and even formaldehyde. It can of course be managed by improving the ventilation, but that often is no easy matter. General considerations of air quality are considered in Part 5.

Heat and cold

Thermal conditions are also largely a function of the ventilation system although commonly, if a problem exists, it is more a local than a general problem, occasioned by some local thermal aberration such as a draught or a local heat source. And although such complaints may appear trivial they can, over a prolonged period, subtly reduce proficiency.

Noise

The problems of noise in the office are not normally considered as a hazard. The combination of printers, duplicating machines, ventilation systems, paper shuffling, telephones, and conversation can readily reach levels of 70–80 dBA, but rarely, if ever, gets to levels that might injure hearing. As a general rule, however, noise levels should not be permitted to exceed 55 dBA in areas where a high level of concentration is required, and 65 dBA in areas for routine tasks.

Other environmental problems

Another physical entity that can on occasion produce problems is the disturbing oscillation that derives from a floating floor, which, as one walks on it can induce resonant vibrations within the balance organ of the middle ear. Again, these problems are not hazardous, and not even incapacitating, but over time they can be sufficient, in susceptible persons, to reduce proficiency. Even more subtle, of course, are the psychological effects of colour and decor. Psychologists and interior decorators have examined and classified these intangibles to the extent that a reasonable consensus has developed on which some aspects of human behaviour may be predicted, at least on a statistical basis, and certainly there is no reason to doubt that a well-illuminated, pleasingly decorated, brightly coloured office is more conducive to productive work than the opposite.

Office layout

As a general rule an ideal office layout should optimize the flow of work between various departments and various people, while at the same time minimizing movement and noise distractions. These contradictory criteria may be difficult to meet in an open office. Galitz (1980) has summarized some of the requirements succintly as follows:

* Keep people close to those with whom they must frequently communicate
* Keep files and other references close to the people who use them
* Keep people with many outside visitors close to the work area entrance
* Keep common destinations (toilets, elevators, photocopy machines, and so on) close together and accessible by direct routes
* Keep work stations away from sources of intermittent sounds and areas of frequent conversation
* Use barriers to block sound transmissions and prevent visual distractions. Barriers should be at least 5 feet high and go to the floor.

* Use efficient sound absorbing materials for ceilings, barriers, and walls (desktop height to 6 feet above the floor)
* Position overhead lights to minimize sound reflection
* Maintain privacy to the extent feasible

For normal privacy allow at least 80 square feet per work station. For confidential privacy allow a minimum of 200 square feet. Open work stations should have at least three walls, with a slightly widened opening, to maintain both territoriality and privacy while still allowing the worker to participate in communication.

Part IV
Physical agents in the work environment

Chapter 12
Vibration

Vibration can occur as part of the natural human environment, as in the low frequency earth rumblings from waterfalls, and high frequency rattles from the wind and so on. As a hazard, however, it has arisen through technological development and is found in association with the use of tools, equipment, and vehicles. The vibration that is so generated is transmitted from the vibrating object to the body surface, and into or through the body tissues.

Parameters of vibration

Vibration may be random and irregular in its occurrence and its characteristics, or it may take the form of a simple harmonic pattern. Even random vibration, however, can be analyzed in terms of its simple harmonic components. Simple harmonic motion is referred to as *sinusoidal* vibration, since its motion can be represented as the displacement of a particle following a sine wave, as demostrated in Figure 12.1.

A sine wave has fundamental characteristics which can be defined in terms of the amplitude and frequency of the waveform. From Figure 12.1 it will be observed that the imaginary particle moves from the zero line to a peak, decreases through zero to reach a lower peak before returning to the zero line. The displacement of that particle may be defined as the *single amplitude,* (the distance from the zero line to a peak), or the *peak to peak* amplitude, sometimes called double amplitude, which is the vertical distance from the top of one peak to the bottom of another, or vice versa. Amplitude is commonly measured in millimetres or inches.

The *period (T)* of a sine wave is the time duration from peak to peak as shown in the x-axis of Figure 12.1. From the period can be defined the *frequency,* which is the reciprocal of the period (1/T), or the number of cycles of the pattern per unit time. Frequency is normally measured in Hertz (Hz), one Hertz being equal to one cycle per second (cps).

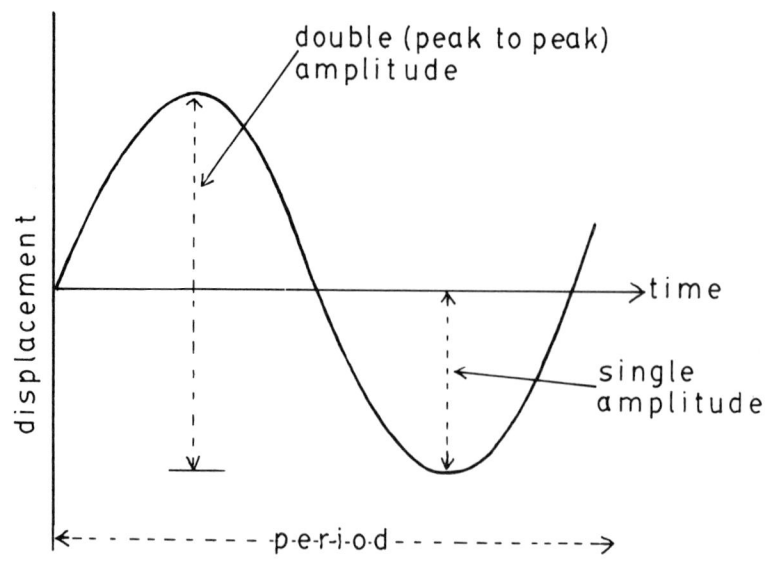

Figure 12.1. Simple harmonic motion.

The imaginary particle has a certain velocity as it completes its cycle. Figure 12.2 illustrates the velocity of the particle. The velocity is the rate of change of displacement. It will be noted that the instantaneous velocity

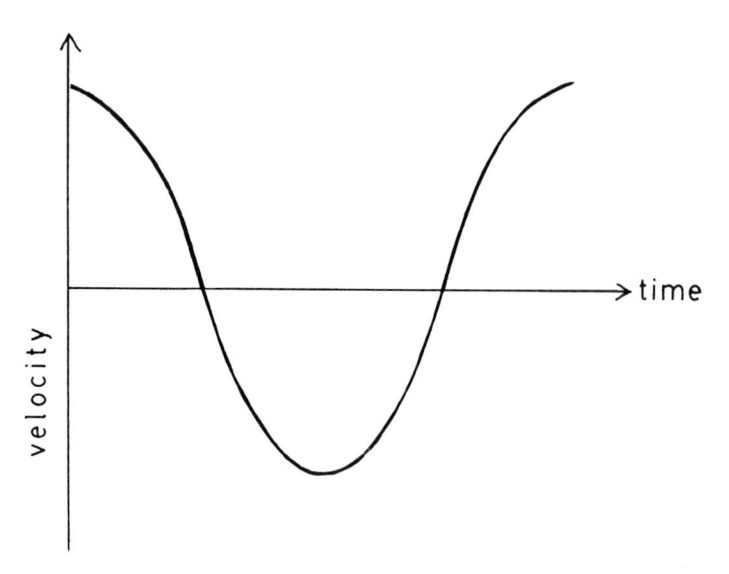

Figure 12.2. Velocity of a particle in Figure 12.1.

is at its peak as the motion begins, reaching zero as it turns at the peak. Thereafter it follows a sinusoidal pattern out of phase with the amplitude pattern. Similarly the acceleration curve, which represents the rate of change of velocity, would follow the same curve shifted still further out of phase.

A fourth parameter is not commonly considered but can be important in examining the human response to vibration at very low frequencies. It is the rate of change of acceleration, sometimes known as the *jerk* or *jolt*. It has the same phase relationship as displacement.

If one were to roll the sine wave curve into a circle, as it were, that circle could be defined in terms of its angular motion. Thus if the angle is given as θ, then the displacement can be shown to be represented by $\sin \theta$, the velocity by $\cos \theta$, and the acceleration curve by $-\sin \theta$. Mathematically then, the velocity (v), acceleration (a), and jerk (jolt) (j) can then be defined as follows:

$$v = 2\pi R f \sin \theta$$
$$a = -4\pi^2 R^2 f^2 \cos \theta$$
$$j = 8\pi^3 R^3 f^3 \sin \theta$$

From the foregoing it will be apparent that the frequency of the vibration is of major significance in determining the parameters of concern, and that in particular the acceleration varies with the square of the frequency. The intensity of vibration is a function of both amplitude and acceleration, but since the acceleration is dependent on the square of the frequency, the importance of frequency in determining the effects will be clearly recognized. At very low frequencies, for example below 5 Hz, the effects of the jerk are of even greater subjective significance, and jerk, of course, is a function of the cube of the frequency. Acceleration is normally expressed in terms of metres per second per second (m/sec^2), or in terms of the gravitional unit 'g', (approximately 10 m/sec^2).

Vibration does not commonly occur in neatly defined sinusoidal form. It is possible by a mathematical analytical method known as Fourier analysis, however, to break up vibration into its sinusoidal components, each of which can be analyzed independently. Alternatively, using a mathematical averaging process involving the root mean square (RMS) value of the sine curve, a power spectral density measure can be determined which gives an overall measure of intensity.

Vibration involves a transfer of energy within a mechanical system, and indeed involves the alternating transfer of energy between potential and kinetic forms, with some loss of that energy due to damping actions of the masses concerned. The human body, however, is not a single rigid mass. It can be considered, indeed, as a complicated system of masses united by springs and dampers. Vibration transmitted into that system meets with an obstruction known as *mechanical impedance*.

Mechanical impedance

Body tissues then, such as bone, muscle, liver and so on, offer varying degress of impedance to the transmission of mechanical oscillatory force, varying with the nature of the tissue and the physiological state of the tissue, for example contraction, relaxation, hydration and so on.

Mechanical impedance is analogous with electrical impedance, and it is convenient to think of it in terms of the electrical analogy. Electrical impedance (Z), as found in the circuit of an alternating current (AC), is akin to resistance in a direct current (DC), although with significant differences. Electromotive force (emf) in an AC current varies sinusoidally from positive to negative. The resulting current is also sinusoidal, but the phase relations are determined by the impedance in the circuit.

Impedance comprises resistance and reactance. The resistance is similar to that in a DC circuit. In mathematical terms it is mathematically real, and it does not affect the phase, as shown in Figure 12.3.

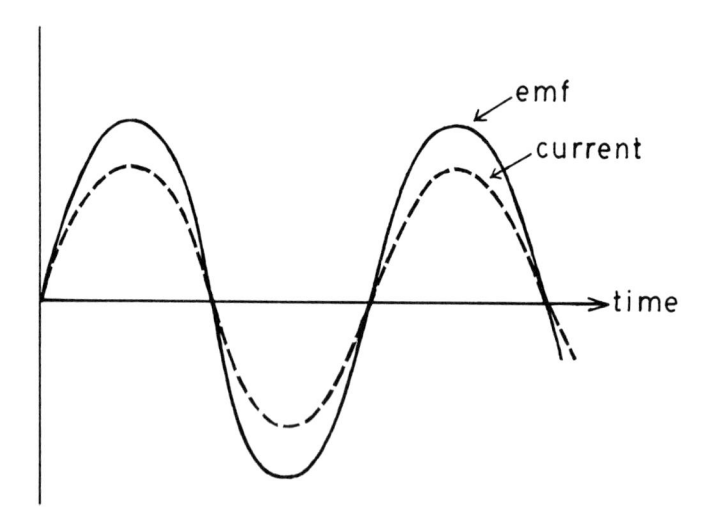

Figure 12.3. Resistance in a direct current.

Reactance comprises capacitative and inductive components and is mathematically imaginary. The capacitative component (X_c) can be shown to vary inversely with the frequency (f) and the capacitance (C), that is $1/2\pi fC$, and causes the current to lead the emf in phase, as shown in Figure 12.4.

Inductive reactance (X_L), on the other hand varies directly with the frequency and the inductance (L), namely $2\pi fL$, and causes the current to lag behind the emf, as shown in Figure 12.5.

Each is clearly a function of frequency and consequently there will occur a frequency in which the reactive component is balanced by the capacitive component, that is:

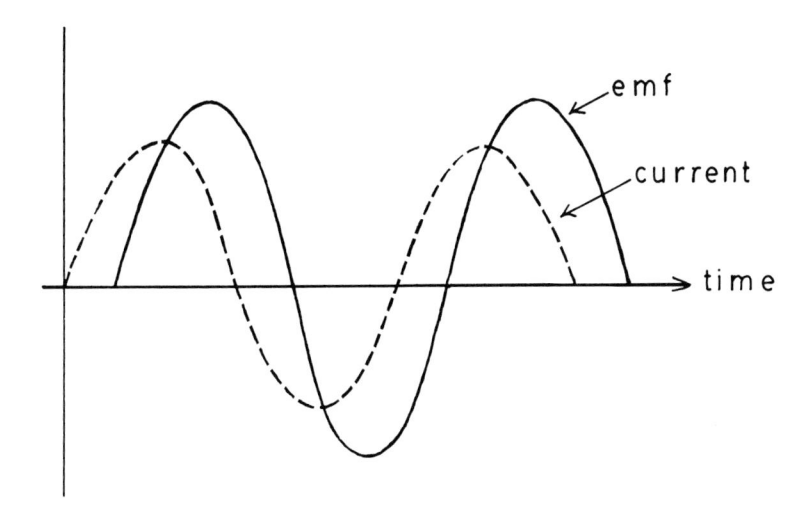

Figure 12.4. Effect of the capacitative component of reactance.

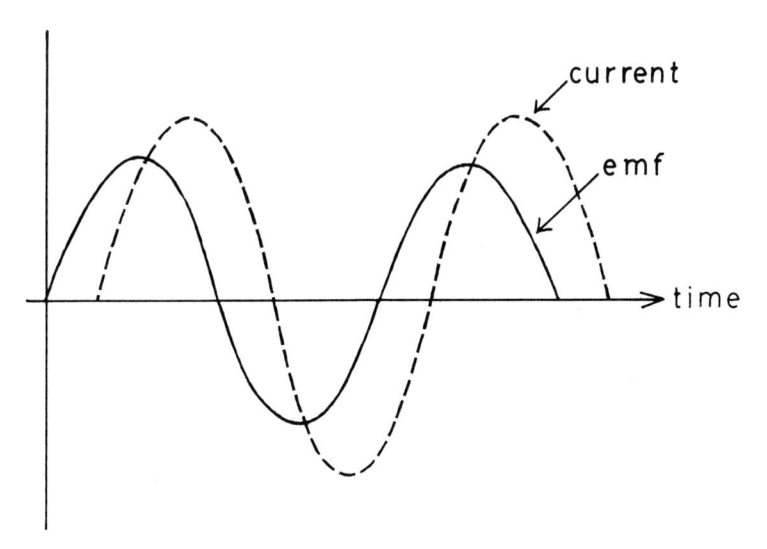

Figure 12.5. Effect of the inductive component of reactance.

$$X_L - X_C = O$$

This frequency is known as the *resonant* frequency (f_o), at which point the energy in the circuit is rhythmically exchanged between X_L and X_C with little loss. The resulting current flow is then obstructed only by the resistance in the circuit. When the resistance is low the oscillations are large.

In a mechanical system the emf is equivalent to the applied force; X_C is equivalent to the elastance of the system, X_L the mass, and current to the velocity. Mechanical impedance is given, then, as the ratio of the applied to the resulting velocity, or $Z = F_{in}/V_{out}$. This relationship is, of course, akin to the relationship defined by Ohm's Law, namely $R = E/I$, where R is resistance, E voltage, and I current. At the resonant frequency the mechanical oscillations are large when the damping is low and less so when the damping is high.

As noted earlier, the human body is not a rigid mass. It is a complex of many masses and it is not surprising that it has many different resonant frequencies. Indeed, all tissues, and all body organs have natural resonant frequencies, some of which can be determined empirically. Because of the immense number of variables it is virtually impossible to calculate the resonant frequency of all tissues and organs, but some can be defined experimentally. Those that have been defined include the following:

Table 12.1. *Human Resonant Frequencies (data from many sources)*

Region	Resonant Frequency (Hz)
Head	20
Chest and abdomen	3
Hips	5, 9
Pelvis	5, 9
Legs	5, 11
Spine	6
Liver	6

It will be noted that the hips, pelvis and legs have primary and secondary resonances, a fact again which is not surprising in view of the complex spring/mass linkages in these regions.

These various resonances, the most of which occur at about 5 Hz, give rise to a very significant whole-body resonance in the range of 5–7 Hertz, depending on the individual. This whole-body resonance, as well as the resonances of individual regions, become very important in determining the biodynamic and subjective responses to vibration.

Terminology

In addition to the terms already introduced, a specialized terminology has arisen to describe the direction in which the vibration is acting on the body. This is shown in Figure 12.6 below, as described by Gell (1961).

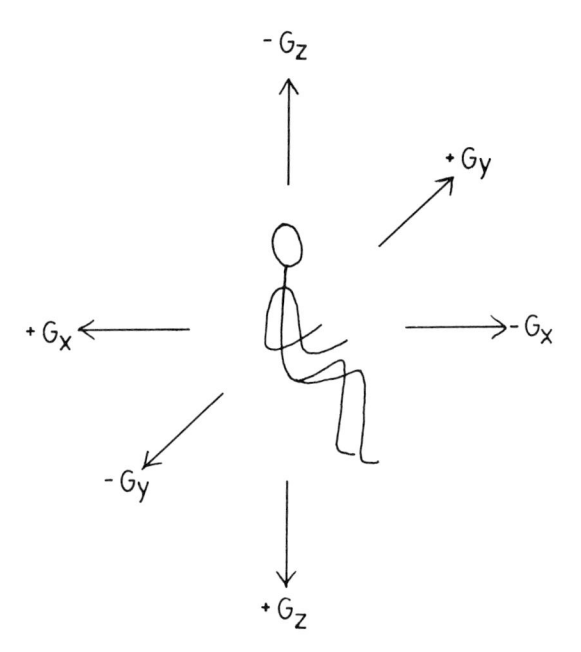

Figure 12.6. Terminology for major areas of human vibration (after Gell, 1961).

Whole-body vibration

Whole-body vibration is also low frequency vibration, and generally occurs at frequencies below 30 Hz. Indeed, as the frequency increases towards and above 100 Hz most of the energy becomes damped out at the surface of the body. Below 100 Hz, and even more so below 30 Hz, the vibration is transmitted into and through the body tissues.

Since the early work of Coermann in Germany in the 1940s, summarized by him in 1962, a number of authors have investigated the effects of whole-body vibration, using widely different devices to apply a controlled vibrational stimulus, or different types of real vehicle or vehicle simulator. These have been summarized by Fraser (1960), Hornick (1973), Kjellberg and Wikstrom, (1985), and Kjellberg *et al.* (1985).

The whole-body effects depend on a number of factors including the physical parameters of the vibration, the posture of the recipient, the presence of restraint and rigid support, as well as personal factors in the individual.

Physical parameters

From the foregoing discussion it will be apparent that the frequency of the vibration, particularly in relation to the natural frequency of whole-body resonance and resonance of tissues and organs is a prime factor in determining effects.

The intensity of the vibration, as determined in particular by the acceleration, is another prime factor. The subjective intensity, however, is modified by the direction in which it is applied to the body. Numerous workers quoted in the summaries above have determined that the most effective human damping is found when the force is applied in the vertical $\pm G$ direction, next best in the fore and aft $\pm G$ direction, and least in the sideways $\pm G$ direction.

Posture

The posture of the subject to whom the force is applied is also significant. The most effective human damping is possible when the subject is standing with the knees bent. In this posture the large muscles of the thighs and back aid in the damping process. The next best is seated in a relaxed position, followed by standing in the erect position, and then seated in the erect position.

Restraint and support

The provision of restraint and/or rigid support is a matter of controversy. The use of a shoulder harness, lap belt, and thigh belt in the seated subject, for example, ties the person closely to the source of the vibration and reduces any differential movement between seat and body. At the same time, provision of restraint and support reduces the capacity of the body to utilize the large muscles of back and thighs to assist in damping the applied vibration. Restraint is probably most useful under conditions of relatively high amplitude and intensity at whole body resonant frequencies, such as may be found for short terms in military and space operations, for example 0.5–2G at 5–7 Hz, or at lesser intensities for longer times such as may be found among drivers of heavy construction or lumbering vehicles. Where the operator can readily maintain his physical equilibrium and is not being heavily bounced within his vehicle very close restraint is not desirable.

Personal factors

In any conditions where subjects are exposed to adverse environments the results of the exposure can be considerably modified by reason of personal factors within the individual, both psychological and physiological. In vibration exposure the physical shape and form of the individual can be significant. The muscular, squat person is less affected by, and more tolerant of vibrational exposure, even at frequencies of whole-body resonance, as compared with the taller lean individual (Fraser et al., 1961). Similarly the more physically fit the subject the greater is his capacity to endure the exposure, as compared with the unfit, and particularly the obese. Perhaps most significant of all is the motivation. Whatever the mechanism, the highly motivated subject is less affected by the exposure than is the poorly motivated.

Effect on performance

In exposure to whole-body vibration the effect on the capacity of a person to perform a task is very largely the result of physical difficulty. While some problem is found at any of the lower frequencies, that is below 100 Hz, the greatest difficulty is found at frequencies less than 30 Hz, and of course the most disrupting disturbance is evident at the whole-body resonance frequencies of 5–7 Hz. The type of problem encountered can be considered as follows:

Interference with fine discrimination

The ability to conduct a task that requires fine discrimination can be grossly interfered with. In different types of exposures, whether on simulators or in actual vehicles, a number of authors have commented on such difficulties as those associated with touch, manipulation of controls and switches, pattern matching, and so on (Buckhout, 1964, Clarke et al., 1965, Fraser et al., 1961, Schmitz et al., 1960).

Tracking

Again, either on simulators or vehicles, tracking of a visual display by some form of control has been found to be difficult under vibration exposure, not only because of the difficulty in maintaining visual contact but also because of the problems of maintaining coordinated control (Catterson et al., 1962, Chaney and Parks, 1964, Fraser et al., 1961).

Visual acuity

As noted in connection with tracking, visual acuity may be a problem. There is loss of fine visual discrimination, and in an operational context significant difficulty has been observed in such tasks as reading dials and maps (Dean et al., 1967, Dennis, 1965, Fraser, 1960, Mozell and White, 1958, Pradko, 1964, Taub, 1964).

Disturbance of speech

Where the vibrational input is sufficiently intense, and particularly at frequencies below 20 Hz, the maintenance of smooth coordinated speech can be difficult (Nixon, 1962). At very high intensities of random vibration embracing the frequencies of whole-body resonance, as in high speed low level flight, the attempt to speak can be frustrated by inability to control breathing (Fraser, 1960).

Reaction time and higher mental processes

The ability to respond rapidly to a stimulus under vibration has been investigated by various authors (e.g. Buckhout, 1964, Holland, 1967, Hornick *et al.*, 1961, Hornick and Lefritz, 1966, Soliday *et al.*, 1965, Schohan *et al.*, 1965). The consensus is that vibration typically does not affect reaction time, or if so to a very slight degree, namely by a few hundredths of a second.

Tasks that primarily involve higher mental processes are not measurably affected. Mental addition, pattern recognition and matching, and navigational behaviour have been shown not to be impaired at frequencies below 20 Hz.

Physiological effects

At the intensities of vibration that could normally be expected in industrial or even military situations the physiological effects of vibration are non-specific and negligible. Any that are observed are probably more associated with the acompanying non-specific stress than the specific characteristics of the vibration. There has been some indication of increase in heart rate, respiration rate, and oxygen consumption (Fraser, 1960, Hornick, 1973, Ziegenruecker and Magid, 1959).

Subjective effects

Subjective effects vary with the intensity and the frequency. The greater the intensity the more unpleasant are the subjective effects, while in addition specific localized effects may occur at frequencies of regional resonance, such as the head (20 Hz), the eyes (60–90 Hz), the hand (30–40 Hz) and so on. In general terms high intensity exposure may arouse annoyance, accompanied by a feeling of insecurity, which in turn may lead to anxiety and even fear. There is generalized bodily discomfort, aggravated at whole-body resonant frequencies, while if regional resonances are stimulated there is discomfort, which may be severe, of the part or parts concerned. Exposure beyond a few minutes to levels above the limits of tolerance leads to a generalized and seemingly disproportionate fatigue.

As the limits of tolerance are reached there may be headache (aggravated at resonant frequencies), sweating, accompanied by nausea and vomiting. At intensities of 1–2 G, at whole-body resonance, dyspnoea (difficulty in breathing) can become marked, accompanied by pain generalized over the chest and torso, or localized to resonant sites (Fraser, 1961).

Tolerance levels

Tolerance levels have been defined by a number of authors. For military and astronautic performance in particular some extreme tolerance levels have been defined, assuming highly motivated, fit, young persons. Magid et al. (1960) defined short time (1 minute and 3 minute) tolerance levels as indicated in Figure 12.7.

In this figure the characteristic decreased tolerance level in the whole-body frequency range will be noted. Virtually all tolerance curves follow this same general pattern.

More generalized long duration tolerance levels are displayed in curves developed by the International Standards Organization (ISO 2631, 1974). This standard provides curves covering exposures of durations from 1 minute to 24 hours, at frequencies ranging from 1 to 100 Hz, and accelerations up to 1 G or more, in each of the three axes of transmission.

The Standard also recognizes three subjective criteria, namely, and firstly, the fatigue-decreased proficiency boundary which, as the name implies, is the level at which proficiency may be decreased by reason of vibration-induced fatigue; secondly, the exposure limit, which is of

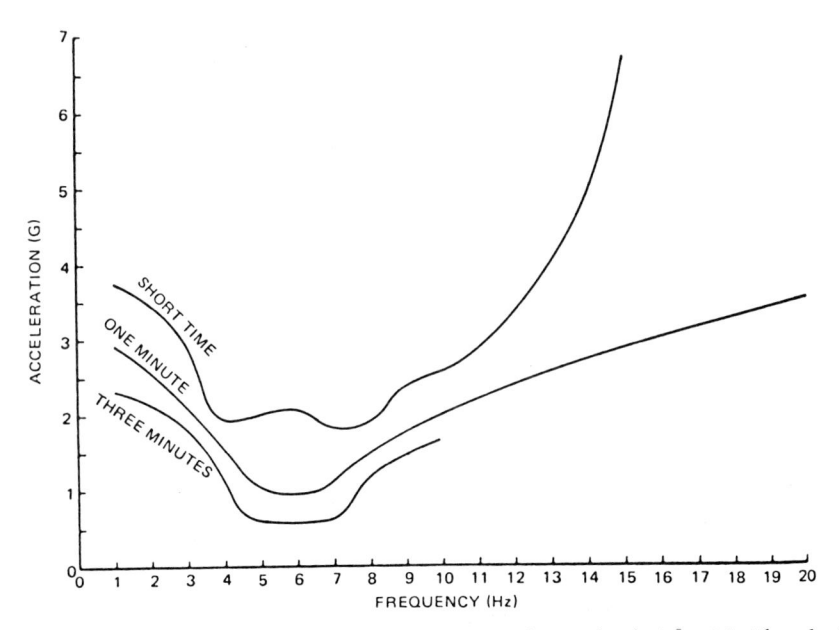

Figure 12.7. Short term, 1 minute and 3 minute tolerance levels (after Magid et al., 1960).

intensity twice as great as the fatigue-decreased proficiency boundary and determines a level which is acceptable for the preservation of health and safety; and thirdly, the reduced comfort boundary which lies at a level intermediate between the fatigue-decreased proficiency boundary and the exposure limit. An example of the type of curve presented, which is by no means all of the guideline and should not be used as such, is shown in Figure 12.8.

Some authors have argued that the ISO standard is not adequate as a guideline. Notable among these is Oborne (1983) who points out that the standard lacks empirical support in a number of different areas, although it is supported in others, and particularly with respect to its recommendations on vibration in the vertical plane. He observes, however, that for recommendations on performance and comfort there is no experimental justification, and the use of the same shape of frequency weighting for different criteria is not experimentally justified. At this time, however, there is little or no reasonable alternative, and while there is no doubt that continued revision is desirable, the ISO curves are still widely used throughout the world.

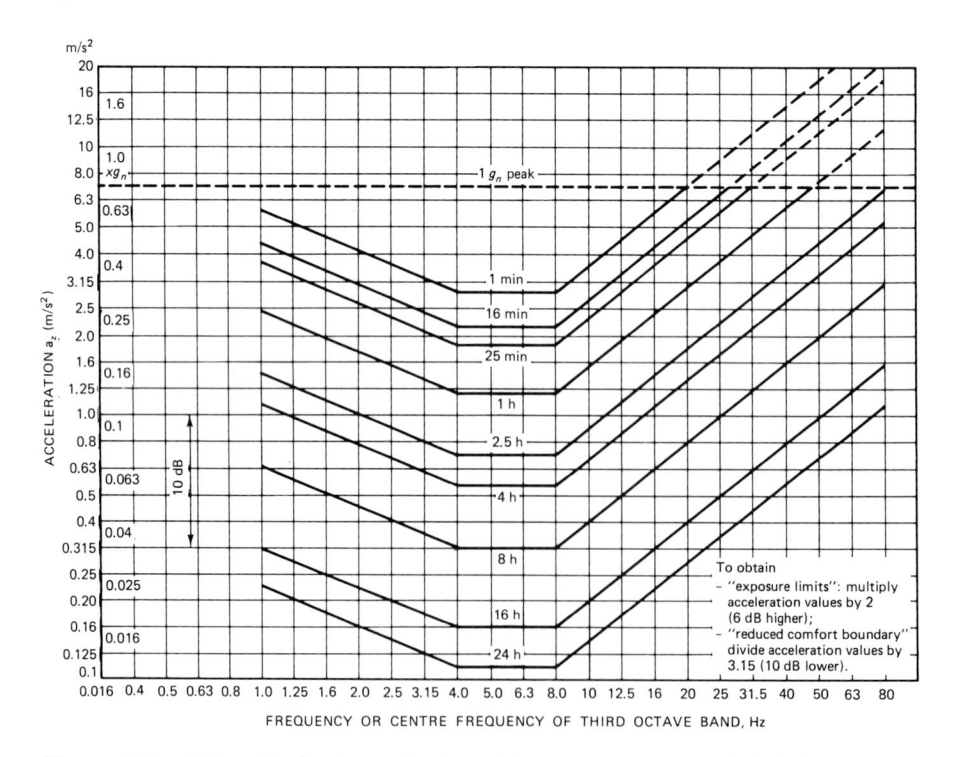

Figure 12.8. ISO guide for the evaluation of human exposure to whole-body vibration (after ISO, 1974).

Protection against whole-body vibration

Since the origins of whole-body vibration are normally inherent in the source, which is commonly a land, water, or air vehicle, protection is not easy except by major engineering change. Clearly the input energy could be reduced and thereby reduce the resulting applied force. Since, however, applications of that energy are what is desired this is not normally a feasible approach. Similarly one could reduce the mass of the vibrating source and thereby apply the same force with reduced acceleration. Again this is not normally a feasible solution.

The most common solution is to isolate the person from the vibrating source by providing an adequate suspension system, either on the vehicle, as in a car, or on the seat, or both where applicable. In this regard it might be noted that mere provision of a cushion on a seat is not necessarily a good solution. As illustrated in Figure 12.9 below a soft cushion is ineffective against heavy vibration and indeed may amplify the acceleration, although it is effective for comfort in low level acceleration or static conditions.

The figure shows that a particular forcing function over a specific time will reach a certain peak and will have an 'impulse' (which is acceleration times time) of a certain value. The force is applied to the mass of the body via a cushion. The cushion initially deflects, storing energy as it does so; the body may then 'bottom out' before finally accelerating in the direction of the vibration vector to hit a still higher peak. The total impulse of each of the two curves will be the same, but the peak of the second may be higher. A suitable cushion in these circumstances should be firm and slightly yielding.

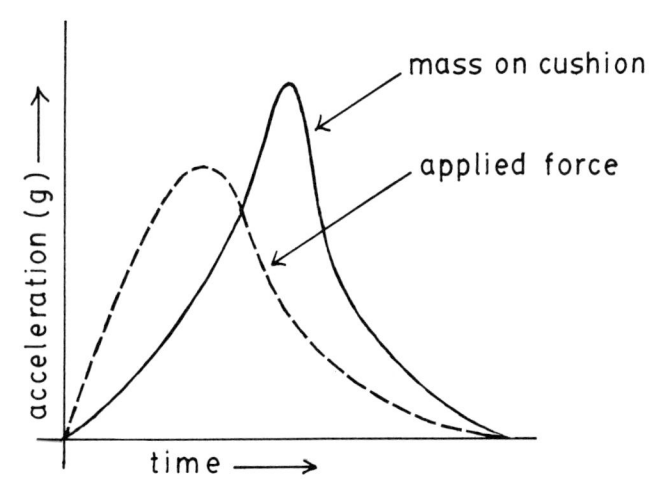

Figure 12.9. Effect of a soft cushion on transmissibility of whole-body vibration.

The possibilities of support and restraint as protection against vibration have already been discussed, along with the importance of ensuring the best posture wherever feasible.

Human response to segmental vibration

Segmental vibration is the term given to vibration of a segment of the body such as the hand/arm/shoulder region. The term is commonly applied to vibrations resulting from the use of hand tools. Undesirable localized vibration tends to occur in impact tools at frequencies of 30–50 Hz, and in oscillatory or rotary tools up to 1000 Hz, although above 150 Hz symptoms are less common. Symptoms are most likely with vibrations between 25 and 150 Hz (Ramsay, 1980).

Relevant history

In 1862 Maurice Raynaud published a medical thesis entitled *De L'Asphyxie Locale de la Gangrene Symetrique des Extremities* in which he described recurrent blanching of the tips of the fingers of persons exposed to cold. Some 90 per cent of his subjects were young females. It was assumed that the condition resulted from intermittent spasm of the small blood vessels supplying the fingers, and that there was some inherent susceptibility amongst the sufferers.

Raynaud's Disease, as it was called, remained something of a medical curiosity, but in 1911 a similar vascular spasm was described in Italy leading to blanching of the fingertips among miners using vibrating pneumatic tools. In 1918, the condition was observed among limestone drillers (Hamilton, 1966, Rothstein, 1918).

Between 1918 and the 1980's the condition was described in many different occupations, including other types of mining, fettling, drilling, riveting, hammering, chiselling, punching, and polishing, among others. By the mid-sixties (Grounds, 1964), it was recognized that one of the most serious problems arose in chain sawing, to the extent that some 50 per cent of chain saw workers in Sweden would ultimately be affected.

Terminology

Since no actual disease process was involved in the occupational occurrences the condition until relatively recently was known as the Raynaud Phenomenon, a term which is still encountered. More recently other terms have been applied, such as traumatic vasospastic disease (TVD), white finger disease (WFD), or vibration induced white finger (VWF).

Development and progression

The condition occurs among workers exposed to manual vibration in the frequency range of about 100 to several hundred Hz with acceleration levels of approximately 10–30 metres per second per second. Below 50 Hz the vibration tends to be transmitted into and through the body; above 250 Hz it tends to be localized to the hand. The critical range is in the order of 50–150 Hz, with a peak of 125 Hz. Acceleration at levels less than 10 metres per second per second is insufficient to cause a problem. Occurrence of the condition is encouraged and its progression aggravated by concurrent exposure to localized cold. It is also probable that there is a genetic susceptibility in those who acquire the condition.

While many of the aetiological and provocative factors are known, and while pathology clearly derives from spasm of the small blood vessels (arterioles) of the fingers, the actual mechanism of causation is unknown. It might be speculated that since neural information governing the signals transmitted to the arterioles via the autonomic nerve fibres is pulse coded in the range of 100–200 Hz, vibrational signals superimposed in that range on the nerve fibres might act to block or interfere with their normal function, resulting in arteriolar spasm.

Several authors have detailed the development and progression of the disorder, notably Taylor *et al.* (1971), and Taylor and Pelmear (1975). After beginning work with a potentially hazardous tool there is generally a latent period of months to years without symptoms, or with occasional very brief episodes of blanching of the tips of the fingers, particularly in cold. Thereafter, there appear initially irregular and infrequent episodes of finger blanching lasting for 5–15 minutes. The blanching is preceded by cyanosis or blueness of the fingertips from stagnation of the blood and may be accompanied by numbness, tingling, stiffness and loss of dexterity. The fingers may be swollen. The phenomenon resolves spontaneously, with initially no lasting effect, and may not recur for months.

Should exposure continue, the condition will recur more and more frequently, and last for increasingly longer durations, perhaps up to one or two hours. There is progressive crippling and ultimately the subject will be unable to work. In the early and moderate stages the conditions will resolve completely on removal from the causative work. In established disease, however, the condition may continue to progress even after the worker is removed from the exposure. In the final stages, fortunately rare, there may even be gangrene with total destruction of the fingertips.

Stages of development

Taylor *et al.* (1974) have outlined a schema for defining the stages of the condition, which allows comparisons and definition of different management and control regimens. In *stage one* there is blanching of one or more fingers but no interference with the capacity to work.

In *stage two* there is repeated blanching of one or more fingers, generally in the winter. There may be slight interference with capacity for activities in the home and surrounds, but no interference in the capacity to continue work. In *stage three* there is extensive blanching affecting all of the fingers in each hand, occurring in summer as well as winter. There is interference with personal, social, and work activities.

Stage four is the final stage. There is extensive blanching occurring frequently in all fingers, both summer and winter. The condition has become so severe that the subject can no longer continue at that type of work and will voluntarily change his occupation.

Prevention and management

Prevention and management of the condition is feasible only to a certain extent. In addition to certain specific actions which will be described it is predicated on the definition of certain guidelines for permissible exposure.

Guidelines for permissible exposure

A number of different guidelines for permissible exposure have been suggested. Three of the most significant are those of Miwa (1968), those of the British Standards Institution (1975), and those of the International Standards Organization (ISO) (1975).

Miwa proposed tolerance levels between 4 and 1000 Hz, for 10 minutes, 1 hour, and 8 hour exposures. Up to 16 Hz the tolerance levels for each duration have equal acceleration levels, but above 16 Hz the limits increase with increasing frequencies.

Taking note of the fact that detailed information on the effects of exposure, and in particular the effects of interrupted exposure, not available in the BSI approach, the next step has been to develop two curves, one for 150 minutes of exposure to different frequencies, and one for 400 minutes. Figure 12.10 below (Hempstock and O'Connor, 1977) shows a number of different curves, including the BSI 400 minute curve.

The ISO approach has been more complex and has taken into account more types of exposure. The ISO curves are shown in Figure 12.11. In promoting the BSI curves, Hempstock and O'Connor argue that there is insufficient evidence to support the ISO curves.

Bearing in mind the guidelines for permissible exposure, there are certain approaches that may be taken for the control and management of vibration induced white finger disease, although control is feasible only

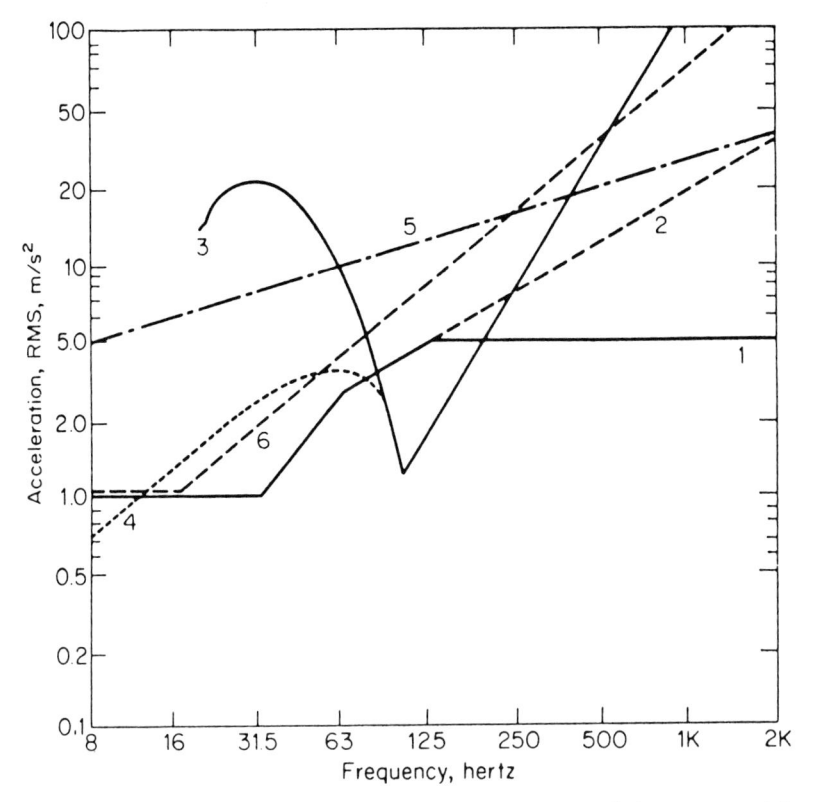

Figure 12.10. Maximum vibration levels for continuous daily exposure (after Hempstock and O'Connor, 1977).

in part. These approaches can be considered as administrative, engineering, and personal protection.

Administrative approaches

Selection and screening: Wherever possible, and sometimes it is neither feasible nor even legally acceptable, a potential worker should undergo a medical examination to determine his fitness to undertake work that could present a vibration hazard. Clearly the medical history is one of the most important aspects of any physical examination. Any worker with a history of previous white finger disease, or any history of spontaneous blanching in the absence of provocative stimuli, should not be accepted.

Unfortunately there is very little in the way of feasible objective test procedure to determine potential occurrence. There are two possible approaches. One is to measure digital blood flow to ensure that the fingertips are getting a reasonable blood distribution. This is not an easy

220   *Physical agents in the work environment*

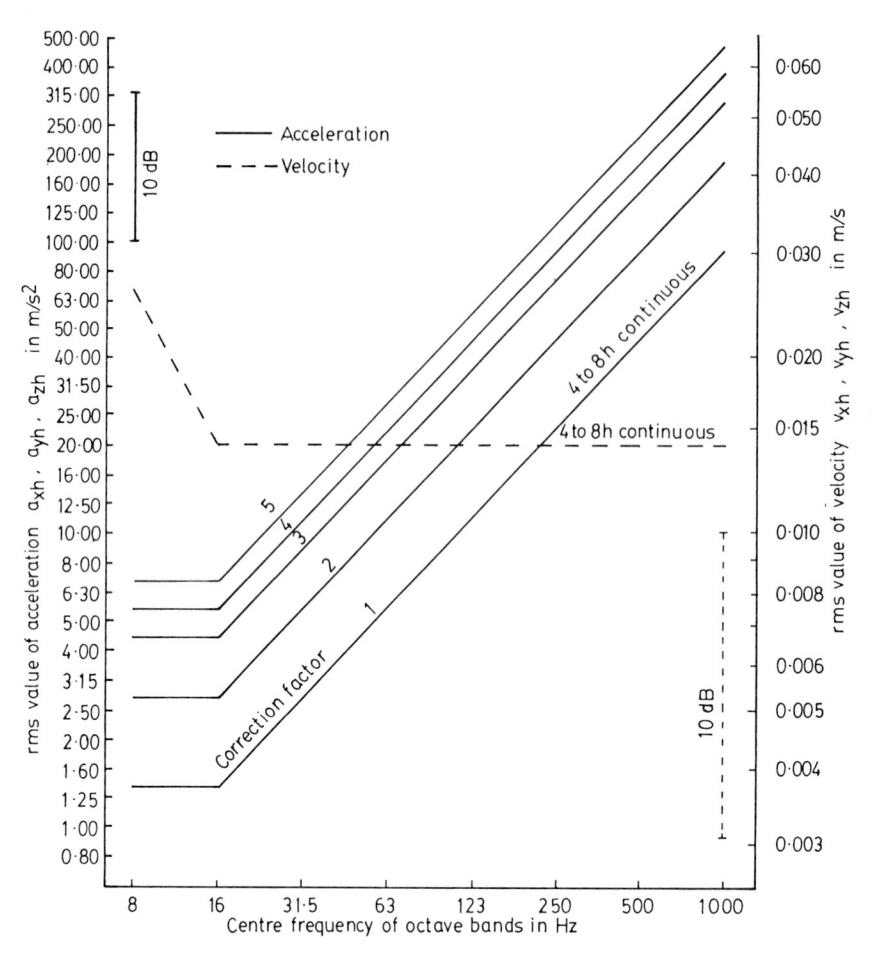

Figure 12.11. Guide for human exposure to vibration transmitted to the hand (after ISO, 1975).

task. It can be done through the medium of digital plethysmography whereby the finger is placed in a device which records the change in volume of the finger with each arterial pulse and provides a basis for calculation of blood flow. Another, and even less direct method, is to measure the change in skin temperatures by skin thermometer, or in heat generation by infrared thermography.

A second approach is to stimulate a possible spasm by a cold provocative test, such as placing the hand in ice water and observing, and timing, the occurrence of blanching.

Removal from exposure: Clearly removal from exposure is indicated where a clear stage one or stage two pattern has been established. Certainly at the stage two level it is desirable to remove the worker from the source of his problem to prevent any further progression, and even at the stage one level it is desirable where feasible. Even if complete removal is not feasible some form of rotation through non-hazardous work is desirable, although it is still argued as to whether work interruptions or separations from the exposure make any ultimate difference to the outcome.

Engineering approaches

The concept of engineering control is based on recognition of the fact that if the vibrational energy is reduced at the source less will be transmitted into the body, and that if the acceleration component is reduced the less will be the generated force.

Reduction at the source: Reduction at the source is generally not very feasible since it is the high energy level of the vibrating tool that is required to undertake the task that the worker is attempting to do. In other words, overall reduction of power at the source reduces the applied acceleration and defeats the purpose of the tool.

One approach to reduction at the source is found in certain chain saws. A chain saw is commonly operated by a single piston internal combustion engine. The action of the single piston creates a high impact vibration which is transmitted through the handle to the hand. Use of a balanced twin-piston engine has had some effect in reducing the vibration, as also has the use of a rotary Wankel engine.

Reduction in transmission: Vibration from the source is transmitted to and through the handle. Reduction in transmission can be achieved by increasing the mass of the handle or by physically isolating the handle. The effect of increasing the mass depends on the Newtonian relationship that force equals mass times acceleration. In other words, by increasing the mass of an accelerating object the resulting force can be achieved with a reduced acceleration. Therefore by increasing the mass of the handle relative to the mass of the tool the same effective force will be attained. Obviously, however, there are practical limits to this approach.

One such technique has been applied to chain saws, which of course are the greatest culprit, by placing the fuel tank in the handle and thereby increasing its mass. Unfortunately as the fuel is used the value of the technique diminishes.

Isolation of the handle is another method of reducing transmission. This technique is based on the concept of mismatching impedances. Every material has its own impedance. Where impedances are matched, transmission will readily take place across a join. Where impedances are different there is obstruction to the transmission. Thus the handle can be isolated by various connecting materials such as cork, rubber, plastic and so on. The isolating material readily succumbs to wear and tear. One of the most useful in this regard is plastic bonded to steel, which provides for both durability and mismatching of impedance.

Personal protection

Personal protective devices are unfortunately not very effective in the range of frequencies associated with vibration induced white finger. Gloves and pads can be useful when the frequency is above 800 Hertz.

Since the onset and severity of the condition are affected by a cold environment, maintenance of warmth can be helpful, both whole-body warmth and local warmth at the site affected.

Chapter 13
Noise and sound

Sound is a specialized form of vibration, specialized because it is perceived by the human ear. It can exist wherever motion occurs in an elastic medium. Normally that elastic medium is air, although it can, for example be water. No sound can be transmitted in a vacuum. The resulting vibrations are transmitted through the air as fluctuations of pressure, referred to as waves of compression and rarefaction. These waves of compression impinge on the eardrum and are transmitted into the ear to generate nerve signals which are perceived as sound.

Parameters of sound

As noted in the discussion of vibration, all vibrations including those of sound are analyzable in terms of their sinusoidal frequency components. With respect to a sound tone, the most significant frequency component is said to be *dominant*. It determines the *pitch* of the tone. Others may be subdominant or supradominant, depending on whether the frequency is lower or higher than the dominant frequency, and are known as *harmonics*. The harmonics are integral multiples of the dominant frequency, that is, for example, twice or three times the dominant frequency, and determine the quality, or *timbre*, of the sound. A tone generated on a piano may have the same dominant frequency as a tone generated on a trumpet, but the harmonics of those tones will allow the person hearing them to distinguish their different qualities.

The *amplitude* of vibration of a sound wave is determined by the displacement of the molecules of the transmitting medium from their place of rest. The energy involved in that displacement determines the intensity, or its subjective equivalent, the *loudness*, of a sound. Subjectively, the intensity as appreciated at the ear does not increase in a linear fashion with the amplitude. The ear is more sensitive at the speech frequencies (approximately 300–3000 Hz).

Noise

Noise has been considered as a non-harmonious assortment of audible vibrations. This, however, is a rather unsatisfactory definition since what is harmonious to one may not be harmonious to another, witness cultural appreciation of music. Most definitions of noise tend to be unsatisfactory since they depend on subjective interpretation. Perhaps the simplest is that noise is unwanted sound. Burrows (1960) took another approach when he defined noise as 'that auditory stimulus or stimuli bearing no informational relationship to the presence or completion of the immediate task'. Clearly, however, acoustic noise is subjectively appreciated; to understand its perception it is necessary to examine the human ear.

The ear

The human ear can be considered as a transducer; that is it converts one form of energy into another. The ear, in fact, is a remarkable example of a complex transducer which takes vibrational energy and converts it by way of mechanical and hydraulic linkages into electro-chemical nerve impulses. From the ear the nerve signal is transmitted to the brain via the auditory nerve whence the signal is processed, codified, and ultimately perceived.

The ear comprises the external ear, the middle ear, and the internal ear. The external ear is made up of the *pinna*, or the projecting portion of the ear, and the *auditory canal*. It acts as a collecting chamber, in the form of an exponential horn, and allows for selective amplification of sounds (about 3–5 dB) in the speech frequencies. The auditory canal terminates in the *ear drum*, or tympanic membrane, which is an elasic membrane separating the external from the middle ear. It can vibrate in response to the applied waves of compression and rarefaction. Its mechanical impedance, that is, its capacity to respond to vibration, favours the speech range.

The *middle ear* is a hollow bony chamber filled with air. It is traversed by a mechanical linkage system comprising three tiny articulated bones, the *ossicles*, attached at one end to the back (or inside) of the eardrum, and at the other terminating by insertion into a tiny hole in the bone, the *oval window*, that opens into the inner ear. The ossicles transmit the signal from the ear drum to the inner ear. They are known, because of the fanciful shapes ascribed to them by the ancient anatomists, as the hammer (malleus), anvil (incus), and stirrup (stapes). Attached to the stapes is a tiny muscle, the *stapedius* muscle, which is always maintained in some contraction, or tone. By changing the level of tone the amount of 'slack' in the transmitting system can be changed in an involuntary manner, in response to a potential or perceived stimulus. Thus, when about to be faced with a very loud noise, such as gun fire, the tone is reduced, thereby reducing the transmitted energy; when listening very intently, on the other hand, the tone is increased, thereby increasing the transmission of faint sounds, or, 'pricking up one's ears', which is no doubt the sensation ascribed to increased contraction of the stapedius.

Since the middle ear is filled with air it is responsive to changes in atmospheric pressure. To allow equalization of pressure there is a tube, the *Eustachian tube*, connecting the middle ear and the back of the throat. The anatomical structure is such that effectively there is a flap valve at the throat end of the tube. This valve allows easy exit of air from the middle ear but does not permit such easy ingress. This phenomenon causes a build-up of pressure across the ear drum when one is descending rapidly from high altitude, or ascending from underwater, which can be relieved to a greater or lesser extent by forcible opening of the valve in the act of swallowing, or the more violent Valsalva manoeuvre in which one attempts forcible exhalation of air against a closed mouth with the nose held firmly between the thumb and forefinger. Failure to relieve the pressure head will cause acute pain or even rupture of the drum. A much lesser pressure differential can also occur on ascent to altitude or descent under water, and can if necessary be relieved in the same manner.

The oval window opens into the internal ear, or *cochlea*, which is Latin for snail and applies to the shape of the internal ear, which vaguely resembles a snail on its side. The cochlea provides the hydraulic linkage in the transmission. It is filled with fluid from base to apex. Internally it is divided by a *basilar membrane* which winds through the turns of the snail to the apex, which it does not quite touch. Thus within the cochlea there is formed a liquid filled channel on one side of the basilar membrane from the oval window through 2½ turns of cochlea to the apex, at which point it connects with the channel on the other side of the basilar membrane which retraces the turns to end at still another window, the *round window*, which opens back into the middle ear.

Thus, a signal enters the external ear, induces vibration of the ear drum which is picked up by the ossicles and transmitted to the inner ear at the oval window. This induces a wave motion in the lymph fluid of the internal ear which passes through the cochlea to the apex and back down again, whence the pressure of the following wave is relieved into the middle ear, and subsequently through the Eustachian tube into the back of the throat. The wave motion in the inner ear induces motion in the basilar membrane. Attached to the basilar membrane are microscopic nerve fibres which are stimulated by its motion. These fibres unite to form the *auditory nerve*. The stimulation pattern, which is complex and not merely a rectilinear representation of the frequency with low frequencies at the base and high at the apex, is interpreted in the brain as sound.

Range of hearing

Human hearing covers a broad range of frequencies, although not so broad as in many animals, nor necessarily in the same band. While the range varies from individual to individual and is markedly reduced at the upper end in the process of aging, the general range is from about 15 Hz to 15 000 Hz. Young people can commonly hear frequencies of over 20 000 Hz. Noise at frequencies less than 15 Hz can be perceived, but it tends to be felt rather than heard, and may be perceived over the whole body.

It is convenient for analytical purposes to divide the operational range into octave bands, an octave occurring when the frequency is doubled. Standard octave bands are as follows, although any octave can be defined. There are of course eight (hence octave) in the convenient hearing range, namely:

Table 13.1. Standard Octave Bands of Sound Frequency

Octave	Frequency range (Hz)
1.	31.5–63
2.	63–125
3.	125–250
4.	250–500
5.	500–1000
6.	1000–2000
7.	2000–4000
8.	4000–8000

Units of measurement

To categorize noise it is necessary to measure it. Since noise is fundamentally a power function then the intensity of noise can be defined as the average power per unit area. In fact, the minimum perceptible sound power was measured empirically many years ago by experimentally exposing a selected group of persons to varying levels of sound and asking them to identify the minimum perceptible. Although the experiment was not well controlled the average level of 10^{-12} W/cm^2 has been accepted as a baseline.

The maximum tolerable level is limited by pain in the ear. It is arbitrarily considered to be 10^{-2} W/cm^2. There is thus an extremely broad range of intensity, complicated by the fact that there is no absolute zero. Alexander Graham Bell met these problems by defining a log scale and a ratio unit, thus:

$$L_W = \log_{10} W_O/W_R \quad \text{Bel}$$

where

L_W = noise intensity in Bel

W_O = intensity observed

W_R = baseline reference, namely: 10^{-12} W/cm²

While this approach overcame the problems, the unit was now too large. Division by ten produced the *decibel*:

$$1 \text{ Bel} = 10 \text{ decibels (dB)}$$
$$= 10 \log W_O/W_R \text{ (dB)}$$

A new problem now arises. While the decibel unit as defined is a unit of power, perception at the ear is based on changes in pressure, or, force per unit area. Since the intensity in pressure units varies as the square of the pressure, then the power unit requires to be multiplied by two to account for the square, in log terms. Thus, the Sound Pressure Level (SPL, or L) is given as follows:

$$L_P = 20 \log_{10} (P_O)/(P_R) \text{ dB}$$

P_R must now, of course, be changed to the equivalent pressure reference, namely 20 micropascals, or 20 μN/m² (0.0002 dynes/cm²).

The foregoing defines a fairly complicated unit, although one rapidly develops a subjective understanding of the intensity of a sound as expressed in decibels, for example:

Table 13.2. *Common Noise Levels in dB*

Condition	Sound pressure level (dB)
pain threshold:	135–140
rock music (near source):	110–115
heavy punch press:	110
lawn mower:	85
average office:	55–60
library:	35 –40
hearing threshold:	0 (by definition)

While one can readily become used to the unit it may be necessary on occasion to convert to the original pressure that gave rise to the decibel value. This also is relatively easy, thus:

What pressure at the ear drum is represented by a sound pressure level (Lp) of 100 dB? Consider the following:

$$20 \log P_O/20 = 100$$

therefore, $\log P_O/20 = 5$

and, $\quad P_O = 10^5 (2 \times 10) = 2 \times 10^6 \ \mu N/m^2$.

Now, if that sound pressure level was reduced by half, what would be the level in dB? Or, in other words, what Lp in dB is represented by a pressure at the ear drum of $10^6 \ \mu N/m^2$?.

$$Lp(dB) = 20 \times \log 10^6/20$$
$$= 20 \times \log 50\ 000$$
$$= 20\ (\log 10\ 000 + \log 5)$$
$$= 20(4+0.7) \ \text{(Note: log 5 = 0.6990)}$$
$$= 94\ dB$$

In other words, using a pressure unit, when you reduce the pressure by half you reduce the L_p by 6 dB (or $20 \times \log_{10} 2$). In a power unit the L_W is reduced by 3 dB (or $10 \times \log_2 2$). To complete the issue, sound power at the source of noise is measured in decibels, reference 10^{-12} W/cm², while noise perceived at the ear is measured in decibels, reference 20 $\mu N/m^2$.

Sound power level and sound pressure level

As noted, the Sound Power Level is a measure in decibels of the power output from a sound source. This sound radiates in all directions, and as it radiates, the pressure fluctuations which it produces will decrease as the available sound power is spread over a larger and larger surface area which takes the form of the surface of a sphere. Thus the relationship between Sound Power Level (L_W) and Sound Pressure Level (L_p) is as follows:

$$L_W = L_P - 10 \log S$$

where S is the area of the sphere ($4\pi r^2$)

If we then consider the Sound Pressure Level at two different distances from the point of origin (point source), namely L_{P_1}, and L_{P_2}, then:

$$L_{P_1} = L_W - 10 \log 4\pi r_1^2 \ \text{and}$$

Therefore,

$$L_{P_2} = L_W - 10 \log 4\pi r_2^2$$

$$L_{P_2} - L_{P_1} = 10 \log \frac{(4\pi r_1^2)}{(4\pi r_2^2)}$$

$$= 10 \log \frac{(r_1^2)}{(r_2^2)}$$

Thus once again it may be seen that if the distance from a point source is doubled, the Sound Pressure Level is reduced by half.

The same principle applies to a line source, but in this case, since we are no longer dealing with a sphere then:

$$L_{P_2} - L_{P_1} = 10 \log \frac{(r_1)}{(r_2)}$$

Thus, if the distance from a line source is doubled then the Sound Pressure Level is reduced by 3 dB.

Subjective loudness

The subjective intensity of sound is not perceived as being equal at all frequencies. The greatest sensitivity, and hence the greatest perceived intensity is found in the speech frequencies. Sensitivity is less both above and below the range. Thus, at lower frequencies the sound pressure level has to be greater to be perceived as having the same intensity. The unit of loudness is the *phon*. The loudness level in phons is the decibel level of a tone of 1000 Hz which is judged to have equal loudness. Fletcher and Munson (1933) developed a set of curves, called *equal loudness contours* which show these subjective relationships. These curves have been re-developed by Robinson and Dadson (1957) and are shown in Figure 13.1.

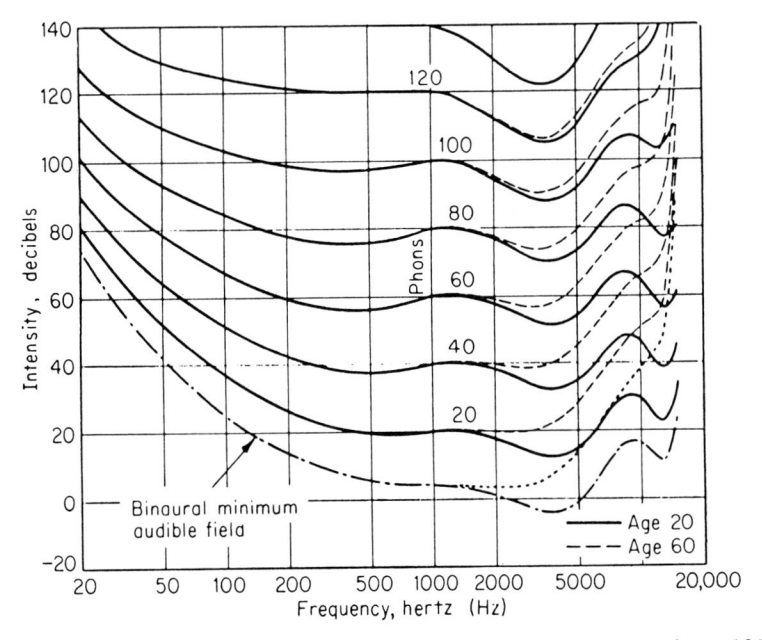

Figure 13.1. Curves of equal loudness (after Robinson and Dadson, 1957, with permission of the Controller of Her Majesty's Stationery Office).

It will be seen that they form a family of curves with the greatest sensitivity at 1000 Hz. Each curve portrays the subjective intensity of a sound at different frequencies as related to the subjective loudness of that sound at 1000 Hz. Thus a sound of 40 dB at 1000 Hz to be perceived as being equally loud at 50 Hz would have to be boosted to about 65 dB.

Still another loudness unit is the sone. While the phon defines the subjective equality of sounds it does not in fact define the relative subjective loudness. One *sone* is defined as the loudness of a 1000 Hz tone at 40 dB, or in other words the loudness of 40 phons. A sound judged to be twice as loud is then 2 sones, while a sound three times as loud would be three sones. Similary, a 30 phon would be 0.5 sones, and a 20 phon, 0.25 sones.

Noise measurement

Noise measurement is conducted by way of a sound pressure level meter. Various different commercial sound pressure level meters are available on the market, but all employ the same principles in their approach to measurement. Sound is sensed via a sensitive microphone, which commonly uses as its sensor a piezo electric crystal which has the property of generating a minute electrical signal when its shape is deformed. The compression and rarefaction waves of the sound act to deform the crystal. The resulting output is converted to a larger voltage change which is used to drive the pointer of a meter, or to change the reading of a digital electronic display. In either case the display is calibrated in decibels. Since the potential level may cover a large decibel range which would tend to clutter a meter it is common practice to present the display in decades, for example a decade range of 50–60 dB, or 80–90, such that the pointer can specify a particular value within a given decade. A switching mechanism permits change to the appropriately expected decade.

To analyze the noise spectrum more closely the meter may incorporate an octave band analyzer (or a fractional octave band analyzer such as half-octave, or one-third octave) which divides the spectrum into octaves or less for selection by switch as required. The output, indeed, may be fed either directly or via tape recorder into a frequency analyzer which will present the output at each frequency within a selected range of frequencies.

Each meter is carefully calibrated before being delivered, but it is essential that the calibration be checked on each occasion before use. Devices are available for the purpose which for example emit a pure tone, normally of 1000 Hz at 40 dB, when in contact with the microphone. The meter can then be adjusted to respond appropriately. In addition, each meter should be regularly returned to the manufacturer for more sophisticated calibration.

The output of the microphone is normally passed through one or more specialized electrical circuits, or weighting networks, before it is displayed on the scale. These networks change the relative output in different frequency ranges before finally displaying it. They are known respectively as the A, B, C, and occasionally D scales. A good standard meter normally possesses a switching mechanism to incorporate any of these scales as required. The A scale applies a weighting to the lower frequencies (below 1000 Hz) to make the output of the meter compensate in broad general form to the 40 phon Fletcher-Munson curve; in other words, the A scale presents sounds approximately as they would appear to the human ear. The B and C scales have less and less weighting respectively. The D scale is a special purpose scale used in acoustical research. The best fidelity of the A scale is found below 55 dB while that of the B scale lies between 55 and 85 dB; the fidelity of the C scale is best above 85 dB.

For industrial noise investigation the A scale is by far the most commonly used; in fact, many common commercial meters have only an A scale. Virtually all noise legislation requires measurements to be conducted using the A scale.

A good meter also has a damping mechanism for the pointer, with a switch labelled 'fast' and 'slow'. With fast response the meter reflects rapid changes in noise output but is difficult to read accurately in a changing noise environment. With the slow response, as is demanded by most legislation, the meter indicates an overall trend.

Other types of meter are also used. An *impulse* meter has a response that is faster than the fast response of the SPL meter and is valuable for recording impact noise. It commonly has a digital display which also incorporates an electronic 'hold' circuit which retains on the display the highest level noise presented to it.

A *personal noise dosimeter* is a meter which is worn on the person of the worker, commonly on his belt with the microphone in his hearing zone. The signal from the microphone is accumulated over a duration of up to eight hours, and eventually displayed as a percentage of a predetermined permissible exposure, for example 110 per cent of a predetermined average level of 85 dB over 8 hours.

Equivalent energy levels

Many noisy environments vary considerably in the noise intensity over a period of time. It is possible to take numerous sound pressure level readings over a given time and express the results as a histogram showing the number of readings at a given level or the percentage of time the readings were at above a given level. (Figure 13.2).

From such a histogram it is possible to develop a cumulative distribution of these noise levels (Figure 13.3) from which can be defined the percentage of time during which a certain levels are exceeded. These are known as the L_N values, for example:

L_{100}: the level exceeded for 100 per cent of the time, or the lowest noise level.

L_{90} : the level exceeded for 90 per cent of the time or the ambient noise level.

L_{50} : the level exceeded for 50 per cent of the time, or the median noise level.

L_{10} : the level exceeded for 10 per cent of the time, or the average level of intrusive noise.

L_1 : the level exceeded for 1 per cent of the time, or the level of highly intrusive noise.

L_0 : the level exceeded for 0 per cent of the time, or the highest noise level (L_{max}).

In order to describe time-varying noise more effectively in a single number still another approach is required. Simple averaging of the sound pressure levels does not express the subjective noisiness. A good correlation with subjective noisiness is found by averaging the energy levels in a sound, which are proportional to the square of the pressure. The resulting average is known as the energy equivalent continuous level, or L_{eq}, which is that constant sound level which has the same energy as a time-varying noise for a specified time duration. It is mathematically defined as:

$$L_{eq} = 10 \log 1/T \int_0^T \frac{P_A(t)}{P_O} \, dt \text{ dBA}$$

where, P_O = standard reference pressure, 20 Pa
$P_A(t)$ = the A-weighted time-varying sound pressure
T = measurement duration.

Let us consider, for example, a machine which was found over a period of one hour to generate sound at the following levels:

    78 dBA for 30 minutes
    80 dBA for 20 minutes
    83 dBA for 10 minutes

$$L_A = 10 \log \left(\frac{P_A}{P_O}\right)^2$$

$$\left(\frac{P_A}{P_O}\right)^2 = \text{antilog} \frac{L_A}{10}$$

Thus, to determine the ratio of the squares of the pressures each sound level must be divided by 10 expressed as an antilog, so:

    $78/10 = 7.8 = $ antilog 6.3×10, for 30 minutes
    $81/10 = 8.1 = $ antilog 1.26×10, for 20 minutes
    $83/10 = 8.3 = $ antilog 2.0×10, for 10 minutes

STATISTICAL DISTRIBUTION

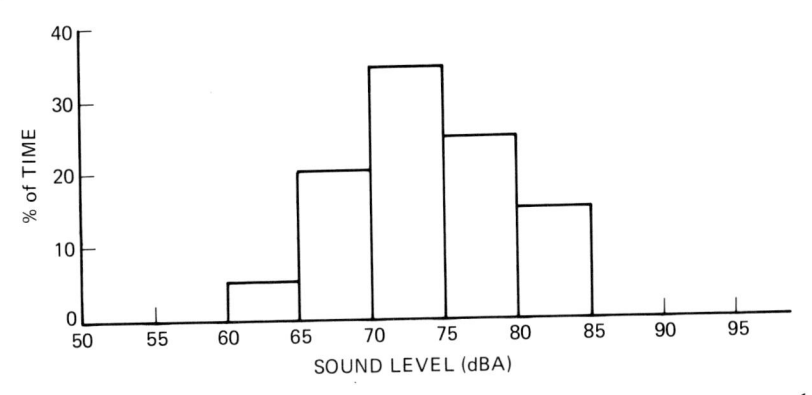

Figure 13.2. Histogram representing per cent of time at which a noise was measured at or above certain levels.

CUMULATIVE DISTRIBUTION

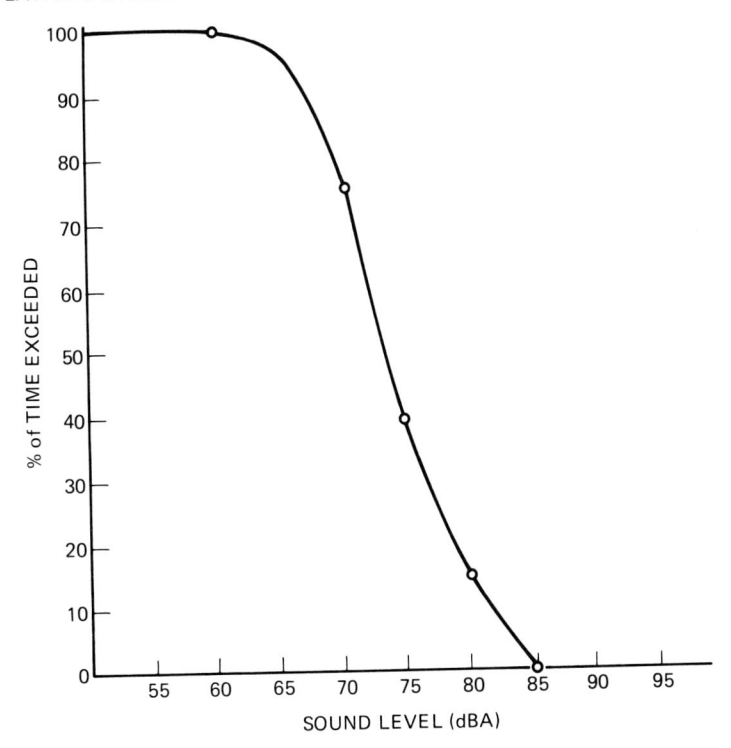

Figure 13.3. Cumulative distribution of per cent of time at which noise exceed defined levels.

Thus,

$$L_{eq} = 10 \log \frac{1}{60} (6.3 \times 10^7 \times 30 + 1.25 \times 10^8 \times 10^8 \times 20 + 2.0 \times 10^8 \times 10)$$

$$= 10 \log \frac{1}{60} (189 \times 10^7 + 25 \times 10^8 + 20 \times 10^8)$$

$$= 10 \log \frac{1}{60} (63.9 \times 10^8)$$

$$= 10 \log (1.06 \times 10)$$

$$= 10 \times 8.03$$

$$= 80.3$$

Thus, the L_{eq} for 1 hour would be 80 dBA in this case. Note that the time duration for an L_{eq} is always specified, and that value will apply only for that length of time. The L_{eq} for a certain time will be decreased when the time interval is lengthened.

Interference with communication

Interference with communication is a problem of vital interest to the communications engineer concerned with the design, for example, of telephone systems. From the point of view of the work place, however, it becomes important when noise on the shop floor is sufficiently great to interfere with receipt of orders, executive discussions, requirements for provision of instruction, emergency activities, and social interaction.

In industry, then, as far as communication noise is concerned, the significant requirement is to conserve, as far as is reasonable, speech intelligibility on or near the shop floor.

Several approaches have been developed towards measurement of communication effectiveness in the working environment. One of these is determination of the Speech Interference Level (SIL), (Beranek, 1947). The SIL is an index of noise level with respect to its potential interference with speech. It is derived by summing and averaging the SPL in each of three octaves within or overlapping the speech range, namely 600–1200 Hz, 1200–2400 Hz, and 2400–4800 Hz. Certain conditional expectations have been established with respect to the values in those ranges, thus:

Table 13.3. Speech Interference Levels (after Beranek, 1947)

SIL (dB)	Condition
45	normal voice at 10 feet
55	normal voice at 3 feet
65	raised voice at 2 feet, shouting at 8 feet
75	very loud voice at 1 foot, shouting at 2–3 feet

When the SIL is 6 dB or more below the average overall speech level, that is when the signal to noise ratio equals 2, then all normal voice communication is audible.

The Preferred-Octave Speech Interference Level (Peterson and Gross, 1978) is the numerical average of the SPL in octave bands with centres at 500, 1000, and 2000 Hz and is used for a similar purpose. Telephone usage, in terms of PSIL, is shown in the table below:

Table 13.4. Preferred-octave Speech Interference Level, (Webster, 1969)

PSIL (dB)	Telephone use
Less than 60	Satisfactory
60–75	Difficult
Above 80	Impossible

Still another approach was developed by Beranek (1978). He defined Noise Criteria (NC) curves by measuring noise levels in a very wide variety of occupational situations such as offices, restaurants, theatres, libraries, discussion rooms, shop floors, and so on, and then related the type of work which could be done to the noise levels measured. These noise curves are then presented in charts governing groups of activities. In use, the noise spectrum of the area under consideration is plotted on the chart and each octave band level is compared with the relevant NC curves to find the one that penetrates to the highest level. The corresponding value of the NC curve is the NC rating for that noise.

A representative NC chart is shown in Figure 13.4. Thus, a location where the noise level lies in the range of NC 40 to NC 50 might be suitable for conferences at a 4–5 foot table with a normal voice at 2 feet, or a raised voice at 3 feet, and with some difficulty in using the telephone. Alternatively it might be appropriate for a restaurant or a large drafting office.

Non-speech signals

Some of the characteristics of non-speech signals have been discussed in Chapter 9 in consideration of displays. In general, the requirements for non-speech signals should meet three sets of criteria, namely, environmental characteristics, ambient noise levels, and the specific needs of the worker.

Environmental characteristics include consideration of the separation between the signal device and the listener, or the maximum range at which the signals must be intelligible, as well as the geometry of the work facility, including the location of walls or other barriers, the location of noise-

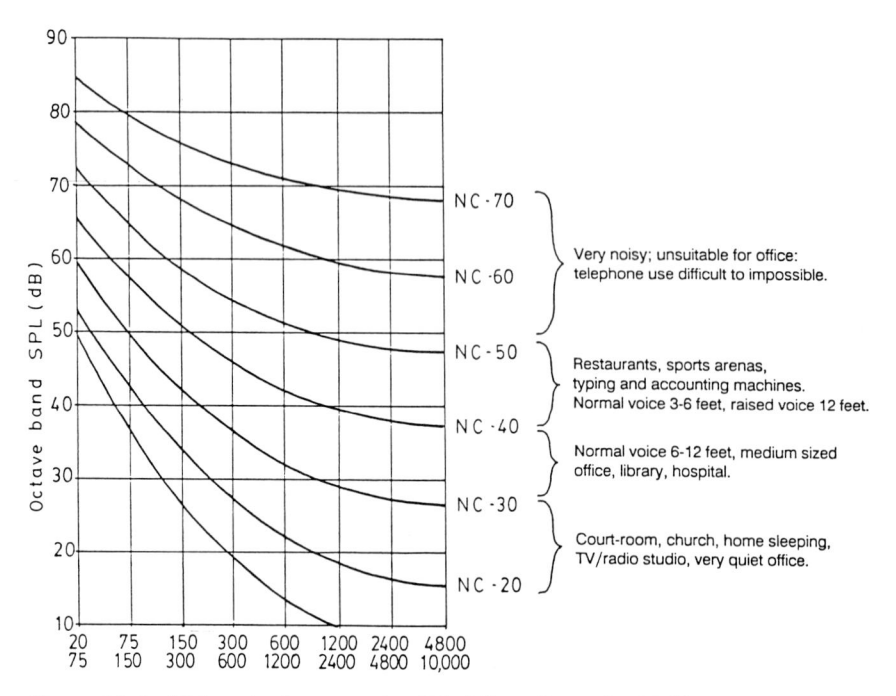

Figure 13.4. Noise criteria curves (modified from Beranek, 1960).

generating sources, the location of the work station in the work place, and so on.

The level of ambient noise within the facility requires consideration. Noise levels must be measured within the identified work locations. Information is also required on the extent of continuity or discontinuity of the time profile over a working shift.

As far as the individual workers are concerned the signal, of course must satisfy the needs of hearing handicapped persons, or persons using hearing protective equipment.

Noise hazard

Noise as a hazard is very largely noise induced hearing loss, sometimes, and indeed more properly, called noise induced permanent threshold shift (NIPTS). Generalized hearing loss, or presbyacusis, tends to be manifest, at least to a slight extent, in all urban dwellers by the age of 30. Whether this loss is normal, in that it is a natural accompaniment of aging, or whether it is the result of lifelong exposure to relatively low noise levels, is still a matter of debate. Some comparative studies of hearing capacity among urban dwellers and nomads in the African arid lands suggest the latter, in that among the Africans it was noted that hearing acuity was preserved to old age.

In the urban dweller old-age hearing loss tends to be observed more commonly at the higher frequencies, and may show itself as an exaggeration of a normal rise in threshold which is found at about 4000 Hz. Thus, in a male, age 65, one could find a 5–10 dB loss at 500 Hz, but a 35–40 dB loss at 4000.

Hearing measurement is conducted by way of audiometry which will be considered in more detail later. In audiometry the frequency of a sound, and the threshold at which it can be heard are recorded on a chart either manually or automatically, until a chart of hearing capacity, or an audiogram, is developed. A typical audiogram for a 35-year old might appear as shown in Figure 13.5.

It will be noted that there is a characteristic loss even in the normal person around 4000 Hz, which is probably determined by characteristics of the hearing mechanism in the inner ear.

Deafness

Deafness, or abnormal hearing loss can be considered as loss of serviceable hearing, that is, a change in threshold of 40 dB or more in the speech frequencies.

Deafness can be *conductive*, that is, interference with the conduction mechanism of the middle ear, or it can be *neural*, or 'nerve' deafness, that is interference with the function of the inner ear, acoustic nerves, and their connections. Conductive deafness occurs by way of disease,

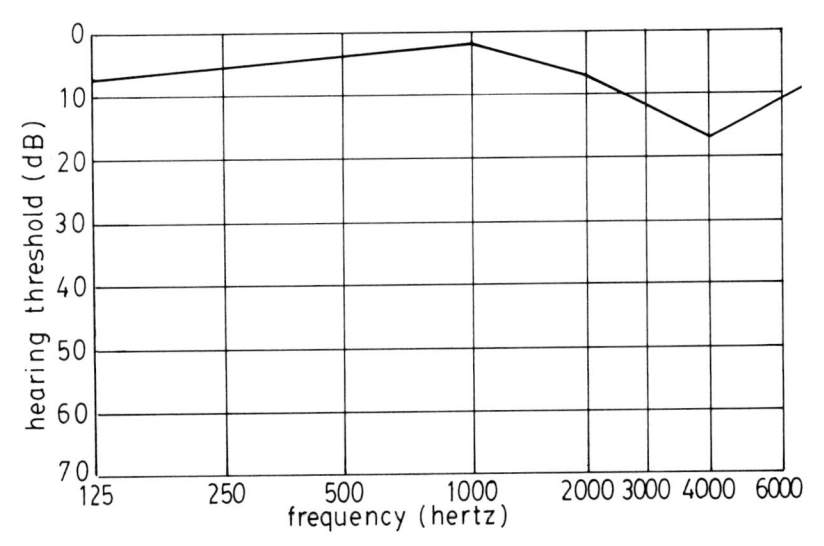

Figure 13.5. Typical audiogram of a young person.

degeneration, or injury to the conductive system such as the clinical condition of otosclerosis, which is an excessive overgrowth of bone that blocks the round and oval windows; or by, for example, a ruptured drum from illness or injury, or damaged ossicles, again from injury.

Nerve deafness occurs by disease or degeneration of the structures in the inner ear and nerve supply. Industrial deafness, except in rare instances of ruptured drum or damaged ossicles, is a form of nerve deafness caused by excessive localized vibration of the basilar membrane of the internal ear induced by noise exposure. The vibration causes swelling and ultimate destruction of the hair cells of the membrane, with loss of capacity to initiate a nerve signal.

The loss is initially short term and is referred to as a temporary threshold shift (TTS). The extent of the shift varies with the intensity and duration of the exposure, as well as with the type of exposure, that is whether it is continuous or impulsive, the latter taking a greater and earlier toll than the former. In addition, the band width of the noise, or the width of the energy spectrum, affects the shift. Where the energy is spread over a wide band the effects are less marked.

A measure of the severity of noise exposure is the permissible end-of-day TTS_2, which is a measure of temporary threshold shift taken two minutes after leaving the noisy work. For the TTS_2 to be acceptable, loss should not exceed 10 dB at 1000 Hz or less, 15 dB at 2000 Hz, and 20 dB at 3000 Hz and above. Should exposure to excessive noise be continued, the TTS will progress to a noise induced permanent threshold shift, in which there is permanent loss of hearing.

There is a great deal of discussion about the relationship between TTS and permanent threshold shift (PTS). The arguments have been summarized by Shaw (1985) in a special report. He notes that there are two concepts, loosely described as the 'total energy theory' and the 'principle of equinocivity', predominant in the field. According to the former, noise induced hearing loss is simply determined by the total amount of A-weighted sound energy received by the ear. Permanent damage is immediate and irreparable though, on a daily basis, generally immeasurably small; the rate at which damage accumulates is proportional to the sound intensity. This means that a doubling of the intensity, corresponding to a 3 dB increase in sound level, can be offset by halving the exposure time. It is implicit in the total energy theory that, by itself, intermittency confers no benefit so far as permanent hearing loss is concerned. The pattern of noise exposure is therefore immaterial. A single two-hour exposure to noise at 96 dBA has precisely the same permanent effect on hearing as four 30-minute bursts of noise at the same level separated by 90-minute periods of quiet.

According to the principle of equinocivity, there is a close relationship between temporary threshold shift measured at the end of the working day and the permanent noise induced threshold shift to be expected after ten or more years of near daily exposure to the same noise. Furthermore, since it is an experimental fact that intermittency generally reduces the amount of TTS produced by exposure to a fixed amount of sound energy, the principle of equinocivity indicates that there will be a similar reduction in the amount of permanent threshold shift produced by many years of near daily exposure with similar intermittency.

Shaw also points out, however, that when the total energy theory and the principle of equinocivity are critically examined it becomes apparent that neither is firmly founded on fundamental knowledge of the microphysiology of hearing, and that while there is strong support for the belief that temporary and permanent noise induced hearing loss are each related to the physiology of the hair cells of the inner ear there is as yet no convincing hypotheses that delineates the relationship between noise exposure and threshold shift.

Damage risk criteria

To prevent or minimize permanent loss of hearing various legislative authorities have introduced laws governing permissible exposure, or what are often referred to as Damage Risk Criteria. These criteria attempt to set up permissible exposure times and exposure levels. While there is still considerable debate on what these levels should actually be all bodies accept the principle. The concept is based on recognition of the fact that recovery from exposure obtained during a week's work should be more or less complete during one or two days, for example a weekend. Thus if a worker starts work on a Monday he accumulates some TTS by Monday night. Some of that is recovered by Tuesday morning. He accumulates more by Tuesday night and so on until Friday night. Normally he is away from his noisy environment for two days at the weekend, and sufficient recovery occurs for him to begin again on Monday. The concept is also predicated, although not always correctly, on the assumption that he has no noisy exposure when he is away from work.

Workers, however, even in the noisiest of environments, are not normally exposed to the same level of noise for a complete working day. Allowance therefore has to be made for intermittency of exposure. Beginning with the work of CHABA, the Committee on Hearing, Acoustics, and Biomechanics of the United States National Academy of Science-National Research Council, summarized by Kryter *et al.*, (1966), it has long been accepted that the maximum unprotected exposure level, as noted above, should be set at 90 dBA for an 8-hour day, 40-hour week. For exposures of less than 8 hours during a working day the sound pressure level should be allowed to increase proportionately to a maximum of 115 dBA, beyond which no exposure would be permitted without hearing protection.

Every time the exposure duration is halved, for example from 8 hours to 4 hours, the allowable sound pressure level is increased by 5 dB. This question of halving has raised considerable discussion. The 5 dB halving is based on the CHABA recommendations to ensure that exposure would produce no more than 10 dB of TTS_2 at 1000 Hz, 15 dB at 2000, and 20 dB at 3000 Hz of exposure. Five dB represents a threefold increase in sound intensity (that is, the square root of 10) which is traded against a halving of the total noise duration.

Shaw (1985) pointed out that not long after publication of the CHABA report it was recognized that there were deficiencies in the procedures used to calculate the contours for intermittent exposure. The assumption that all exposure to noise below 89 dBA could be ignored was untenable, and equally disturbing was the discovery that intermittent exposure to high frequency high intensity noise often produced delayed recovery from temporary threshold shift even though the amount was moderate. Hence the magnitude of TTS_2 could no longer be accepted as a reliable indicator of potential long term hazard.

It became apparent that the equal energy principle as expressed by the 3 dB rule, where doubling the intensity (that is, increasing by 3 dB in terms of A-weighted exposure) is accompanied by halving the exposure, is more closely related to the occurrence of permanent noise induced hearing loss. A-weighted exposure ($E_{A,T}$) is defined by the International Standards Organization (ISO 1999, 1984) as the value of the time integral of the squared A-weighted sound pressure level over a specified time period (T) or event, is given by the equation:

$$E_{A,T} = \int_{t_1}^{t_2} p_A^2(t)\, dt$$

where P_A is the instantaneous sound pressure starting at t_1 and ending at t_2. The period T measured in seconds is usually chosen to cover a whole day of occupational exposure to noise (commonly 8 hours, or 28 800 seconds) or a longer period that is to be specified, for example a working week.

Using this concept it is further concluded that in the measurement and specification of sound exposure no distinction should be made between impulsive noise and other types of noise. Steady, intermittent, varying and impulsive noise should all be included in a comprehensive measurement of the time integral of the squared A-weighted sound pressure in accordance with the International Standard.

Non-auditory effects of noise

It is difficult to show direct cause and effect in physiological changes from exposure to common industrial noise levels, even in the range of

80–90 dB for up to ten years or so, since most such effects are generalized and might be the result of many different causes. Anticaglia and Cohen (1969), have outlined some of these. They include increase in the concentration of corticosteroids, gastrointestinal pathology, electrolyte imbalance, changes in pulse rate, blood pressure, and heart rhythm, headache, and so on, as well as fatigue and loss of vigilance.

At very high intensity, for example above 130 dB even for a few seconds, marked effects may be observed. These occur partly by the direct effects of the high frequency vibration on the surface and depths of the body, and partly because the fluid that transmits the signal to the basilar membrane of the inner ear is physically connected with, and indeed is the same fluid as, the fluid in the labyrinth, or balance organ. Thus excessive motion of the fluid in the cochlea of the inner ear induces false signals within the balance organ which can give rise to dizziness, nausea, vomiting and disorientation or vertigo. These sensations may be accompanied by distracting and very uncomfortable vibration of the teeth, the sinuses, and chest, as the sound induces motion at resonant frequencies.

Hearing conservation

The fundamental objective of a hearing conservation programme is to maintain, or conserve, the hearing of the workers. The reasons are both humanitarian and economic; humanitarian since obviously impaired hearing is a disability, and economic since a good acoustic environment is a safer and more productive environment, and furthermore a bad acoustic environment can lead to financial penalization by workers' compensation or insurance institutions.

A hearing conservation programme can be considered as having seven steps, each of which will be examined in turn. These are:

* define the problem
* evaluate and monitor employee hearing status
* control noise at the source
* reduce noise exposure
* provide hearing protection
* maintain adequate records
* conduct noise safety education programme

Problem definition

The first step in problem definition is to recognize that there is a problem. Having recognized that there is a problem the extent and severity has to be defined. This process requires a noise survey.

Noise survey

A noise survey begins with discussion with representatives from management, employees, and health and safety personnel in an attempt to determine the extent, severity, and likely cause of the noise problem. It includes evaluation of health statistics and available medical information to establish the existence and severity of any hearing impairment in employees, as well as evaluation of possible sources and the noise specifications of equipment that might be responsible. A map of the plant area should be made or obtained, indicating the work areas, the equipment (and particularly the noisy equipment) in each area, and the number of personnel exposed in these areas.

On completion of the preliminaries, measurements of noise level should be made throughout the plant, concentrating on those areas where noise is believed to be a potential hazard. A broad overall survey can be made with a simple noise meter, with a single weighting on the A-scale, although a more sophisticated study may be necessary, using a complex meter with a built-in octave-band frequency analyzer. It might indeed, where a major problem is anticipated, be desirable to use the services of an acoustic technician or engineer with tape recorder and computer-mediated frequency analyzer to provide a detailed analysis of the noise spectra in the areas of significance.

A broad overall survey with simple instrumentation, however, may well be adequate. The data acquired should then be recorded on the previously prepared map and converted to 'isobels' by joining all the points of equal noise level in noise contour lines. The isobel map provides a readily interpreted graphic picture showing where the chief source of noise are found. In conjunction with the information on employee distribution in the plant the map also indicates where primary attention should be directed.

Personal noise dosimetry

It is, however, not only necessary to identify noise sources, as described above, but also to quantify the noise exposure experienced by the employee. A noise dosimeter is used for this purpose. A noise dosimeter is carried on the person. It comprises a microphone, located at the ear, the signal from which is accumulated within the dosimeter worn at the belt. The dosimeter is set with a baseline at some predetermined noise level, for example 85 or 90 dB, and thereafter records the dose accumulated over a period of time, for example 8 hours, as a percentage greater or less than the pre-set value. Dosimetry taken on different workers in different areas of the plant provides a good indication of the noise actually being received by the worker.

Hearing status

Having defined the environment, it is necessary to determine the hearing status of the employees. Ideally of course each employee should have a physical examination by a doctor or nurse, including hearing evaluation, prior to his first day of employment, with further hearing testing at times of job transfer, or reassignment, as well as annually and just prior to termination or retirement. This, however, is not always feasible in small companies. It is highly desirable, if not essential, that a worker who is going to work in a noisy environment should have his hearing examined before he begins such work. This examination serves two purposes. Firstly, it allows management to steer the worker into other work should it be found that his hearing is already impaired and could be further damaged by continued exposure, and secondly, it provides a baseline reference should the employee claim at a later date that his hearing was damaged by exposure to noise during his employment.

Audiometry

The process of hearing evaluation is called audiometry. The audiometer, of which there is a variety of types – manual, semi-automatic, and computer mediated – is a device which determines the threshold of hearing of a subject at selected frequencies, commonly 500, 1000, 2000, 4000, 6000, and 8000 Hz, through a range from 0 dB or less to as much as 60–70 dB or more, if required. Excessive noise exposure produces a threshold shift which is easily identified by the test. The resulting audiogram, which is a graph or print-out of hearing at the respective frequencies, requires skilled interpretation. Audiometry is best conducted in a specially designed sound-proof booth, of which several are commercially available. Failing a booth, the test can be conducted in a sheltered room with noise level no greater than 60 dB. The testing should be conducted by a trained and skilled technician or nurse. Certain institutions and commercial organizations provide full audiometric services.

Interpretation of audiograms is a skilled matter. Interpretation requirements vary from authority to authority, but a common protocol is that if preplacement testing shows an average loss of more than 15 dB at 500, 1000, and 2000 Hz, or if there is any unusual irregularity – particularly an abrupt loss beginning at 2000 Hz – the potential employee should be referred for specialist examination. In general, those accepted for employment in noisy areas should not have impairment in either ear exceeding 20 dB in the 500–2000 Hz range, 30 dB in the 3000 Hz area, and 40 dB at 4000 Hz. There are some work situations of course where 'complete' hearing loss may constitute an advantage, although many people have scruples about hiring even 'totally deaf' persons to work in excessively noisy areas.

Follow-up testing should be conducted three months after the initial placement test. Should any change observed at that time be less than 15 dB then tests can be repeated at yearly intervals, unless there is long and continuous exposure to levels above 100 dB or the worker complains of hearing loss or tinnitus (noises in the ear). If the loss is greater than 15 dB the worker should be removed from the noisy environment and not returned until his hearing has recovered or he has received other medical approval. Audiometry should also be undertaken not less than 12 hours and not more than 24 hours after a worker has been unexpectedly exposed to noise well in excess of his usual environment. Normally workers should be removed from a noisy environment for at least 12 hours prior to routine audiometry.

It is essential to record carefully all audiometric findings. Commercial forms are available for the purpose. Records should include demographic data on the subject or subjects, as well as history relevant to previous noise exposure both occupational and leisure (for example, shooting, flying, motor cycling, injury). The type of audiometry (for example, manual, semi-automatic, computer mediated) should be recorded along with the actual equipment used and the conditions under which it was used (for example, sound-proof room). The threshold for each ear at all recorded frequencies should be stated, either in graphic or columnar form. The form should be signed, showing the qualifications of the tester, and dated accordingly.

Noise control at source

The third major step in a hearing conservation programme is reduction of noise at the source. This process normally requires the services of an acoustic engineer, but the techniques can be summarized here. There are two basic approaches, namely prevention and attenuation.

Prevention

Wherever possible the design for a plant and its equipment should begin with consideration of a non-noisy environment. The engineer should consider what his processes are going to be and what equipment he is going to use in the light of potential worker exposure. He should so place his equipment to minimize noise exposure to the extent feasible, both in terms of position and in vibration-limiting mounting. In addition, he should select his equipment with a knowledge of its noise specifications to reduce to the extent possible the noise generated.

Attenuation

Attenuation refers to the process of reducing noise levels to the extent feasible. Noise reduction by means of attenuation can employ either the principle of *isolation*, (of the source) or that of *insulation*, with the former taking advantage of the inverse square law as applied to reduction of noise in decibels at the ear, while the latter is predicated on the capacity for certain materials to absorb acoustical energy. With respect to the reverse square law, it will be recognized that the reduction in noise varies with square of the distance of the noise from the listener. Thus if you double the distance you reduce the noise four times.

Isolation is a simple straightforward procedure by which noisy items of equipment are physically separated from workers exposed to that noise. While simple in principle it may be difficult to achieve in practice, and requires careful consideration of where in the plant work is going to be done in relation to locations where the largest number of persons are going to be found. The ultimate isolation is found with automated or semi-automated equipment which requires no human intervention except for occasional tending and is located in a sound-proofed room away from human contact.

Insulation may be employed in three different ways, and together they comprise the vast majority of noise reduction applications in working environments. They comprise:

(a) insulation of the noise generator itself, for example, with an enclosure as mentioned above, or by insulation of specific noise sources on a piece of equipment such as an exhaust duct, or the pump set of a hydraulic press.
(b) insertion of noise barrier insulation between the noise generator and nearby work stations to block the transmission of noise between the source and the worker's ears. Barriers can be free standing and movable where required, or they can be built in to plant architecture. Personal protective equipment, which will be considered later, is of course a form of barrier.
(c) use of acoustically absorbent materials on reflective surfaces such as floors, ceilings, and walls.

Types of insulative materials are illustrated in Table 13.5. Almost any substance can provide some form of acoustical energy attenuation, but special materials are available for application of the above three insulation treatments. More definitive material should be sought in an engineering reference book, such as the *Handbook of Noise Control* by C. M. Harris, published by McGraw-Hill, New York and elsewhere, or the *Industrial Noise Manual* of the American Industrial Hygiene Association.

Table 13.5. Energy absorption capacities of common noise insulating materials (derived from Handbook of Noise Control)

Substance or treatment	Insulation efficiency dB reduction
Plastered concrete block	44
Single brick wall	37
Asbestos board (¼ in) in steel frame	31
Single door	21–29
Double door	30–39
Single glazed window	20–24
Double glazed window	24–28
Surrounding source with absorbent casing	20–40

Reduction of reverberation

The problem of a noisy environment is often complicated by the fact that some of the total energy reaching human ears is reflected from walls, ceiling, and floor of the work area containing the noisy source or sources. This noise reverberates from surface to surface until damped out. The time it takes to do so is referred as *reverberation time* and is a function of several factors, namely the volume of the space in which the noise is reverberating, the total area of the reflecting surfaces, and the surface absorbency.

The coefficient of absorption of various materials is shown in Table 13.6, below:

Table 13.6. Coefficient of absorption of various materials (derived from Handbook of Noise Control)

Material	Coefficient
Perforated double aluminium tile with ¾" glass wool filling — suspended	0.90
Unplastered 3" wood slabs on hard backing	0.80
Hardboard, 10% perforated with 1" fibreglass filling — suspended	0.75
Fibreboard tile, ¼" soft wood, perforated on battens with 1" air space	0.70
Expanded polystyrene tile, ¼" with 1" air space	0.70
Plywood panel, ¾", 15% perforated, on 2" fibreglass insulation	0.60
Metal and wood	0.50
Hard and soft concrete	0.20

Specific engineering controls

For detailed engineering control options it would be wise to discuss the aforementioned Handbook, but some broad general approaches can be considered here.

Maintenance

Worn, loose or unbalanced parts contribute greatly to the noisiness of a piece of equipment. Care should be taken to ensure that all equipment is kept in a good state of maintenance, and that machines are well lubricated. Dull, improperly shaped machine cutting tools are also responsible for increasing noise levels as they are driven through protesting metal. Cutting tools should be shaped properly for the purpose and kept sharp.

Substitution of machines

It is not uncommon in a metal working plant to find unsuitable machines being used to effect a particular purpose — the classic use of a sledge hammer to drive a tack. It is generally better, for example, to use larger, slower machines in place of smaller, faster and noisier. In order to work faster there may also be a tendency to use a single die to shape a piece of metal with heavy force (and excessive noise), rather than several dies and less force. Sometimes, indeed a noisy press is used when a hammer might suffice. The driving force for the heavy equipment may also be an unnecessary culprit. A hydraulic press for example is less noisy than an equivalent mechanical press, and, curiously, the old fashioned belt drives tend to be less noisy than gears.

Substitution of processes

Impact riveting of metals is a very noisy process. Compression methods can be used in suitable situations in place of impact riveting, and indeed the riveting itself can be replaced with welding, which, although still noisy, is commonly less so. Similarly hot working is generally less noisy than cold working, and pressing is less so than rolling or forging, where appropriate.

Reduction of vibration

Much of what is appreciated by the ear as noise originates in some form of vibration of the equipment or the piece being processed. The vibrational energy is of course also a function of the applied force. Consequently if the force is reduced, the vibration will also be reduced. As noted before, sometimes an unnecessarily powerful machine is used to accomplish a task which could be accomplished with much less force.

In addition to being transmitted through the air, the vibrational forces may be transmitted through the floor or ground. In particular where large forces are required it is desirable to isolate the equipment from the transmitting medium, by, for example, building it on a separate foundation. This would normally require the services of an engineer.

It is common in a metal working plant to see large sheets of metal being unrolled from a coil and passed into a press for further processing. In its passage to the machine the sheet metal can undergo noisy vibration. This should be damped out by applying damping to the surface of the metal, or by providing additional support for the sheets in their passage, or even, if appropriate, using a stiffer material.

Since the vibration is a function of the mass of the material as well as the acceleration, it is possible to reduce a vibrating source by increasing the mass of the vibrating members, or, for that matter, to change the size to change the resulting resonant frequency.

Reduction of exposure duration

Another approach to the minimization of hazard lies in reduction of the time during which the workers is exposed to noise. This is an administrative rather than an engineering procedure. Should there be one particularly noisy process which cannot be properly controlled by other means, then an employee can be assigned to that particular job for only a proportion of his shift, working in a less noisy area for the remainder. This procedure of job rotation is a valuable means for controlling noise that cannot be feasibly controlled in any other way, although it often creates logistic difficulties in its implementation.

Noise exposure is additive, and it must be remembered in calculating an employee's exposure that he spent different portions of the day under different noise conditions.

Personal protective equipment

Provision of personal protective equipment should be considered the last resort in protecting personnel against hearing loss, to be used for short time special occasions or when no other method is feasible. In an emergency, of course, it is always possible, and quite effective, to stick one's fingers in one's ears, but little work can be done under these circumstances. Personal protective equipment can be inserted into the ears or it can be applied outside the ears, or both.

Soft inserts

Probably the most common, and the least useful, protection is undertaken by pushing cotton or waxed cotton balls into the ears. In some respects this is worse than useless since it provides an illusion of protection. In fact, under the best circumstances, this approach gives about 3–5 dB attenuation. While this could conceivably be useful at levels of 88–90 dB its affect at 100 dB would be worthless. Significantly better than cotton are commercial soft plugs (for example, made by the Bilsom company) of glass wool surrounded by thin plastic sheeting which are comfortable and easily inserted into the ear. In the same category are soft, firm, mouldable plugs which also are readily inserted into the ear and cling closely to its contours once inserted.

Hard plugs

A variety of commercial organizations manufacture ear plugs or inserts of different sizes and shapes, made of firm plastic or rubber which are pre-moulded to fit the ear. They may vary in size and shape but the size and shape are relatively unimportant in comparison with the fitting. They must fit tightly. A loosely fitting plug loses much of its protection. The plug should be fitted with one arm across the head holding back the pinna of the opposite ear while the other hand inserts the plug into the ear with a screwing motion in a slightly forward direction until it is uncomfortably tight. The best plugs, although also the most expensive, are custom moulded for the individual. Good plugs, property fitted, can give, under controlled conditions, about 15–25 dB attenuation, with the higher attenuation being found at frequencies below 1000 Hz.

Plugs are simple, convenient, easy to carry, and effective. Some are joined with a cord that allows them to be slung across the neck when not in use. Some are provided with a small carrying case. They have two disadvantages, however. It is difficult for a supervisor to tell if they are being properly worn, and they are uncomfortable for prolonged wear. Because of the discomfort there is a tendency for the worker not to wear them or to wear them so loosely that they are relatively ineffective. Indeed, there is a tendency for some workers, particularly the older ones, to reject any protection, claiming that the noise does not bother them. In fact, it may not, since they already may be partially deaf in the frequency range of exposure. Their condition, however, can be aggravated by continued exposure. Consequently, the wearing of protective equipment should be enforced where it is necessary.

Ear muffs (ear defenders)

A variety of ear muffs is also available from commercial organizations, some with rubber or synthetic foam pads, some with glycerine-filled pads. In general, the more expensive the muff the better is the quality. Again attention must be directed to the fit. The skull bone surrounding the external ear is irregular in shape. No air gaps should exist between the pad and the skull when the muff is worn. A properly worn muff is even more effective than ear plugs at the higher frequencies, although not much if at all better at frequencies below 1000 Hz. Between 1000 and 5000 Hz, however, it can provide attenuation under controlled circumstances in the range of 20–40 dB. While more comfortable for prolonged periods than ear plugs they are of course more clumsy and may be disliked by workers just as much as ear plugs. A muff worn round the neck serves no useful purpose; when needed its use must be enforced.

Muff and plug combination

The best ordinary protection lies in the combination of ear muffs and plugs. A communication system can also be built into the ear muffs which can be heard despite the ear plugs. The combination of muffs and plugs provides for good protection below and above 100 Hz, with a range in laboratory conditions of 24–45 dB, depending upon the frequency.

Helmet

In unusual circumstances, where noise is abnormally excessive, for example in a jet engine test cell where the noise level might be 140 dB, a noise protective helmet can be worn. A suitable helmet can provide protection up to 50–60 dB. It must be remembered, however, that above 130 dB the non-auditory effects of noise become clearly manifest. These include dizziness, pain in the chest, teeth, sinuses, and so on. In addition the noise is now transmitted directly through the bone of the skull to the ear, or even through the sternum to the skull and ear. These latter circumstances, of course, are exceptional.

Records

In all cases it is necessary to ensure that records be kept of hearing conservation and findings. These records are necessary not only to guide future actions, but also to act as a data source if some claim or other problem should arise on some future occasion. In this connection there are two basic types of record, namely, physical plant records, and personnel records.

Physical plant records

The physical plant records refer to the findings during the surveys of the physical plant. The information should include the following:

* grid maps or plans of plant
* measuring equipment data, e.g. types and serial numbers of microphones, sound level meters, and so on
* corrections for measured values, such as cable type, temperature, calibration, and so on.
* location and description of survey area, including nature of

work and noise sources, dimensions, time pattern of noise, location of test apparatus, personnel exposed, and so on.
* sound levels measured (scale, meter response, and so on)

Records should also be kept of the noise control measures instituted, including engineering and personal protection.

Personnel records

If medical records are kept for each employee as part of routine personnel records, perhaps in a medical clinic, then the audiometry record should become part of the general medical record. Otherwise separate audiometric records should be kept for each employee. It should be remembered that audiometric records are confidential and that audiometry, under many legislations, can only be conducted, and records maintained, with the informed consent of the employee. It is wise to obtain that consent in writing.

Noise hazard education

For a hearing conservation programme to be effective, employees must be informed about the hazards of noise exposure, Any educational programme that is developed should be repeated, perhaps with modifications, at regular intervals. It is only through repeated discussion and continued emphasis that employees come to appreciate the seriousness of the problem to which they are exposed. Formal programmes should be backed up with on-the-job supervision and example.

While an appropriate programme can be presented by any well-informed and experienced person it may be found more convenient to permit and encourage employees to attend the special programmes developed for the purpose by safety and accident prevention organizations and other institutions. Topics to be discussed should include the nature of noise, the effects of noise on humans, the various types of exposure that can occur, the proper action to protect oneself from such exposure, and the engineering methods by which noise can be controlled. It should be remembered that under many legislations a worker may refuse to work under certain conditions where he is likely to endanger himself or another worker. It is only where both management and worker are suitably informed that these provisions can be properly applied and countered.

Chapter 14
Heat gain and heat tolerance

Introduction

From the thermal point of view the body can be considered as a heat engine. That is, it is a device which burns fuel in oxygen to release energy. The fuel normally is in the form of carbohydrate, although fat or even protein can be broken down into constituents for use as fuel in the absence of carbohydrate. The fuel is derived from ingested foodstuffs. The oxygen, of course is derived from the air through the gas exchange process of respiration. The chemical processes involved in the conversion of carbohydrates and fat to energy and heat are collectively referred to as energy metabolism.

Heat production

All body activities produce heat. The liver, heart, brain, and so on generate large amounts of heat. The heat production of the skeletal muscle, however, accounts for 20 to 30 per cent of all the body's heat production (Guyton, 1984), even at rest. During severe exercise the heat produced by the muscles can rise for about a minute to as great as 40 times that produced by all the remaining tissues together. Thus the degree of muscular activity is one of the most important means by which the body regulates its temperature.

The industrial worker, of course, may not only be undertaking severe exercise, but also may be acquiring heat from the environment, particularly in certain hot industries, or when working outside in hot climates.

Heat loss

Heat is lost from the body in three ways, namely, radiation, conduction, and evaporation. About 60 per cent of body heat in a nude person at rest in a thermoneutral environment (21°C) is lost by radiation to the environment, provided that the environment is cooler, while about 18 per cent is lost by conduction to the air, and about 3 per cent to contact objects such as chair and floor (Guyton, 1984). The heat is conducted to the air and is carried away by air convection.

Sweat glands are continuously secreting small quantities of watery sweat which diffuses through the skin and is evaporated with loss of heat. Some 22 per cent of body heat at rest in a neutral environment is lost by evaporation, again assuming that the relative humidity of the air is less than that of the skin.

Metabolic heat

For continued survival the body core requires to be maintained at a critical temperature of 37°C ± 1 which is achieved by balancing metabolic heat gain against any environmental heat loss. Metabolic heat is measurable in kilocalories (kcal). Each kilocalorie is 1000 times a calorie (cal). Metabolic heat can also be measured in joules (1 cal equals 4.184 J).

While direct measurement of metabolic heat is possible, it is very difficult. The measurement is made in a whole-body calorimeter which is an environmentally controlled tank into which a subject can be placed while measures are taken of the accruing change in temperature as his body heat is generated. The values obtained are converted into measures of heat generated. The process is impractical for routine or regular use.

It is fortunate, however, that there is a simple and linear relationship between heat generated and oxygen consumed. One litre of oxygen consumed is approximately equivalent to the generation of 5 kcal of heat (approximately 20 000 J). The measurement is accomplished by allowing the subject to breathe from a respirometer (see Chapter 17) containing pure oxygen in a closed system. Exhaled carbon dioxide is absorbed by soda lime in the closed system. From the volume of oxygen used in a given time the metabolic rate can be expressed in terms of kcal per square meter of body surface per hour (average body surface is 1.9 square metres). The measurement is normally done in a laboratory, but less precise approximations can be undertaken using portable equipment (see Energy Cost of Work, Chapter 3).

Thermal regulation

Heat balance is essential for survival. Complex physiological mechanisms exist to ensure that the core temperature (that is, the internal temperature) remains within the critical range of optimum. In this connection the core is contrasted with the surface. The surface skin temperature can vary quite widely, and indeed the skin and surface tissues can be sacrificed in frostbite and burns while the core is preserved. There is a need, therefore, for heat exchange and thermal regulation. Thermal regulation is a classic example of a feedback control system which has heat sensors, a data processing mechanism, and an effector mechanism for making the necessary physiological adjustments.

Heat sensors

Physiological heat sensors exist in two forms and two sets of locations, namely superficial, or skin, and deep, or core. The superficial sensors are free nerve endings in the skin which do not sense temperature specifically, but sense the direction of heat flow, that is, loss or gain. Some histologists (that is, scientists concerned with the form and function of body cells) suggest that there are special microscopic organs, called the end-bulbs of Krause, at the end of dedicated nerve fibres that are responsible for the heat sensing process. Others consider that any superficial free nerve ending will respond appropriately.

The deep sensors exist in the depths of the brain near the thermal control centre. There may be others elsewhere in the inner confines of the body. The deep sensors monitor the core temperature.

Control nerve

The control centre is found in the base of the brain in an area known as the hypothalamus. The hypothalamus is in that primitive region of the brain which is responsible for the coordination and control of functions necessary for basic survival. The hypothalamus coordinates the activities of two subsystems, namely the autonomic nervous system and the endocrine system.

The autonomic nervous system has control centres in the brain for the control of involuntary functions such as respiration, circulation of the blood, blood pressure, and so on. The endocrine system comprises a group of glands dispersed throughout the body, such as the thyroid gland, the adrenal gland, and others, which secrete chemicals called hormones into the blood stream. These chemicals act to increase or decrease the activities of body cells and organs, and thereby operate as an auxiliary chemical control system.

The heat control centre coordinates some of the activities of the vasomotor centre. The vasomotor centre is responsible for adjusting the calibre of small blood vessels and thereby controls the blood flow to tissues and organs. The blood acts as a heat exchanger. The heat control centre also coordinates the 'shivering' centre which in turn coordinates involuntary actions of skeletal muscles in repeated spasmodic contractions which generate heat, or at the opposite extreme, slacken the tone to reduce the rate of muscle metabolism.

Effector mechanisms

The autonomic and endocrine systems provide a varying response to the environment, as driven by change or threat of change in the core temperature. There are five well defined types of response that can be executed. These are as follows:

1. *Redistribution of blood flow*

The blood vessels in the body exist in two networks, namely a superficial network in the skin and a deep network in the underlying tissues and organs. Blood flow can be directed locally and regionally by constriction or dilation of very small arterial blood vessels called arterioles. These arterioles have circular muscular walls. When the muscle contracts at the behest of the control centre the calibre is reduced to the extent that the supply to the region it serves can be cut off. When relaxed the arterioles dilate and open up the supply. Thus in cold conditions the superficial blood supply is reduced to the point of cut off. This diverts circulation to the deep network. In hot conditions the flow is diverted to the surface for easier heat exchange with the environment. Should the surface blood flow be cut off for a prolonged period, as in extreme cold, there is a danger of depriving the tissues of gas exchange and nutrients, a condition which will lead to potential or actual tissue damage. Thus, at that point, even in continued cold the hypothalamus will instruct the arterioles to re-open the circulation. The circulating blood will now become chilled and reduce the core temperature. When the core temperature falls to the critical level the control centre will instruct the arterioles to close off the circulation and sacrifice the tissues, which will now suffer from frostbite.

2. *Change in sweat production*

As noted, the evaporation of sweat is one of the main ways in which the body maintains its heat balance. Sweat is a salt-containing transudate from the blood plasma, which is secreted by specialized glands in the skin. There are two kinds of sweat glands—appocrine, located on the forehead, the back, the palms of the hands, and the armpits, and eccrine, which are found in other parts of the body. While at all times some sweat glands are secreting, the appocrine secrete a protein-containing sweat in response to stressful situations. The protein may break down and create an odour. The eccrine secrete a water sweat in response to heat. Evaporation of the sweat, via the latent heat of water, removes heat from the skin and produces chilling, provided that the water vapour pressure at the skin surface is greater than the water vapour pressure in the ambient air.

The maximum continued production rate in a well hydrated person is about 1 kg (or 1 litre) per hour (Blockley, 1965). Some degree of sweating, to the extent of 650 ml per day is always occurring. The maximum short-term production is about 10–15 litres over a six hour period (1.5 to 2.5 litres per hour approximately). This would provide a cooling rate of more than 1200 watts. It is of interest to note that since the latent heat of water is approximately 600 cal/gm then the evaporation sweat will require some 6000–9000 kcal, or in other words, the theoretical limit of heat dissipation will be 6000–9000 kcal which is approximately the heat output of lumberjacks in a cold climate.

3. *Change in muscle tone*

It was pointed out above that muscle metabolism is a major source of body heat. All skeletal muscle is held in a mild degree of contraction, or tone, in the conscious individual. The maintenance of tone is involuntary, mediated by the autonomic nervous system. Increased tone requires a higher metabolic rate and decreased tone a lower. In response to environmental stimuli the tone will be increased in cold conditions and reduced in hot conditions, thereby increasing or reducing the heat generated. These activities are coordinated by the control centre.

Shivering is a special form of this increase in tone which results from a kind of oscillation in the neuronal circuitry which controls the stretch reflex of a muscle. The stretch reflex occurs when a muscle fibre is slightly stretched, for example when a knee tendon is tapped with a hammer. The stretch induces contraction in the muscle, mediated through the spinal cord. In cold the circuitry becomes more sensitive. Thus when a minor movement occurs in a muscle it contracts; the resulting action induces movement in an antagonizing muscle, that is one operating to move a joint in the opposite direction. This movement induces further contraction in the first muscle, and so on until shivering ensues. Shivering can increase the rate of heat production by 300 to 400 per cent (Guyton, 1984). Subjectively, increased tone is associated with alertness and desire for activity which in turn increases muscle metabolism.

4. *Change in voluntary activity*

As noted, just as change in tone is associated with increase or decrease in the generation of metabolic heat so is change in voluntary muscle activity. Thus, in cold conditions one tends to seek activity, for example jumping up and down, and in hot conditions one is lethargic.

5. *Change in metabolic rate*

Certain animals, such as bears, possess the capacity to hibernate. In hibernation all their involuntary activities fall to a very low level, including metabolic rate. The phenomenon is induced by the approaching winter and allows them to survive without food or water in the cold. Man cannot hibernate, but among certain primitive peoples, notably the Australian aborigines and the natives of Tierra del Fuego, South America, there appears to be a capacity to reduce their basal metabolic rate and involuntary functions at night in the cold, to the extent that they can sleep unprotected in tolerable comfort in environmental conditions down to freezing.

The basal metabolic rate can be reduced artificially by refrigerating the circulating blood. This is done under anaesthesia in an arterial bypass for special kinds of surgery, for example during heart surgery where the circulation is going to be cut off for a period. Reducing the basal metabolic rate reduces the requirement for oxygen which is carried by the blood. It is a dangerous procedure since reduction in the core temperature below a few degrees means almost certain death without special precautions.

While man cannot hibernate, he has another possible response. Under conditions of cold for several weeks the thyroid gland, which is the endocrine gland in the neck responsible for the hormonal control of metabolic rate, is induced by the control centre in the hypothalamus to produce larger quantities of thyroid hormone. Over a period of several weeks this increase in circulating hormone can raise the metabolic rate some 20 to 30 per cent.

Bearing the foregoing in mind, the input-output relationships of the control system can be expressed by a diagram, see Figure 14.1.

Body temperatures

Core temperature

As already noted, the internal activities of homeostasis contrive to keep the internal body temperature within a narrow range centred on 37°C (98.6°F). This is known as the core temperature and it is maintained as long as the body remains in thermal equilibrium. While it is not feasible to obtain a true reading of core temperature a close approximation can be obtained by way of the rectal temperature, measured by a mercury thermometer or an electric probe. Changes in core temperature will affect metabolic rates, and vice versa. At temperatures above 98°F there is an approximate 7 per cent increase in metabolic rate per 1°F increase in internal body temperature with a similar decrease below 98°F. As temperature cools to unacceptable levels the metabolic rate slows to a dangerous extent in the condition of hypothermia, and as the temperature increases to unacceptable levels the metabolic rate rises to dangerous levels as in heat stroke (see later).

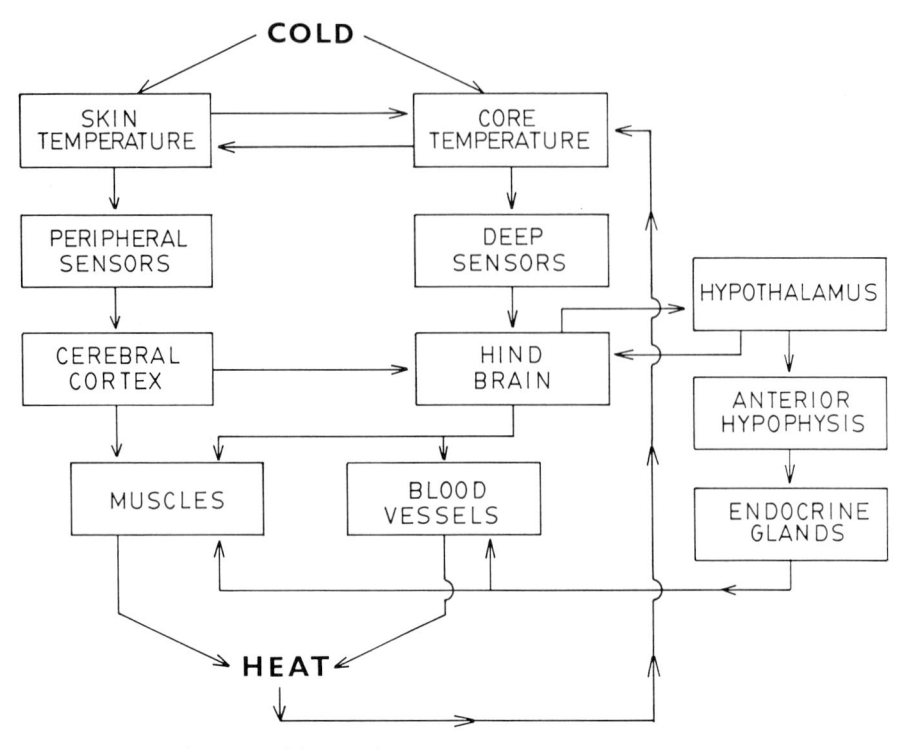

Figure 14.1. Schematic of human heat control system.

Skin temperature

Unlike the core temperature, which remains constant in thermal equilibrium, the skin temperature varies significantly with the ambient and normally remains several degrees below core temperature.

The mean skin temperature can be estimated by taking the weighted sum of skin temperatures over various parts of the body as follows (Berenson and Robertson, 1973):

Table 14.1. Estimation of mean skin temperature

$$T(mean) = 0.12T(back) + 0.12T(chest) + 0.12T(abdomen) + 0.14T(arm)$$
$$+ 0.19T(thigh) + 0.13T(leg)$$
$$= 0.05T(hand) + 0.07T(head) + 0.06T(foot)$$

Mean body temperature

The mean body temperature is derived from the weighted sum of the rectal (core) and mean skin temperatures, thus (Burton, 1946):

$$T(body) = 0.67T(rectal) + 0.33T(skin)$$

Heat balance

From the foregoing it will be apparent that to maintain homeostasis it is essential to ensure a balance between heat input and heat output. There is therefore, a need for heat exchange. Heat exchange in the body is accomplished in much the same way as in any mechanical system, except that there is emphasis on one particular physical phenomenon, namely the latent heat of evaporation. Conduction as a mechanism is minimal except under water. In water, however, conduction as a mode of heat loss can cause serious problems unless the subject is provided with effective insulation. Convection can be a significant means of both heat loss and heat gain; its effect is increased by air movement. Radiation, to and from the surrounds, for example radiation from the sun or an industrial heat source, or from the body to the ambient environment, is also significant. Evaporation of sweat, however, provided that the water vapour pressure of the boundary layer surrounding the skin is greater than that of the ambient atmosphere, remains the single most significant mechanism of heat loss.

The relationships of input and output are established in the Balance Equation, as follows:

$$M = (E \pm C \pm R) = 0, \text{ in equilibrium}$$

where,

M = metabolic heat
E = evaporation
C = convection

and conduction is ignored.

In hot conditions the evaporative requirements can, therefore, be defined by:

$$E_{req} = M \pm C \pm R$$

where E_{req} is the required evaporation, M is measured or estimated (see Chapter 3), and C and R are calculated from thermodynamic equations, or derived from nomograms (see later).

Since heat balance is so dependent on evaporative loss it is not surprising that the extent of any heat stress can be evaluated by examining the rate of sweat loss from the body. Loss of water through the skin is influenced by the vapour pressure gradient, the skin temperature and the barometric pressure (Blockley, 1964). A high skin temperature is related to a high diffusion loss, while diffusion is also increased as barometric pressure is lowered. Blockley has also shown that the more a subject maintains his hydration by drinking water the lower is his core temperature under working and environmental conditions that would otherwise tend to raise it. Blockley concludes that the failure to replace completely the water lost in sweat, hour by hour, leads to elevation of body temperature and excessive physiological strain. Unfortunately, thirst is not a reliable indicator of water loss.

Temperature measurement

Ambient temperature for environmental and physiological purposes is commonly measured by dry-bulb, wet-bulb, and globe thermometers. Electronic devices, commonly using a thermocouple as a base, are also available. The *dry-bulb thermometer*, as the name would imply is an ordinary mercury thermometer the bulb of which is maintained in a dry state. It provides a mean temperature without accounting for the effects of evaporation.

The *wet-bulb thermometer* is a mercury thermometer the bulb of which is wrapped with a woven cotton wick. Before use the wick is soaked in water. The temperature is measured while the wick is wet to allow for the cooling effects occasioned by the evaporation of the water. Sometimes the wet-bulb and dry-bulb thermometers are mounted together on a wooden or metal frame like a football rattle to form a *sling psychrometer*. In use, the sling is whirled rapidly for about half a minute to allow the full cooling effects to take place before the temperatures are read.

The *aspirating thermometer* comprises a wet-bulb and a dry-bulb thermometer mounted in tubes which shield the thermometers from radiant heat. Air is drawn through the tubes by a motor driven fan.

The *globe thermometer* comprises a matt black hollow copper sphere of 15 cm diameter at the centre of which is a dry-bulb thermometer. The thermometer provides a reading of the effects of radiant heat. It should be allowed to equilibrate for about thirty minutes before reading.

Wind velocity is measured by a *vane anemometer,* in which the rate of rotation of the low friction mounted vane is calibrated in terms of air velocity. Velocity also can be measured electronically, commonly from the temperature change occurring in a heated wire exposed to the air flow.

In use, these devices are commonly mounted together on a stand at approximately chest height to expose them to more or less the same conditions. A miniaturized device, (the Wibget), is also available, containing, integrated into one electronically displayed meter, the readings of dry-bulb, wet-bulb, and globe temperature.

Thermal indices

Because of the influence of air movement, radiation, ambient vapour pressure, and so on, ambient temperature alone is inadequate to describe the factors within the environment that give rise to heat stress. Various attempts have been made to combine different factors into one number or index.

Heat stress index

One of the most physiologically oriented, but at the same time one of the most mathematically complex to derive, is the Heat Stress Index (HSI) developed by Belding and Hatch (1955). It is defined as the ratio of the evaporation rate required to maintain heat balance to the maximum sweat rate safely attainable over long periods. The measures are in litres of sweat per hour, and the index is expressed thus:

$$HSI = \frac{E_{req}}{E_{max}} \times 100$$

where,

E_{req} = evaporation required
E_{max} = maximum evaporation feasible under the applicable conditions.

These conditions are as follows:

(a) body heat storage does not exceed the limit represented by a skin temperature of 35°C, and

(b) E_{max} does not exceed 1 litre/hr, the equivalent of 400 kcal/hr.

The values for the required and maximum rates of evaporation were originally derived from the appropriate equations for heat conduction and radiation, along with measures or estimates of metabolic heat. This can indeed be done but it is a long and laborious process. Instead the HSI can be evaluated by the use of a nomogram, such as is shown in Figure 14.2 below.

To use the nomogram it is necessary to obtain values for the metabolic rate, which can be derived from tables or actual measurement, as well as measures of wet bulb, dry bulb, globe temperature, and air velocity. Instructions for utilizing these measures are given in the figure.

The significance of different levels of HSI is shown in Table 14.2.

P4SR index

The P4SR index utilizes the phenomenon of sweating in a different manner. The index, which refers to the *predicted four-hour sweat rate,* uses the rate of sweating as a measure of stress under conditions which are hot enough to cause sweating (Smith, 1955). The index was derived on the basis of empirical evidence following which a nomogram was prepared indicating the probable amount of sweat that would be generated over a 4-hour period by fit acclimatized men sitting in shorts. The nomogram Figure 14.3) is entered via the dry-bulb (or globe) temperature and the wet-bulb temperature. A line drawn between the two indicates the predicted sweat rate. The index, however, has considerable limitations,

262 *Physical agents in the work environment*

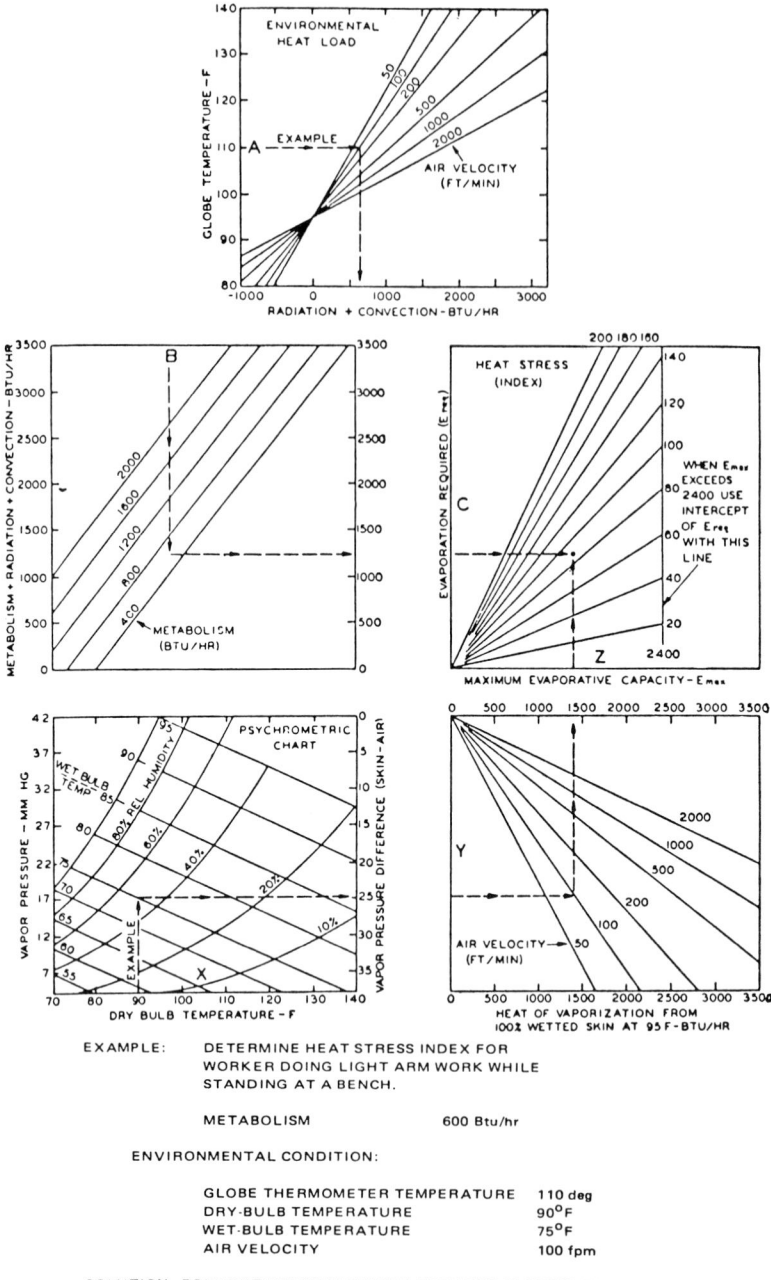

EXAMPLE: DETERMINE HEAT STRESS INDEX FOR WORKER DOING LIGHT ARM WORK WHILE STANDING AT A BENCH.

METABOLISM 600 Btu/hr

ENVIRONMENTAL CONDITION:

GLOBE THERMOMETER TEMPERATURE 110 deg
DRY-BULB TEMPERATURE 90°F
WET-BULB TEMPERATURE 75°F
AIR VELOCITY 100 fpm

SOLUTION: FOLLOW THE BROKEN LINES FROM THE GLOBE THERMOMETER AND FROM DRY-BULB TEMPERATURE TO THEIR INTERACTION ON ABOVE DIAGRAM C TO READ A STRESS INDEX OF 90.

Figure 14.2. Heat stress index nomogram (reprinted with permission from the ASHRAE, Guide and Data Book—Fundamentals and Equipment for 1965 and 1966).

Table 14.2. *Physiological and work limitations with increasing levels of the Heat Stress Index (after ASHRAE, 1965)*

HSI	Physiological and Work Implications
0	no thermal strain
10–39	mild to moderate strain; subtle to substantial decrements in performance of intellectual function, dexterity or alertness; little decrement in physical work unless ability is marginal.
40–69	severe heat strain, threat to health unless physiologically fit; acclimatization required, medical selection desirable; some decrement in physical work capacity; unsuitable for activities requiring sustained mental effort.
70–99	very severe heat strain; tolerable only to a few. Selection after (a) medical examination (b) job trial after acclimatization. Ensure adequate water and salt intake, ameliorization of work conditions desirable and may decrease health hazard.
100	maximum strain tolerated daily by fit, acclimatized and motivated persons.

and is really only applicable under the conditions in which the data were originally derived. It can be used, however, for comparing differing environments in terms of their sweat-generating potential.

ASHRAE comfort index

The American Society for Heating, Refrigerating, and Air Conditioning Engineers (ASHRAE) developed a comfort index based on much experimental work. This index, the Generalized Comfort Chart, takes into account four factors, namely dry bulb (operative) temperature, relative humidity, dew-point temperature, and the seasonable effects of clothing, under light, mainly sedentary, activity. The chart is shown in Figure 14.4 below. Related charts are available for other conditions.

The psychrometric chart demonstrates that a given dry-bulb temperature and an associated relative humidity (or wet-bulb temperature) can be specified by a point on the chart. Thus a series of points derived from different combinations of wet-bulb and dry-bulb temperatures will be found to lie on the same relative humidity line. Recognition of this concept led to the development of a measure of environmental comfort known as the Effective Temperature (ET) scale.

Effective temperature scale

The original effective temperature scale was developed by inviting subjects to compare the thermal sensations invoked in passing from one thermally controlled room with a given combination of heat and humidity to another with a different combination. The resulting sensations were

264 *Physical agents in the work environment*

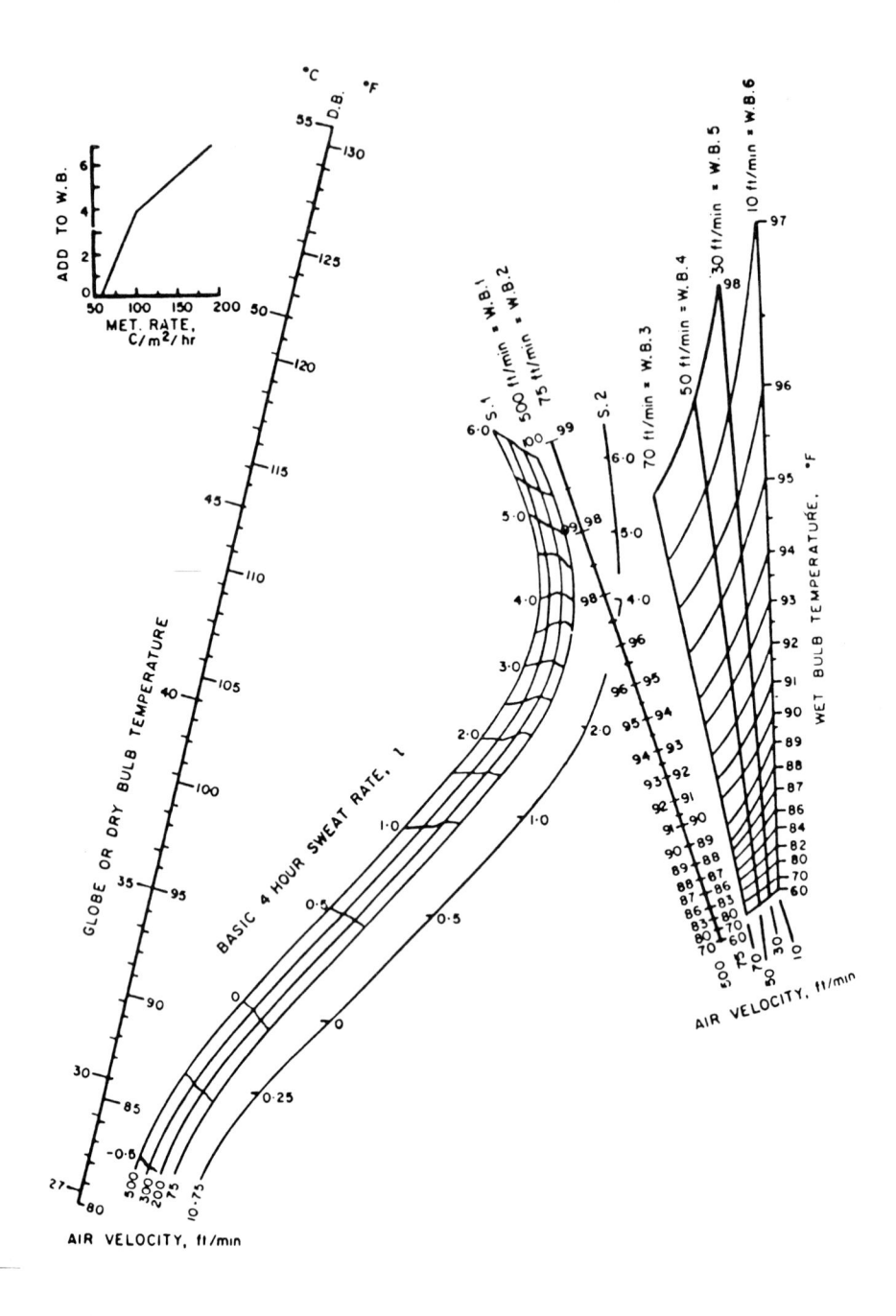

Figure 14.3. Nomogram for the prediction of 4-hour sweat rate (P4SR) (after Macpherson, 1960).

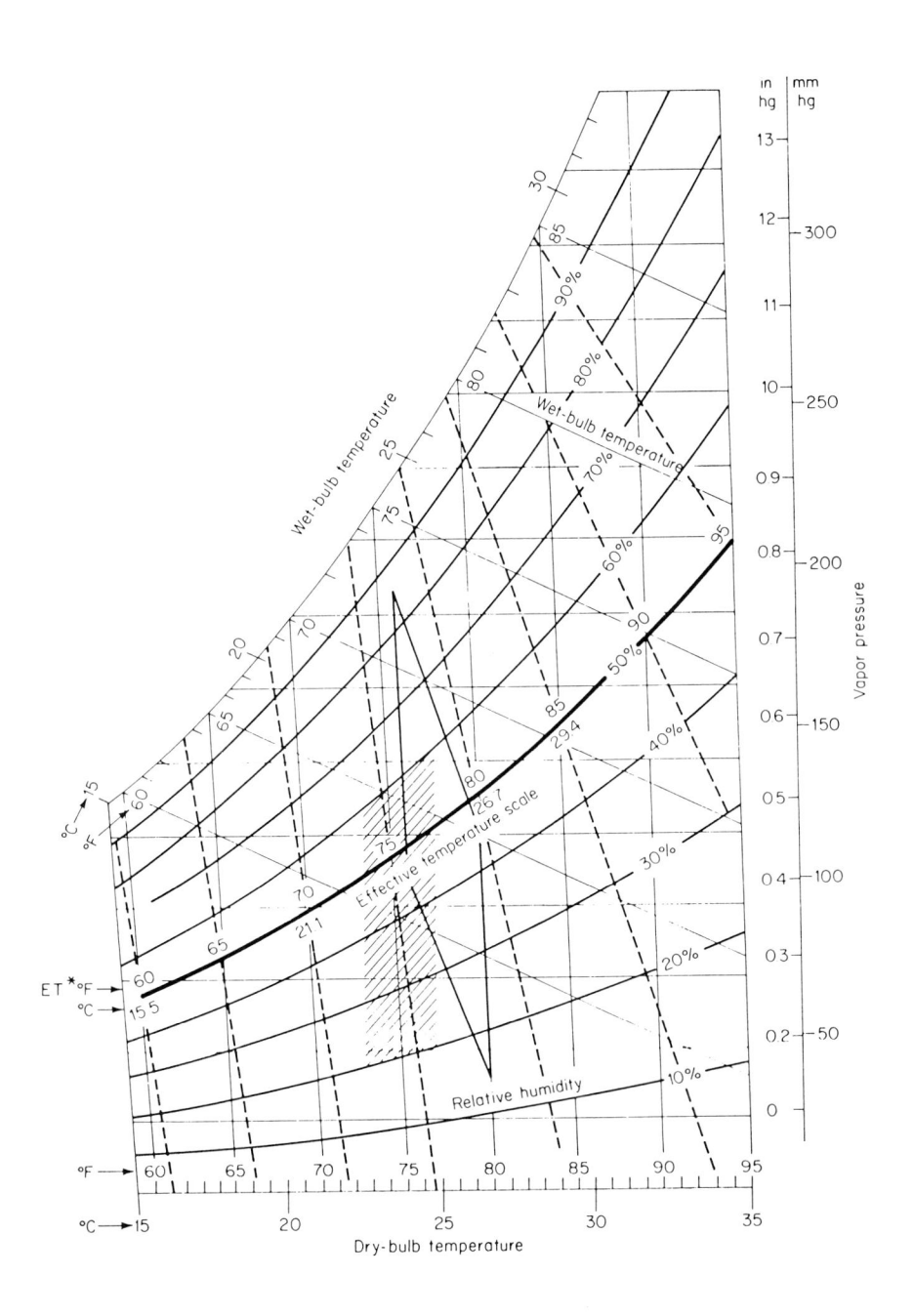

Figure 14.4. Psychometric chart (after ASHRAE, 1977).

equated with sensations produced by a combination of some dry-bulb temperature in 100 per cent relative humidity. Relative humidity, of course, is a function of vapour pressure and dry-bulb temperature.

It was found that the use of 100 per cent relative humidity as a standard tended to overemphasize the effects of humidity in cool neutral conditions, and underemphasize the effects in warm conditions. Accordingly ASHRAE developed another scale, referred to as the New Effective Temperature scale (ET*) which, by setting the relative humidity standard at 50 per cent rather than 100 per cent, takes into account the great part played by the evaporation of sweat in human heat regulation. Thus, by holding the relative humidity constant at 50 per cent, the scale can demonstrate the effectiveness of higher or lower temperatures on modifying the evaporative process, and thus demonstrate the relative role of skin 'wettedness' in determining comfort. Any given level of 'wettedness' can be produced by different combinations of dry-bulb temperature and vapour pressure; the combinations that produce the same level of 'wettedness' have the same ET* value.

The ET* scale applies in conditions of low air movement and minimal heat radiation, where the subjects are lightly clothed and doing sedentary work.

Oxford index

The Oxford, or WD, index (Leithead and Lind, 1964) is a simple combination of wet-bulb and dry-bulb temperatures which takes into account the subjectively more intolerable effects of high relative humidity. It is given as follows:

$$WD = 0.85 \; WB + 0.15 \; DB$$

where,

$$WB = \text{wet-bulb temperature}$$
$$DB = \text{dry-bulb temperature}$$

Wet-bulb globe temperature (WBGT)

The Oxford index does not take into account the effects of radiant heat which, in many circumstances, can be very significant. Coming into increasingly more common use, and sponsored by the American Conference of Government Industrial Hygienists, is the WBGT index which includes the dry-bulb temperature, the relative humidity (or vapour pressure) as well as the mean radiant temperature and the air velocity (Azer and Hsu, 1977). It is given as follows:

Outdoors with solar load:
$$WBGT = 0.7 \; NWB + 0.2 \; GT + 0.1 \; DB$$

Indoors or outdoors with no solar load:
WBGT = 0.7 NWB + 0.3 GT

where,
NWB = natural wet-bulb temperature
DB = dry-bulb temperature
GT = globe temperature

The instruments used comprise a dry-bulb thermometer, a natural wet-bulb thermometer, and a globe thermometer. These should be mounted on a stand to allow free airflow; the wet-bulb and globe thermometer should not be shaded. ACGIH, whose recommendations are widely accepted throughout the world, recommend the following Threshold Limit Values for heat exposure as determined by WBGT measurements:

Table 14.3. Heat exposure Threshold Limit Values (ACGIH, 1986-1987)

Work/Rest Regimen	Light	Work Load Moderate	Heavy
Continuous work	30.0	26.7	25.0
75% work/25% rest, each hour	30.6	28.0	25.9
50% work/50% rest, each hour	31.4	29.4	27.9
25% work/75% rest, each hour	32.2	31.1	30.0

Thermal tolerance

While broad tolerance levels can be established for tolerance to thermal conditions, tolerance is culturally affected. It must be recognized that a person brought up in ambient cold tends to be more tolerant of cold and less of heat, and vice versa. Tolerance is also determined in part by habits and behaviour. In North America, where heated or air-conditioned houses and buildings tend to be more common than in Europe the acceptable comfort levels are similarly different. Thus for North America the comfort level, at rest, nude, is given as 23°C ± 2.5 and 50 per cent relative humidity, whereas for Northern Europe the level is 20°C ± 2.5.

Adaptation

A limited adaptation to thermal conditions can occur. Over and above the physiological changes described, such as redistribution of blood flow, some longer lasting changes can occur of which activation of large numbers of non-active sweat glands latent in the skin is most significant. Adaptation to heat in fact is directly related to an increased capacity to sweat (Blockley, 1964).

Adaptation normally occurs within 4-7 days. It is acquired more rapidly with concurrent activity, which in turn increases the metabolic heat generation and acts as an increased stimulus. Once adaptation has occurred it is maintained at a high level for about two weeks even when the adapted person returns to a cold climate. Thereafter it diminishes slowly over a period of about two months.

Heat storage

When a person is exposed to extended intolerable hot conditions where the heat input is greater the heat output then heat storage occurs. The resulting effects go by a variety of names but it is convenient to classify them under the terms *heat exhaustion* and *heat stroke*. Heat exhaustion is uncomfortable and temporarily disabling, but heat stroke is dangerous and potentially fatal. Uncontrolled heat exhaustion, however, can lead to heat stroke.

Heat exhaustion

Heat exhaustion tends to occur following exposure of an unacclimatized person to temperatures in the range of 38°C to 60°C for periods of 1 to 8 hours. The occurrence and severity depends on a variety of other factors such as the associated humidity, the work level, the clothing worn, the ambient air velocity, and so on, while in addition there is a wide range of individual susceptibility from person to person.

The condition is insidious in onset, beginning as a generalized discomfort with some emotional irritability. As it progresses there is an increasing loss in the ability to concentrate, with subsequent decrement in performance, and/or will to work. Increasing sweating is apparent, along with fatigue and lassitude. Because of the increase in sweating there is a substantial loss of body fluids and of the salts dissolved in sweat. The resulting dehydration causes severe thirst, while the loss of salt produces muscle weakness and may cause muscle cramps. The skin is warm and moist to the touch, a fact which assists in discriminating heat exhaustion from the critically dangerous condition of heat stroke. The core, or rectal, temperature is commonly raised, but not more than one or two degrees.

Heat exhaustion may be accompanied by heat collapse. In heat collapse an excessive amount of blood may be pooled in the superficial veins. If this pooling is sufficient to reduce the return of blood to the heart below a critical level, then the heart will be unable to pump sufficient blood to the brain to maintain its function. The effect will be rapidly increasing weakness and dizziness, culminating in a faint similar to that found when the circulation is pooled in the peripheral veins during, for example, prolonged standing at attention, particularly in hot conditions.

Heat exhaustion can be potentially serious in that it can lead to the peripheral circulatory failure of fainting, and even to heat stroke, but it is not intrinsically dangerous. In heat exhaustion the heat control system is strained but not damaged. If the individual is removed from the hot conditions into a cool environment and given fluids to replace his fluid loss he will normally recover with no damage. In fact, if he is well hydrated, with an adequate salt balance before he begins his exposure, he will tolerate it much better. The question of giving extra salt before exposure or during recovery is controversial and will be discussed later.

Heat stroke

Heat stroke is a serious medical emergency, and indeed can be fatal. It may occur on direct exposure to temperatures in the range of 50°C to 80°C approximately, for a duration of as little as an hour, or at lower temperatures for a longer duration. As with heat exhaustion, its occurrence is dependent on, or aggravated by, other factors such as humidity, clothing, air velocity, work level, and so on. As has already been noted, it may occur as a progression from heat exhaustion.

When it begins the occurrence is sudden, with convulsions and unconsciousness, leading in some situations to death within a few hours. Characteristically the sweating ceases and the skin temperature rises to 40°C to 50°C or more. The core temperature will rise to 45°C to 50°C or even higher. Normal core temperature is 37°C. These dramatic changes are the result of failure of the thermal regulation centre and damage to the control system from the overheating.

Management of the condition, other than the immediate emergency response, should be the responsibility of qualified medical personnel. It calls for urgent specialized procedures such as rapid cooling, by, for example, wrapping the subject in cool wet sheets, or even packing him in ice. The latter approach is considered by some to be too extreme, although it may become necessary if other attempts at cooling fail. Refrigeration of the blood via an artificial circulation has even been attempted. It is vital at the same time to ensure replacement of body fluids and electrolytes by intravenous solution, and to give whatever other medical support is necessary.

Mortality is high. Even if the patient recovers he is likely to demonstrate some permanent brain damage, and at the very least will have a susceptibility to recurrence if exposed again in a hot environment.

Heat pulses and skin pain

The term heat pulse is applied to a short duration exposure to extreme heat, for example during entry to an annealing oven or a brick kiln. In general, the pain threshold is reached when the skin reaches a temperature of 45°C. A skin temperature of 46°C is intolerably painful (Berenson and Robertson, 1973).

The normal tolerable maximum exposure at work for an acclimatized person wearing ordinary clothing, gloves, and a face mask is in the range of 200°C to 230°C for a minute or so (Blockley, 1964). Employees doing such work are generally self-selected. They have a high natural threshold of tolerances of heat pulses and commonly have a dry skin with reduced sweating capacity. The water on moist skin may boil and cause scalds. Under laboratory conditions, with trained and highly motivated subjects, tolerance can be much higher.

Prevention and control of heat stress

The prevention and control of heat stress can be considered under two headings, namely engineering and administrative.

Engineering control

Reduction in ambient temperature of source

Since the occurrence of heat storage or heat pulses results from excessive external heat then reduction in the heat gradient will reduce the potential for harm. Since external heat production, however, is commonly an integral part of the process under consideration this is rarely feasible.

Reduction in conductive gain

Conductive gain is normally a feature of skin contact. Reduction in conductive gain is commonly achieved by placing insulative material, such as gloves, on the skin. The question of protective equipment will be considered later.

Reduction in radiative gain

The most common cause of heat gain leading to heat storage, whether from the sun or from an industrial source, is radiation. Reduction in radiative gain can be achieved by shielding the person from the radiating source. This shielding can be achieved by covering the surface with a layer of radiation-absorbing material such as polyester foam or other suitable plastic material, or by erecting a barrier of radiation-absorbing material between the radiating surface and the worker, or by wearing heat-reflective clothing, as for example an aluminized jacked or coverall. The latter approach is not recommended for extended exposures or for heavy work since the protective clothing tends to reduce or prevent heat loss from evaporation of sweat. Some effective barriers in the work place are aluminized reflective curtains, panels of wood or fabric covered metal, or even curtains of falling water.

One of the most effective ways of reducing radiant heat gain is to increase the distance between the source and the worker. Radiation of heat varies inversely with the square of the distance travelled. Hence, by doubling the distance between the source of heat and the recipient the intensity at the recipient can be reduced by a factor of four.

Increase to convective loss

Heat will be removed by normal convective motion of the ambient air. Air motion can be increased by increasing the natural movement through the use of fans and ventilation systems. These are considered in more detail in Chapter 24. In moderate conditions, however, the optimum air flow is in the range of 100–150 cubic feet per minute (2.83–4.25 cu.m/min).

Administrative control

In addition to engineering control, or where engineering control is not feasible, or where its resources have been exhausted, or where, for example, exposure might be occasional and/or for short durations, then administrative approaches to control have to be considered. A variety of approaches can be made depending on circumstances. These will be considered below.

Worker selection

Where it is known that workers will be exposed to excessive heat, and where it is feasible or acceptable to assign pre-selected workers to a known hot job, then consideration should be given to developing a pre-selection procedure. This should be based on a sound knowledge of the environmental conditions under which men will have to work, the energy costs of the various task involved, and the possible heat tolerance or intolerance of the individuals comprising the labour force. An analysis of such procedures, with specific reference to hot conditions in deep underground mines, has been undertaken by Strydom (1980) from whose work the following is derived. He notes that there are four major physiological parameters to be considered in selection, namely individual maximal oxygen intake, the effects of age, body size, body composition and surface area, along with race and sex differences. He recommends that each of these be considered in developing a selection procedure, thus:

Maximal oxygen intake

It is recognized (Astrand and Rodahl, 1970) that heat tolerant men have a significantly greater oxygen intake capacity than those who are heat intolerant. Similarly men with high work capacities in heat show more favourable heart rate and core temperature responses. Wyndham and his colleagues (1967), quoted by Strydom (1975), recommend that as a heat tolerant selection procedure workers for excessively hot jobs should be given a maximal oxygen intake capacity test (see Chapter 3). Those with high capacity should be allocated to hard work (oxygen cost 1.4 l/min), those with low capacity should be given light work (oxygen cost 0.6 l/min), and those in between should be assigned to moderately hard work (oxygen cost 1 l/min).

Age

Older persons are more susceptible to heat than younger, partly because they secrete sweat less readily, partly because of lower maximum oxygen capacity, and partly because of a readier absorption of radiant heat since their blood vessels are closer to the surface of the skin. In one study (Strydom, 1971), it was shown that although men above 40 years of age represented only 8–10 per cent of the work force they accounted for 30 per cent of the heat illnesses over a period of 5 years.

Body size and surface area

In general, the larger well developed person has a higher work capacity than the smaller, provided that the size is not disproportionately fat. A significant parameter is the ratio of the body surface area to the body mass (weight). Where the ratio is high, as in lean linear persons, there is a greater capacity to dissipate heat. Where the ratio is low, as in small fat persons, there is a greater ability to conserve heat against a cold environment; hence the lean Masai warrior of central Africa and the stocky well covered Eskimo of Northern Canada.

Race and sex

Provided that the workers are adequately acclimatized, any differences that might accrue to age and sex would appear to be eliminated (Strydom and Wyndham, 1963), although males have a better sweat rate response to heat stress, and a higher maximal oxygen capacity than females.

Other factors being equal, there is then no specific reason for selecting males rather than females for work in heat on the basis of sex response alone.

Work load

The heavier the work load the greater is the heat production. To ensure heat balance it is therefore necessary either to reduce the ambient heat, or reduce the load. The latter is commonly done by the worker by slowing down the rate of work or by increasing the number of unofficial work breaks.

At temperatures between 24°C and 40°C, each 1°C rise in ambient temperature increases the physiological stress by 1 per cent of the maximal work capacity. This is accompanied by increased sweating, increased heart rate, increased respiration rate, increased fatigue, and reduction in performance (Astrand and Rodahl, 1970). The work load of different tasks has already been discussed in Chapter 3.

Work practices

Cool break areas should be provided for workers in hot environments who are taking formal breaks. Care should be taken to ensure that the cool areas are not sufficiently cold as to be uncomfortable, nor should there be obvious high air velocity. The temperature should be in the range of 20–30°C, with an air movement of about 100–150 cfm.

When a worker is being introduced to a known hot job, or when a worker previously experienced in hot work is being returned after a lay-off of a week or more, care should be taken to ensure that he is put through a period of acclimatization for about seven days or so before returning to full activity.

Water and salt

The thirst mechanism is not adequate to encourage make-up of fluids lost in sweating. Consequently in hot conditions it is desirable for the worker to drink frequently, perhaps every half hour, small amounts of water in the range of 200–300 ml in order to maintain water balance. Excess of water will not be harmful. Water deficit eliminates the beneficial effects of acclimatization. Even in relatively cool conditions a water deficit has been shown to have a deleterious effect on performance (Strydom, 1980).

The desirability of ingesting extra salt during hot work has raised a lot of controversy. Until recent years, and in fact still today, it has been common practice to encourage workers in hot jobs to swallow salt tablets during their heat exposure. Excess salt, however, is in fact an irritant. It generates an osmotic pressure differential across body tissues, for example the mucosal lining of the stomach, which causes intracellular fluids to be withdrawn from the tissues and thereby set up an intracellular fluid deficit. If salt has to be replaced the best procedure is to restore salt losses at mealtimes, and if it has to be replaced during heat stress it should be taken in fluid form, and preferably in concentrations not exceeding 0.3 per cent salt.

Clothing and protective equipment

Any type of clothing provides some protection from heat. Clothing, however, also insulates the body and obstructs the loss of metabolic heat, consequently adding to heat load in hot environments.

Pre-chilled clothing can be used to assist in maintenance of heat balance. The prime example of this technique is use of the ice vest (Kamon, 1980). The ice vest is a jacket or vest containing pockets in which are placed packages of ice or frozen gel. Wearing of the vest allows longer exposure to hot environments without gross increase in core temperature. The duration of increased exposure will of course depend on the nature of the work and the ambient heat level.

Still another approach is the use of air-cooled or water-cooled suits, each of which, however, requires a trailing hose which limits movement and access to awkward areas.

In regions of high heat where whole body protection is required, the vortex suit (Raven, Dodson, and Davis, 1979) is an alternative for whole-body cooling. It uses a different principle for the distribution of air, but still requires to be attached to an air supply system. For comfort, the air flow rate in clothing should be 0.8 to 1 m^3/min at 27°C. A flow rate of 0.3 m^3/min or less will give rise to heat stress. Flow rates greater than 1 m^3/min will cause uncomfortable cooling (Blockley, 1964).

Radiant heat protection

A relatively simple approach to shielding from radiant heat is the use of insulated and reflective clothing surfaced with aluminized mylar which is a reflective material that lends itself readily to clothing manufacture. Like any protective equipment, of course, it is clumsy to wear, which limits its usefulness.

Some radiant heat situations can be controlled by insulating the radiant surface by covering it with a nonreflective heat absorbing material, or by the erection of barriers between the worker and the source, such as mylar reflective curtains, wooden or other panels, or, in appropriate places, water curtains.

As noted earlier, distance is an effective barrier since the radiant heat level diminishes according to the square of the distance.

Chapter 15
Heat loss and cold tolerance

Introduction

Heat loss occurs when the balance equation is disturbed such that more heat is radiated from the body than is generated or acquired within it. It is the opposite of heat storage.

The occurrence and extent of heat loss varies with the ambient temperature, along with the duration of exposure, the wind velocity, the available insulation, natural and acquired, and the amount of heat generated with exercise.

At 18°C, discomfort and the initial stages of cooling will be manifest in 4-8 hours, depending on other factors, in the unclothed body. Below 18°C to about 10°C, unclothed, there will be progressive chilling, shivering, and discomfort after a few hours, depending again on the ancillary factors. This will be accompanied by numbness and fatigue. With reduction in skin temperature of the hands there will be an increasing decrement in the performance of manual tasks within 5 minutes to 1 hour. (Trumball, 1956).

Continued exposure below the level of physiological compensation, for example, more than 4 hours unclothed at less than about 15°C, will give rise to heat loss accompanied by reduction in core temperature. A fall in body core temperature below 35°C will strain the thermal control centre, while a fall below 28°C is criticl for survival (ASHRAE, 1977).

Windchill

The concept of windchill was originated by Paul Siple in World War II (Siple and Passell, 1945), and was not originally intended as an index of subjective comfort. The numbers generated are based on the time taken for water to freeze under different conditions of temperature and wind velocity. Subjectively the effect of wind velocity will greatly increase the sensation of cold associated with a given temperature. The greater the wind velocity the greater is the subjective coldness. The actual scale is presented in terms of kcal/m²/hr, but subjective terms have been attached to various ranges of heat loss, recognizing that a given range can be derived from various combinations of temperature and heat loss. For example, a temperature of −45°C with air movement of 0.1 mph has the same windchill value as a temperature of −10°C with an air movement of 5 mph. A nomogram derived by Blockley (1964) from the work of Siple and Passell is shown as figure 15.1

Since water readily conducts heat away from the body, tolerance in cold water is less than in air. Little experimental work has been done on this matter, but from the work of Hall and his associates (1953) and by others, it would appear that the voluntary tolerance to a water temperature of 5°C is about 15 minutes, to 10°C is about 30 minutes, and to 15°C about 1 hour. Beyond those limits, continued exposure is

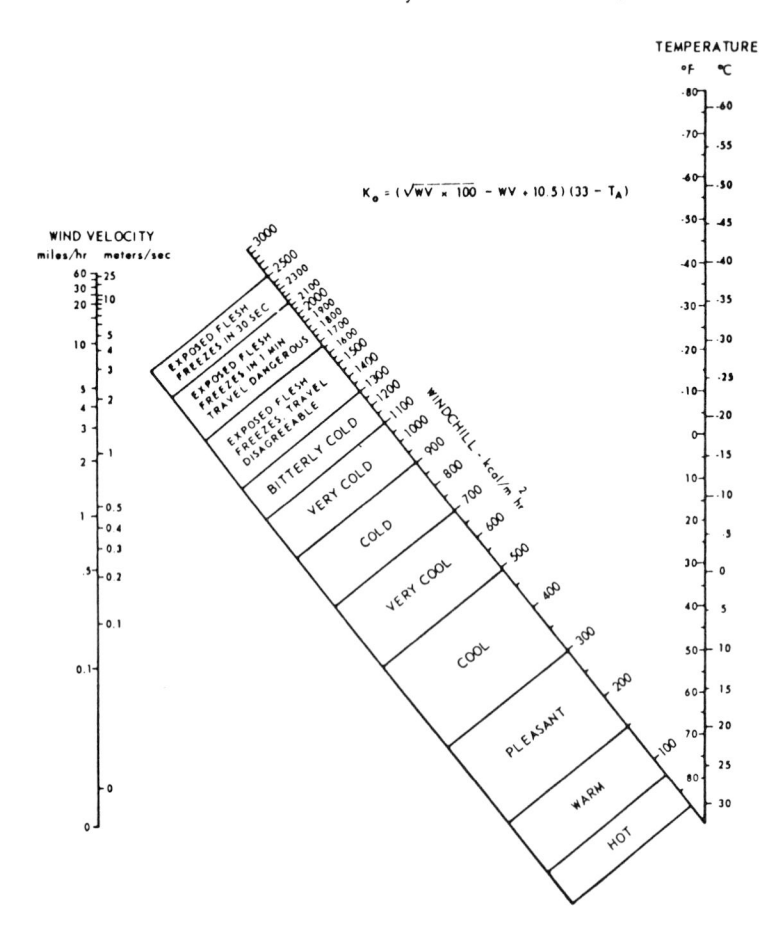

Figure 15.1. Windchill nomogram (after Blockley, 1964, derived from Siple and Passell, 1945).

likely to result in eventual unconsciousness and death from drowning or hypothermia.

Performance in cold

Exposure to cold is accompanied by clumsiness and loss of dexterity. In this connection, cooling of the hands is of more significance than cooling of the body. The loss in dexterity is accompanied by a loss in touch sensitivity. If the core temperature remains adequate, loss in dexterity will begin to appear with hand skin temperatures of about 15–20°C, while loss in tactile sensitivity begins with a hand skin temperature of about 8°C (Poulton, 1970), quoted by Konz (1979).

Konz (1979) also notes that at a core temperature of 35°C dexterity is reduced to the point where one cannot open a jacknife or light a match. Below 35°C the mind becomes confused, and about 32°C there is a loss of consciousness. At 26°C death will occur from heart failure deriving from hypothermia.

Cold protection

Cold protection is effected by reducing conductive and radiative heat loss, that is, by way of insulation. Insulation can be defined as the ratio of the skin/air temperature difference to the rate of heat flow, thus:

$$I = \frac{T_{skin} - T_{air} \,°C}{kcal/m^2/hr}$$

where,
I = insulation
T = temperature (°C)
Heat flow is in $kcal/m^2/hr$

From the above relationship Burton (1946) derived a measure of clothing insulation which he termed the *clo*. One clo is found when the above equation is solved to permit a heat flow of 1 $kcal/m^2/hr$, thus:

$$1 \text{ clo} = \frac{0.18 \, C}{1 \, kcal/m^2/hr}$$

In other words, 1 clo is the resistance to heat flow which will permit the passage of 1 kcal m^2/hr for a temperature gradient of 0.18°C.

From the conditions of the experimentation, 1 clo is sufficient insulation for comfort in a lightly clothed person in conditions where there is:

(a) a resting subject (metabolic rate = 50 $kcal/m^2/hr$)
(b) an environmental temperature of 21°C (that is, thermoneutral)

(c) a relative humidity of 50 per cent
(d) air movement of 5m/min

In practical terms 1 clo is also equivalent to the insulation provided by a light business suit and underwear, or a light overcoat. A sleeveless blouse, with light cotton skirt and underwear represents 0.3 clo, while long lightweight trousers and a light open neck short-sleeved shirt provide 0.5 clo.

There are in fact three components to the insulation of a clothed person. These are the insulation of the tissues themselves, the insulation provided by the air envelope around the skin, and the insulation of clothing. Each is considered below.

Tissue insulation

Since fat is a poor conductor of heat, the tissue insulation depends to a considerable extent on the proportion of fat in the tissues. It is also dependent on the physiological status of the blood vessels, and in particular the extent of constriction or dilation. The fat content, of course, cannot be manipulated in the short term, and the calibre of the blood vessels responds physiologially to the environmental demand and cannot feasibly be altered to meet special requirements. The actual insulation value of the superficial tissues varies from about 0.15 to 0.80 clo.

Boundary layer

The air envelope around the skin, or the boundary layer, provides significant protection provided it remains undisturbed. A significant function of clothing is to keep the boundary layer undisturbed, and also, since air itself is a poor conductor, to trap layers of air between layers of clothing. Even in the unclothed person at rest, however, the boundary layer remains more or less undisturbed. A boundary layer of about 0.3 cm will provide insulation of about 0.5 clo, rising to 0.7 clo at about 2.5 cm.

Mechanism of clothing insulation

Clothing as an insulating material, as opposed to a windbreak, acts essentially by trapping air in the weave. The same thickness of clothing gives the same insulation, namely, 1.5 clo/cm, provided that the textile fibres are of such a nature that they can trap air, and that the material has the property of 'loft', that is, expansible fibres. In addition, the material must not be dense enough to provide conduction.

In the design of cold protective clothing certain virtually incompatible requirements must be met. Firstly, it must provide thermal insulation. Secondly, it must be vapour permeable to allow evaporation of sweat.

However, it must also be impermeable to air in order to maintain the essential boundary layer; and it also must be impermeable to water to avoid conductive chilling and discomfort from wetness.

The foregoing requirements cannot normally be met by a single material. In practice there is commonly a compromise with a triple shell material. The outer shell is a durable fine weave material which is permeable to water vapour which can slowly penetrate the interstices, but relatively impermeable to wind, as well as being water resistant.

The middle shell is made of an elastic, expansile, high loft material. Down has been a traditional filler material, but it is probably no better than wool, kapok, or synthetic materials so fabricated as to meet the physical requirements. The inner shell requires to be readily permeable to water vapour and relatively impermeable to wind in order to provide still another barrier. It does not, however, require to be water resistant. Recent technological developments in synthetic materials such as the proprietary material Gortex, which is permeable to water vapour and permits the evaporation of sweat, allow for a simpler approach to clothing structure.

A triple layer garment, however, still requires to have adjustable ventilation to allow for variations in windchill and in the heat generated by work. Overheating within a cold-protective garment can give rise not only to paradoxical hyperthermia, but the sweat induced by the heating will eventually chill, causing wetness and discomfort.

The garment, of course, also has to be functional and allow relative ease of donning, doffing, movement, seating, seeing, and body hygiene. It must be possible to undertake manual activities while wearing it. Because of bulk, the insulation value of a garment is limited to about 6 clo. Consequently it is normally better to wear several removable layers than to attempt complete protection with one garment.

Chapter 16
Vision, illumination and visual hazards

Vision

Vision is the most widely used human sensory modality. All other senses defer to vision in perception of the environment, a phenomenon which can give rise to strange illusions. For instance, when an air pilot is deprived of vision outside the cockpit in cloud or darkness, and fails to integrate information that can be derived from instruments and non-visual senses, various kinds of disorientation can occur.

A classic example is the condition sometimes known as the 'leans'. A solo pilot is flying in cloud or darkness, occasionally glancing at his instruments. At a rate that is below his threshold of perception of radial acceleration one wing begins to dip slightly and continues to do so until the pilot, commonly by checking his instruments, observes that he is flying tilted. He quickly straightens his wings to the level position, imparting in so doing a perceptible acceleration to his labyrinth sensing system. The aircraft is now flying straight and level but the pilot, because of the imparted acceleration, and because he has no visual reference, feels he is tilted. Consequently he corrects the apparent tilt by leaning his body to one side, and continues to fly in the leaning position, intellectually knowing that he is leaning to one side but subjectively being unwilling to straighten. He maintains this tilted position until he receives an exterior visual reference at which point he is immediately conscious subjectively of his position and is happy to straighten himself.

The human visual system thus provides a person with:

* information on the nature of his/her environment
* information with respect to his/her location and position within that environment
* information to assist in the monitoring and control of the motions of his/her body and its members in the environment
* information that will allow the person to locate and/or track a static or moving 'target' within the visual field in terms of visual angle and range

Structure and function of the eye

The eye is an almost spherical structure largely filled with a gelatinous clear fluid. The tough fibrous outer coating become a clear round window in front—the *cornea*. Behind the cornea lies a retractable diaphragm of very fine muscle fibres—the *iris*. The iris encloses a hole of variable size—the *pupil*. Behind the pupil lies a firm but flexible *crystalline lens*, convex back and front, and capable of being adjusted in curvature.

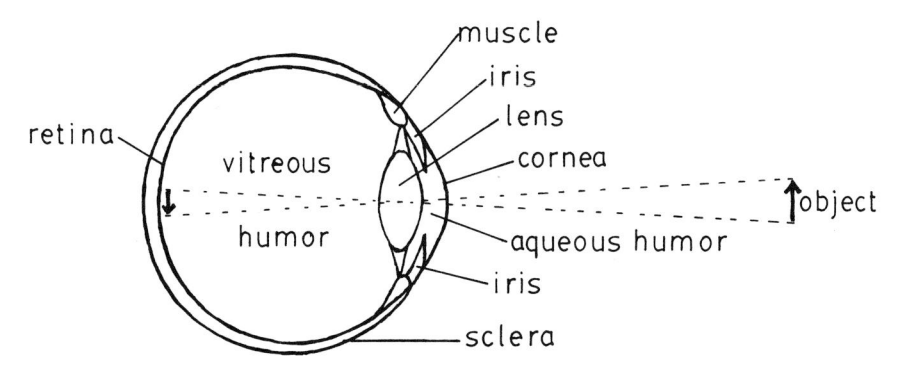

Figure 16.1. Schematic drawing of the eye.

Light enters through the cornea, passes through the pupil, is refracted by the lens and impinges on the lining of the inside of the eye—the *retina*—with reversal of the image. Light stimulates specialized nerve cells in the retina, fibres of which leave the retina near its centre via the *optic nerve*, which passes to the brain, where the stimuli are processed, integrated with memory and other stimuli and ultimately perceived as vision.

Retina

The retina is made up of several microscopic layers of cells and nerve fibres. Light first passes through a layer of nerve fibres and several layers of cells before reaching the light sensitive layer—the *layer of rods and cones*. These are nerve cells so called because of their fanciful microscopic shape. Rods are highly sensitive to light but not discriminative in detail. Cones are much less sensitive to light but acutely discriminative in detail; they are also discriminative for colour, whereas rods are insensitive to colour. Cones predominate in a small area at the centre of the retina specialized for observation of acute detail—the *fovea*—and diminish in number away from the fovea. As cones decrease in numbers rods increase. In direct vision, the centre of gaze is normally in line with the fovea, hence in low illumination vision may be better off centre. Since cones have a high stimulus threshold, night vision is not sensitive to colour.

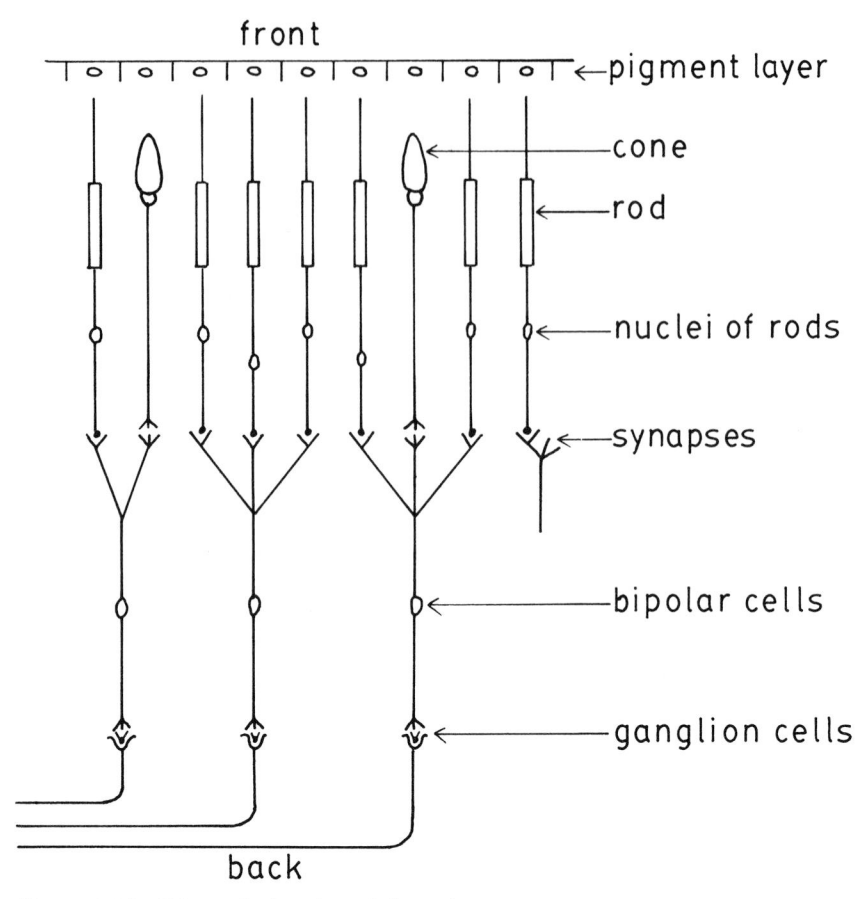

Figure 16.2. Schematic drawing of the retina.

The optical system of the eye

The optical system of the eye comprises the cornea, and the crystalline lens, aided by the iris. Light from the exterior is brought to focus on the retina. The refractive power of the eye, like the refractive power of any lens system, is measured in *diopters*. One diopter is the reciprocal of the focal length of a lens. The focal length is given by the Lens Maker's Equation which states that:

$$1/f = (n-1)(1/R_1 + 1/R_2)$$

where

    f = focal length
    n = index of refraction
    R_1 = radius (m) of the incident curvature of the lens
    R_2 = radius (m) of exit curvature

Note that:

$$1/f = 1/f_1 + 1/f_2 + \ldots + 1/f_{12}$$

The index of refraction is specific to any transparent material and is an indicator of the extent to which light is bent as it passes from one medium to another.

On diopter (D), then, is the refractive power required to converge parallel rays at a point 1 metre from the lens. Because it is defined as a reciprocal, 2 diopters is the power required to converge at 50 cm.

The total refractive power of the normal young eye is 59 diopters, a variable number up to 18 D being derived from the lens, and a fixed 41 D from the cornea itself. In children, who have strong intrinsic muscles in the eye and a very elastic lens, the variable refractive power can rise to as much as 32 diopters. Refractive power is considered to be positive for convex lenses, for example, + 1 D, and negative for concave lenses, for example, − 1 D.

Light rays passing through a convex lens are caused to converge and focus on a point. Light rays from a distant source can be considered as being effectively parallel and focus on a point a fixed distance from the lens. In the normal eye they focus at a point on the retina. Light rays from a point a short distance from the lens are divergent as they reach the lens and focus on a point further back. In a camera this problem is met either by adjusting the distance between the lens and the film to compensate, or by the addition of a stronger lens to focus on the appropriate plane of the film.

Accommodation

In the eye, of course, the distance between the lens and the retina is fixed, and hence the focal length cannot be varied by adjusting the distance; nor can a natural additional lens be added. To focus light from a source near the lens on to the retina, the lens undergoes the process of accommodation, that is, the curvature of the lens is increased to increase its power. Conversely, as the source moves further back from the lens, the curvature is allowed to decrease to reduce its power. This feat can be accomplished because of the structure and suspension of the lens.

The lens is an ovoid body held suspended by ligaments. Small intrinsic muscles are attached to these ligaments in such a manner that when contracted they relax the 'pull' applied to these ligaments. The elastic lens, stretched tightly by the ligaments, is thus allowed to relax and assume a greater natural curvature, which in turn increases its refractive power. Unfortunately, the elasticity of the lens is diminished by the process of aging, and hence older persons progressively lose their capacity to focus on close objects, to the extent that they frequently require additional lenses in the form of glasses for reading and other close work.

Visual stimulus

Light does not directly stimulate the nerve cells in the retina. The rods contain a chemical pigment which is sensitive to light, just as the silver compound in a photographic film is sensitive to light. The pigment is purplish in colour and is known as rhodopsin, or visual purple. In the presence of light the rhodopsin is rapidly bleached into lumi-rhodopsin, and then, under the influence of enzymes, into meta-rhodopsin, and finally it is broken down into two components, one being retinene, which is similar to vitamin A, and the other a protein called scotopsin. It is this process that acts as the stimulus to the retinal nerve cells. The process is reversible, and as the light stimulus is removed a batch of visual purple will return to its original state.

Light and dark adaptation

The nature of the visual stimulus gives rise to the phenomena of light and dark adaptation. On exposure to darkness from bright light, or vice versa, the individual is temporarily blinded. In bright light, the blindness is related to a sudden bleaching of all available visual purple. Once bleaching has occurred several minutes are required even for partial recovery. Sensitivity increases on a log scale, 1000 times in 10 minutes, 10 000 times in 20 minutes, and 100 000 times or more in 1 hour (Guyton, 1984).

In going from bright into dark, the blindness occurs since the available visual purple has been largely used in the brightness and is inadequate to meet the greater demands in the dark.

Design for dark adaptation

There are various situations, particularly in military conditions, where it is necessary to maintain dark adaptation in a light environment. In World War II, for example, before the extensive use of radar, night flying pilots were required to identify their airborne targets visually, and had to maintain their dark adaptation while waiting in lighted rooms before flying. Even today, operators on the bridge of a ship require to maintain their night vision in order to move from the lighted bridge to the exterior and back.

To meet these requirements, and other situations where night vision is necessary in the presence of light, advantage is taken of the fact that although the cones of the retina are sensitive to all light, the rods are only slightly sensitive to red light. Thus by lighting with red filters sufficient light is provided for use by the cones, while the rods are preserved for night vision. Thus, for example, the pilots noted above were fitted with red goggles, and the bridges of naval ships, including

Perception of colour

When white light is passed through a prism the light is refracted to the extent determined by the wavelength of each of its components and presents the classic rainbow colours as follows:

Table 16.1. Wavelengths in millimicrons, and perceived colours of refracted white light ght

Red	> 600 millimicrons
Orange	580
Yellow	550
Green	500
Blue	450
Indigo	400
Violet	< 400

Helmholtz theory

The most common explanation advanced to account for the perception of colour was developed by the nineteenth-century German physicist, physiologist, and mathematician Hermann Helmholtz, who propounded the trichromatic, or three-colour, theory on the basis that all colours can be synthesized from the three primary colours, red, blue, and yellow. Other theories have been developed, and no theory is entirely satisfactory.

The trichromatic theory depends on the fact that while the rods do not discriminate colour, the cones demonstrate selective sensitivity for different wavelengths. While all cones respond to light some respond more actively to different wave bands, thus:

Table 16.2. Sensitivity of cones to colour

Type of cone	Range of response (millimicrons)	Predominant Sensitivity	Other Sensitivity
Green	450–675	green	blue, yellow, orange, red
Blue	400–500	blue	violet, green
Red	< 400–< 600	red	orange, yellow

The combination of stimuli received at the retina is converted into nerve signals and transmitted to the brain via the optic nerve where it is processed and ultimately perceived as colour.

Physically, colour can be defined in terms of three characteristics, namely, dominant wavelength, discussed above, saturation, and luminance. Psychologically, or subjectively, the dominant wavelength is seen in terms of *hue*. The characteristic of *saturation* determines the degree of difference from a similar tone of grey, and the luminance is perceived as *brightness*. The Munsell Color Company (1929) represented these characteristics in the form of a double cone (Figure 16.3) with the intensity of the hue and saturation being demonstrated by the length of the diameters. They are thus able to define colours by their position on the cone.

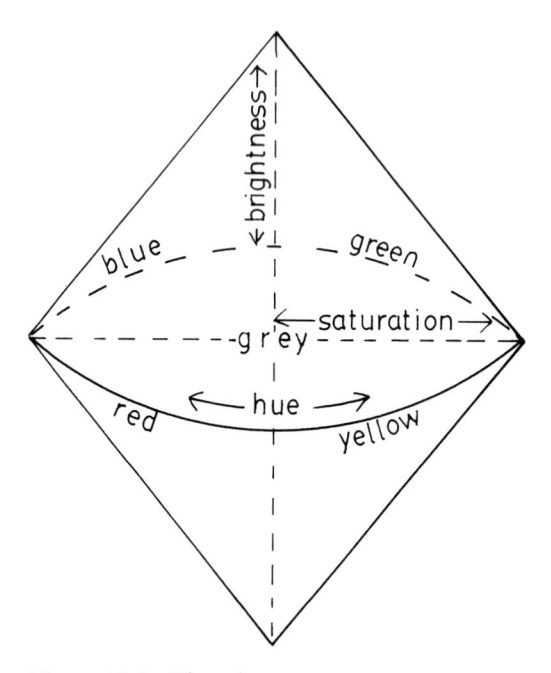

Figure 16.3. The colour cone.

Colour blindness

In the developing cell the genes which carry the genetic code for the cones are carried linked to the female sex chromosomes. Two such chromosomes are carried by the female while only one is carried by the male. Consequently should these genes be lacking in whole or in part there is a greater likelihood for colour blindness to appear in the male. Some 8 per cent of males in fact are colour blind. Colour blindness commonly takes the form of red-green blindness, or in other words where other clues are lacking it may be difficult or impossible for the colour blind subject to discriminate red from green, a condition which can cause problems in the differentiation of signal lights. Less common is blue-white blindness. Total colour blindness is extremely rare.

Depth of vision

Depth of vision, or stereoscopic vision, depends in part, but only in part, on binocular vision. A person with two normal eyes will see an apparent three-dimensional view of an object up to a range of about 6–7 metres. His perception will be aided by other visual clues which are not dependent on the possession of normal binocular vision. These clues will be examined later.

Stereoscopic vision

Because the eyes are separated each eye receives a slightly different view of a near subject. Consequently the image formed on the retina by the right eye is different from that formed by the left. Furthermore, the light approaching from the right side of the body is sensed by the left side of each retina, and that from the left is sensed by the right. Each set of left and right stimuli is carried independently to the brain. In the brain a complex processing activity takes place such that the brain blends the two images and perceives the differences with the help of stored memory in terms of three dimensions.

Other facets in depth perception

Besides the phenomenon of stereoscopic vision there are other factors which unite to assist in providing information on relative depth of field and on the relative position of objects within that field. Some of these result from the physiological function of the visual system, and some from the psychology of perception itself. Some of these cues may appear trivial and obvious by themselves, but when taken in context and integrated with information from memory and all other cues they become an important part of a total picture. They are discussed below:

Convergence: Each eye is held by ligaments in a bony cavity known as the orbit, but in addition small muscles are so attached to each eye as to permit it to be freely moved through 360 degrees in the frontal plane on contraction of the appropriate muscles. In particular, each eye can be rotated inwards to converge on close objects. The extent of this convergence is sensed by nerve sensors in the muscles and provides an indication of the distance at which the object is located in front of the eye. This information is good up to about 7 metres.

Linear perspective: Parallel lines subjectively converge and are said to meet at an infinite distance in front of the viewer. The extent of that visual convergence in the light of a reasoned understanding of actual distances give an indication of the distance of objects in front of the viewer.

Resolution: The closer an object is to the viewer the greater is the resolution of detail. Therefore if two known objects are in front of the viewer the object showing the greater visual detail is closesr. The phenomenon of visual resolution provides a significant cue to a pilot landing an aircraft under visual flight rules. He unconsciously observes the resolution of the surface of the runway as he approaches it and thereby receives an indication of how close he is to it.

Interposition: If one object lies in front of another then it must be closer than the other.

Previous knowledge: Previous knowledge of the environment and objects within it is significant in many of these ancillary cues. In particular, if two known objects subtend the same visual angle and retinal size then the smaller known object must be closer. For example if a church steeple and a darning needle are both seen in the same visual field and are both the same apparent size then the needle must be closer than the steeple.

Contrast and 'induced colour' phenomena

Contrast, or more properly, luminance contrast or brightness contrast, refers to the difference in luminance between an object and its surrounds, or between specific features of an object and their background on the object. It is given by the following relationship:

$$\text{Contrast} = \frac{B_1 - B_2}{B_1} \times 100$$

where

B_1 = brighter of two contrasting areas
B_2 = darker of two contrasting areas

The contrast between black and white or between two complementary colours changes the quality of the perception of these images. Thus, for example, black print on a white ground looks blacker than black on a grey ground or a coloured ground. However, because of other interference, black is easier to read on a white ground than white on a black ground. Complementary colours tend to enhance each other's brightness. Blue looks more blue on a yellow background than on any other colour, while looking at a blue patch on one site reduces the sensitivity to blue elsewhere. Looking at red increases the sensitivity to green. A shade of grey placed alongside a colour will take on the tone of its complementary colour.

Observations of these 'induced colour' phenomena led Edwin Land of the Polaroid Corporation to develop his technique of generating colours with the use of only two colours instead of the three demanded by the Helmholtz theory. Some dramatic artistic effects can be similarly achieved by placing various colours in juxtaposition with one another while confusing the eye with complex graphic patterns.

After images are a related phenomenon. Prolonged gazing at a bright object followed by shutting the eyes or gazing at a black ground will cause projection of an image of the object. If the object is black on a white ground the image will be white on a black ground and vice versa. If the object is coloured the image will be presented in the complementary colours. It is suggested that this phenomenon occurs because of fatigue of the colour receptors due to prolonged viewing which then allows the colour receptors for the complementary colours to take over.

Illumination

Light can be defined as radiant energy that is capable of exciting the retina and producing a visual sensation (IES Nomenclature Committee, 1979). It is, of course, part of the radiant energy spectrum which includes all forms of radiant energy from cosmic rays to radio waves, and even power transmission. Thus it can be considered as visually evaluated radiant energy. It has both physical characteristics and psychological connotations.

Measurement of light

Light can be measured in terms of four characteristics, namely, luminous intensity, luminous flux, illuminance, and luminance, each of which defines some special characteristic. They will be considered in turn.

Luminous intensity

Luminous intensity, sometimes known as *candlepower* is the intensity of a theoretical point source of light. At one time it was literally measured by the amount of light emitted from a standard candle. The unit of measure is the *candela* (cd) which is defined as an international standard, namely the luminous intensity of $\frac{1}{600\,000}$ m^2 of projected area of a black body radiator operating at the temperature of solidification of platinum (2047°K).

Luminous flux

The luminous flux is the rate at which light is emitted from such a source, that is the amount of light per unit time. The unit is the *lumen* (lm). The luminous flux from a 1 cd source is 12.57 lm.

Illuminance

Light from a point source is distributed equally all around and consequently can be considered as falling on the inside of a sphere. The amount of light striking a point on the inside surface of that sphere is the *illuminance,* or illumination. It is measured in terms of luminous flux per unit area, namely lumens per square foot, or lumens per square metre. The unit of lumens per square foot is the *footcandle* (fc), whereas the unit of lumens per square metre is the *lux* (lx). The latter is the international unit. One foot candle is equivalent to 10.76 lux; for convenience it is often considered that 1 fc and 10 lux are interchangeable.

Luminance

Objects are seen because of the light reflected from, or leaving, their surfaces. Luminance is the term given to the amount of light leaving a surface, either by reflection or being emitted from, say, a lighted panel. When the amount of light is measured in lumens (that is, luminous flux) and the area is in square feet, the unit of measurement of luminance is the *footlambert* (fL). When the amount of light is measured in candelas (that is, luminous intensity) and the area in square metres, the unit of measurement is in candelas per square metre. One fL is equivalent to 3.43 cd/m. The ratio of the amount of light actually reflecting from a surface to the amount of light striking that surface is the *reflectance*.

Typical reflectances are presented in the following table (Konz, 1979):

Table 16.3. Typical reflectances of various surfaces (Konz, 1979)

Object	Reflectance %
Mirrored glass	80–90
White matt paint	75–90
Porcelain enamel	60–90
Aluminium paint	60–70
Newsprint, concrete	55
Dull brass, dull copper	35
Cast and galvanized iron	25
Good quality printer's ink	15
Black paint	3–5

Industrial lighting

Lighting in industry may be either natural, that is, from sunlight, or artificial. Sunlight can be provided by windows or by skylights, or both, but natural lighting can present some unwanted problems (Hopkinson, 1972). Thus, unless the window is intended to present a view, or a feeling of subjective openness, it is generally neither adequate nor cost-effective except in a small room. Otherwise light from windows commonly has to be supported by artificial light sources, and heat loss either from radiation or ventilation through the window can become unacceptable. Natural light is also difficult to control in terms of brightness or dullness, while excess sun may give rise to intolerable heat. Windows also admit noise.

Artificial lighting in industry can be provided by incandescent filament bulbs, or by fluorescent tubes. There are various types of both. Konz (1979) points out that cost is one of the most significant factors in determining the selection of lighting. The cost of electricity comprises about 90 per cent of the total illumination cost. Consequently high intensity discharge lamps which provide at least 60 lumens/Watt tend to be favoured over incandescent lamps which provide only about 15 lumens/Watt. There are various types of high intensity discharge lamps, each with different characteristics. These include the common fluorescent lamp, which produces a whitish light and has an output of 65 lumens/Watt for a one metre lamp, and 72 lumens/Watt for a 2.5 metre lamp. Sodium vapour lamps are even more efficient with an output of over 100 lumens/Watt; mercury vapour lamps are less efficient than fluorescent. Sodium lamps have the disadvantage of having a yellowish light, while mercury vapour have a bluish light. Since their light is emitted in very narrow wave bands colour discrimination is very difficult when objects are so lit.

Looking at lighting in another manner Grandjean (1980) defined four distinct light sources with different characteristics. These are the direct radiant sources, the mixed direct and indirect, the opalescent globes, and the indirect.

Direct radiant

Direct radiant sources generate a cone of light, of which 90 per cent is directed towards the target. These lights throw hard shadows with a contrast between light and shadow of at least one to ten. They are commonly used in offices, shop windows, show cases and in switch rooms, but because of the glare produced by the excess contrast they are not recommended for use as working lights unless the ambient illumination is high enough to relieve the contrast.

Mixed direct and indirect

Mixed direct and indirect lighting occurs when a translucent shade is used to modify the light so that about 40 per cent is diffused in all directions while only about 60 per cent falls directly on walls and ceilings. This lighting throws softer shadows than does direct lighting, with diffuse edges, and tends to be used as general illumination in the home, in retail shops, and in offices, where the emphasis is on general comfortable lighting. It is suitable for moderately close work, such as writing, but not for precision work.

Opalescent globes

Opalescent globes, and other fixtures providing equidirectional distribution of light disperse the light equally in all directions with minimal shadow. They may, however, act as bright sources and consequently can cause glare. Thus, while they are suitable for storerooms, corridors, entrance halls, vestibules, lavatories, and other areas where the demands for visual discrimination are low, they should not be used for work rooms, offices, or living rooms.

Indirect lighting

Indirect fixtures are designed to direct 90 per cent or more of their light on to walls and ceilings which reflect it back according to their reflectance capacity. Reflectance is improved by the use of light coloured walls and ceilings. Because it is non-directional, the light is diffuse with minimal shadows. It has the great advantage of producing no glare, but if used in a workroom it should normally be supplemented by additional direct lighting at the work area. It is considered very suitable for displays, salesrooms, and any place where the eye is drawn to the walls.

Lighting intensity for tasks

The work of the great vision psychologist, Dr H. Blackwell, discussed extensively by McCormick and Sanders (1982), showed amongst other matters that the visual needs of a task are determined by the illumination of the task, the contrast, the size of the object to be viewed, and the time available for viewing. Since the size of the object and the time available are intrinsic to the task, only the illumination and contrast are available for modification.

On the basis of the work of Blackwell and others, and also out of practical experience, the Illuminating Engineering Society (IES) of the United States (IES, 1981), as well as various other professional bodies and standards institutions, such as the British Illuminating Engineering Society, and the German DIN, have developed recommendations for the level of lighting required for difference circumstances. The British standards, which are based on the accomplishment of 90 per cent of required performance, tend to be about 50 per cent lower in lux value than those of the United States. (Konz, 1979).

All the recommendations are very comprehensive, categorizing each type of many different tasks according to the lighting required under different circumstances, and commonly including different specified levels of contrast. It is not feasible to reproduce them here; the original documents should be consulted as required. Suffice it to say that in general about 2000 lux with good contrast is required for precision work, 1000 for general office work, and 100 for non-work areas.

Grandjean (1980) points out that too high a level of illumination can also be undesirable. Anything above 1000 lux can increase shadows, contrast, and glare, thereby aggravating the problem it is designed to relieve. He suggests that the overall lighting level for open plan offices is in the range of 400–850 lux. In this connection it is becoming increasingly common to maintain a moderate level of overall illumination, supplementing it as required with local working lights at the place where they are needed. Grandjean again suggests the following guidelines:

Table 16.4. Guidelines for working lights and general illumination (after Grandjean, 1980)

Working lights	General illumination
500 lx	150 lx
1000 lx	300 lx

Role of contrast in illumination

Blackwell (1963) noted that contrast and illumination are each significant in the ability to see an object. For example if the illumination is kept constant at 700 lux, a visual performance equivalent to that with an illumination of 1900 lux can be achieved by increasing the contrast from 0.60 to 0.68.

Contrast can be used to enhance visual capability. To detect shape, for example, it is desirable to increase the contrast of the task versus the background, while to detect surface characteristics the contrast between the task and the background should be reduced.

Aging changes contrast requirements. Blackwell and Blackwell (1968) showed that as age increased a greater contrast was required between object and background for satisfactory performance. Using the age group of 20–25 as a norm the contrast requirements for older person must be multiplied by the following factors:

Table 16.5. Contrast requirements for effective vision in older persons

Age group	Factor
40–49	1.17
50–64	1.58
65+	2.66

Luminance ratios

While the contrast between an object and its background is important in determining its visibility, it is also important to maintain certain levels of relative brightness between different areas in the visible environment. The luminance ratio is the term given to the ratio of the luminances of a given area and the other areas in its vicinity. On the basis of theoretical work and practical experience certain recommended ratios have been established. These are shown in the following Table developed by McCormick and Sanders (1982) from the IES Lighting Handbook, 1972:

Table 16.6. *Recommended maximum luminance ratio for various conditions (after McCormick and Sanders, 1982, from IES, 1972)*

Areas	Recommended maximum luminance ratio — Office	Industrial
Task and adjacent surroundings	3:1	
Task and adjacent darker surroundings		3:1
Task and adjacent lighter surroundings		1:3
Task and more remote darker surfaces	5:1	10:1
Task and more remote lighter surfaces	1:5	1:10
Luminaires (or windows, etc.) and surfaces adjacent to them		20:1
Anywhere within normal field of view		40:1

Reflectance

Part of the brightness within a room environment comes by reason of reflectance from the surfaces within that room including the furniture and equipment. Reflectance from different types of surface has already been discussed. To ensure appropriate distribution of light, however, there are certain desirable reflectances that should be achieved from the walls, ceiling, and so on. Some of these are recommended in the following table:

Table 16.7. *Recommended reflectances for room and furniture surfaces (American National Standard Practice for Office Lighting, 1973)*

Surface	Reflectance %
Ceilings	70–90
Walls	40–60
Machines and equipment	30–50
Furniture	25–45
Floors	20–40

Glare

Glare occurs when there is an interfering brightness in the visual field. *Direct glare* occurs when there is an excessively bright source directly in the visual field: *reflected*, or *specular* glare occurs when the bright light is reflected from surfaces to the eyes.

Terminology

The IES Nomenclature Committee (1979) have categorized three types of glare, namely, (a) *discomfort glare*, which produces discomfort, but does not necessarily interfere with performance or visibility; (b) *disability glare*, which reduces visual performance and visibility and may be accompanied by discomfort: and (c) *blinding glare* which continues to produce blindness for some time after the origin has been removed. In industry, while both disability glare and blinding glare can occur, the main problem lies with discomfort glare.

Effects of discomfort glare

Luckiesh and Moss (1927–1932) in the early General Electric experiments showed that glare has a significant effect on visual effectiveness directly related to the angle of the glare source to the line of site. The visual effectiveness diminished from 58 per cent with the source at an angle of 40 degrees from the line of sight, to 16 per cent with the angle at 5 degrees. Proportional differences occurred within that range. This was confirmed by Bennett and his colleagues (1977) who demonstrated that the effects of glare are reduced not only by increasing the angle from the line of sight, but also by increasing the background luminance and reducing the size of the glare source.

The extent of predicted glare from a luminaire can be determined by calculating the *discomfort glare rating* (DGR), a procedure developed by the Illuminating Engineering Society of the United States (IES Lighting Handbook, 1981). The DGR is an index of potential discomfort that might arise from luminaires in a given environment. It is based on the room size and shape, the room reflectances, the illumination, the nature of the luminaires, their number and location, the field luminance, the observer location and line of sight, and the type of equipment and furniture. The calculation is tedious and the Handbook should be consulted for the technique.

Unfortunately not all glare comes from luminaires. Much comes from windows. These, too, have to be considered when assessing the amount of glare to which workers are exposed.

Reduction of discomfort glare

The approach to the reduction of glare depends on the origin of that glare, whether from luminaires, windows, or reflections from surfaces.

Where the glare is primarily direct from luminaires, these should be selected with a low DGR and/or reduced in intensity wherever feasible. If it is necessary to retain brightness it may be possible to replace one luminaire of relatively high intensity with several of lower intensity. Indirect lighting is normally preferable to direct. Lights should be located off the line of sight to the extent possible, either by raising them above, or to the side. As noted in the earlier discussion also, if the ambient luminance is increased around the glare source the direct glare is perceived as being less. Failing all other approaches, or perhaps ancillary to them, the glare source can be shielded by placing a shield around it, or between the source and the worker, or a polarizing lens on it, or if necessary by providing the worker with a peaked cap or visor.

Where the glare is from windows, one of the approaches, where feasible, is to rotate the work area at right angles to the window so that the glare is no longer direct. Alternatively, a canopy can be placed outside the window. If the situation is recognized in the building stage, the windows can be built above the line of sight to the extent that is necessary, while preferably not destroying any view, and finally, if the source cannot be removed, the windows can be shielded by shades, or venetian blinds, horizontal or vertical.

When the glare is from surfaces, it is first of all important to reduce the source of the glare to the extent feasible, whether from luminaires or windows, and to orient the work to minimize the reflection. Thirdly, a good level of overall illumination should be provided to reduce the difference between the reflection, and the ambient environment. In addition, the surfaces themselves should be selected or treated to minimize reflectability. Surfaces should be matt and non-glossy, avoiding bright metal, shiny paper, glass, and smooth plastic.

Visual Hazards

Visual hazards can be categorized as non-penetrating mechanical, penetrating mechanical, corrosive chemical, and burning. Hazards also exist from ionizing radiation and laser exposure, but will not be considered here.

Non-penetrating mechanical

Non-penetrating mechanical hazards occur by impingement of irritating particulates on the corneal surface, often under the eyelid. While they cause discomfort, and may cause temporary disability, non-penetrating hazards are normally not so dangerous as those in other categories.

Agents

The agents comprise dusts and other particulates, commonly specific to the tasks or to the materials being processed. They include:

(a) Wood—commonly sawdust from cutting and processing.
(b) Metal—all metals involved in industrial processing, from grinding, cutting, and polishing.
(c) Other minerals—including sand and stone, from similar processing, including crushing.
(d) Grain particles—including hull fragments from grain handling, and dust from milling.

Effects

The effects result from lodging of a particle on the cornea or conjunctiva. This causes discomfort, irritation, pain, reflex blinking and tearing. These latter two phenomena are a physiological attempt to remove the particle. Should the tears and blinking be inadequate to remove the particle, the eyes should be washed in water for some 10–15 minutes if necessary. If this fails to remove the particle it should be removed manually by a competent person such as a nurse or a doctor. If it is not removed it may give rise to infection and cause a painful and occasionally disabling conjunctivitis. Corneal scratches may also occur with resulting scarring which can interfere with vision.

Prevention

Prevention is simple, and comprises the mandatory use of protective goggles or spectacles whenever there is a possibility of exposure, combined with good working procedures and good housekeeping to reduce the occurrence of particulates. Specifications for goggles and spectacles are available in safety catalogues and manuals.

Penetrating mechanical

Penetrating mechanical hazards comprise those objects which have sufficient mass and acceleration to penetrate or deform some portion of the eye. The effects can be serious and disabling.

Agents

There is a wide variety of possible agents. One major subset comprises a group of items capable of penetrating in a dart-like manner, such as metal cuttings or sharp fragments, wood slivers, and sharp edges of mineral substances, while the other comprises more blunt objects which damage by impact.

Effects

Collisions with moving objects, or falling or striking fixed objects, may produce laceration and/or bruising of the eyeball, eyelids and associated structures. Long term results can produce scarring of the cornea and damage to the internal structures of the eye. Injuries of this nature require skilled surgical or ophthalmological treatment.

Prevention

Again prevention is simple. It includes the wearing of protective goggles and/or spectacles as well as the prevention of possible injury by the removal of potential hazards, or the erection of barriers and protective devices around them.

Corrosive chemical

Any corrosive liquid, vapour, or gas, when in contact with the eye can cause irritation and destruction of the eye surfaces.

Agents

The agents are varied. They include numerous organic and inorganic chemicals which will be considered in the chapters on industrial toxicology.

Effects

Corrosive liquid, vapour, or gas can cause etching, ablation and destruction of the surface and subcutaneous tissues of the eyes, the eyelids, and the conjunctiva, with severe irritation and pain, blinking and tearing. The results may be permanent scarring and loss of vision. Prompt flushing with water for 10–15 minutes can dilute the agent and reduce the adverse long-term effects.

Prevention

Prevention requires anti-splash goggles and related protective gear, even to the extent of full-face respiratory protection, hood and protective clothing. These matters are considered later.

Burns and heating agents

Burns of eye tissues can occur to the cornea, conjunctiva, and retina, as part of generalized burning of the face, or as a specific result of trauma to the eye from one or more of the agents noted below.

Agents

Burns of the surface structures of the eye, and the deeper structures if the intensity is great enough, commonly occur as part of generalized burns to the face from flame and burning materials.

Burns can also occur from exposure to infrared radiation, from non-fire sources, for example in glass blowing. High intensity sources of ultraviolet radiation, such as sun simulators, and less intense sources such as are found in arc welding, can cause radiation burns affecting the conjunctiva (conjunctivitis) and the cornea (keratitis) with the usual signs of severe irritation, sometimes followed scarring and disability. If the intensity is great enough retinal damage can occur with severe effects.

Effects

The effects are similar to those of the corrosive agents. They may leave permanent scarring in the cornea or retina with interference with vision.

Prevention

Prevention comprises any and all measures that will limit exposure to the hazardous agents. These include removal of the potential hazard and provision of protective equipment. In particular, welders should always wear goggles which have the appropriate shielding factors as defined in safety catalogues and manuals. Bates (quoted by Zenz, 1975) suggests that as far as ultraviolet radiation from welding is concerned, safe exposure without goggles is no more than 20 seconds at 7 feet, and 17 minutes at 50 feet.

Chapter 17
The atmosphere, respiration and barometric pressure

Atmosphere

The nature of the earth's atmosphere has evolved over countless millennia, as the primaeval gases and steam gave way to the atmosphere we know to-day. It is of course, maintained in its location by the pull of gravity, which in turn generates the pressure that the atmosphere exerts upon the earth's surface and its inhabitants.

The atmosphere is a mixture of gases comprising approximately 21 per cent oxygen and 79 per cent nitrogen. The oxygen is continually being consumed by living creatures and in fire, and continually being regenerated by plant life. In addition to the oxygen and nitrogen, the atmosphere contains small quantities of various normally inert gases, namely argon, neon, krypton, xenon, and helium, the latter being in the largest quantity. There are also small quantities of carbon dioxide, largely produced by living metabolism, as well as water vapour, and, of course, sundry chemical pollutants.

The atmospheric pressure at sea level averages 1.013×10 N/m^2, or 1.013 millibars, or 760 mm Hg (29.92 in.), or 14.7 psi, dependent on which measurement scale is being used. The preferred units of the International System are newtons per square metre, but in human physiology the normal unit is millimeters of mercury (Hg).

With increase in altitude above sea level there is a progressive decrease in barometric pressure, although, as will be discussed later, the relationship is not rectilinear. In other words, doubling the altitude does not mean halving the pressure. In fact, at 18 000 feet (approximately 5500 metres) the barometric pressure is equal to ½ an atmosphere, or 380 mm Hg, while ¼ atmosphere is found at 23 000 feet (approximately 7000 metres), or 195 mm Hg.

Similarly at 33 feet (10 metres) of sea water the pressure exerted includes an additional one atmosphere provided by the depth of the water, such that the total barometric pressure is then two atmospheres. Each additional 33 feet (10 metres) adds another atmosphere of barometric pressure.

Respiration

The presence of barometric pressure both permits and is necessary for human respiration. Human respiration comprises external and internal respiration.

External respiration

External respiration is the term given to the gas exchange that takes place between the blood in the lungs and the exterior atmosphere. It comprises both inspiration and expiration.

Inspiration

Inspiration is the active process of breathing. It requires muscle exertion to expand the chest. This is accomplished by coordinated contraction of the diaphragm and the intercostal muscles. The diaphragm is a large dome-shaped muscle which separates the chest from the abdomen. The effect of contraction of the diaphragam is to lower and flatten it and thus increase the vertical dimension of the chest.

The intercostal muscles lie between the ribs and are so arranged that a coordinated contraction will cause the ribs to lift and expand upwards and outwards. The total effect of contraction of ribs and diaphragm is to produce a marked increase in chest volume. According to Boyle's Law, the product of a given pressure and volume is constant, so long as the absolute temperature remains constant, thus:

$$P_1 V_1 = P_2 V_2, \text{ when T is constant}$$

Thus, when the volume of the chest increases, the pressure within the chest cavity must fall. Air then enters through the respiratory passages to restore the balance.

Expiration

Expiration, on the other hand, is normally a passive process. It relies on the elastic qualities of the chest structure and its tissues. After an inspiration the muscles responsible cease to contract. There is a momentary hold and then the chest collapses elastically until the original volume is restored. The internal and external pressures thus become equilibrated as the air exits through the respiratory passages. The process can be assisted by active expiration in which the intercostal muscles act in a reverse manner to force the air out, assisted by the muscles of the neck and shoulders where necessary. This process is much more fatiguing than normal expiration. In addition, forced expiration does not invoke the normal physiological control mechanisms which ensure that the rate and depth of breathing, and the accompanying loss of carbon dioxide, are compatible with the need to ensure physiological acid-base balance. The latter is to a large extent controlled by the amount of carbon dioxide that is exhaled.

Under certain conditions of breathing artificially pressurized oxygen by aviators in emergency conditions, active expiration becomes necessary against an input pressure. This type of breathing may lead to disorders of breathing rhythm and resulting disorders of acid-base balance. These matters will be discussed in the next chapter.

Route of entry of air

During breathing, air normally enters and exits via the nose. When the nose is blocked, or when the demand is great, as in panting, the nose can be by-passed and the mouth used as a portal. This, however, is much less satisfactory since air taken through the nose is passed to the sinuses for heat exchange and saturation with water vapour. The sinuses comprise five bony cavities in the skull, each opening into the nasal cavity. They are lined with mucous membrane rich in blood vessels, and are saturated with water vapour. In certain duct diseases, notably byssinosis, or cotton disease, the sinuses can become choked.

From the mouth and nose the air passes through the glottis (throat) to the trachea (windpipe). The trachea is a wide-bore tube of elastic tissue supported by rings of cartilage. The rings maintain the trachea open during inspiration when the accompanying reduction in pressure would otherwise cause it to collapse. The trachea passes down the chest cavity and divides into two bronchi, (singular: bronchus), each of which goes to one lung. The initial portions of the bronchi have the remnants of cartilaginous rings, but the bronchi are largely made of strong muscle.

The bronchi divide and subdivide, gradually changing in structure to become smaller, less thick-walled, and less muscular. Ultimately the muscle wall disappears and the bronchi become thin-walled broncheoles. These in turn divide and subdivide to become microscopic respiratory broncheoles, which end in terminal alveoli (singular: alveolus).

The alveoli are microscopic, very thin-walled, cauliflower-like structures, each of which is surrounded by a network of capillaries. The capillaries are the microscopic blood vessels which bring oxygen-poor venous blood to the alveoli where gas exchange takes place. The venous blood gives off carbon dioxide into the alveoli and receives oxygen in return. The oxygen-rich blood is now passed through the pulmonary circulation to the systemic arterial circulation.

Lung volumes and capacities

Fundamental measurements of volumes and capacities are made by way of a spirometer. A spirometer comprises an empty cylindrical drum floating within a tank of water. An air tube passes through the water into the drum in such a manner that when air is blown through the tube into the drum the drum rises out of the water to a height proportional to the volume of air added. The drum is connected via a pulley system to a counterbalancing weight. A recording device attached to the connecting system records the upward and downward motion of the drum on recording paper fastened around a second drum which rotates at a constant rate. Less cumbersome techniques, making use of an electronic device known as a plethysmograph, are in common use in industry and elsewhere for the same purpose.

By breathing, or blowing air in a standardized manner into the spirometer, various lung volumes and capacities can be defined. Measurement of these volumes and capacities, and also various other dynamic volumes, become important in determining the effects and relative disabilities incurred in diseases of the lungs, and particularly in evaluating the effects of industrial dust diseases. They are illustrated in the following Figure:

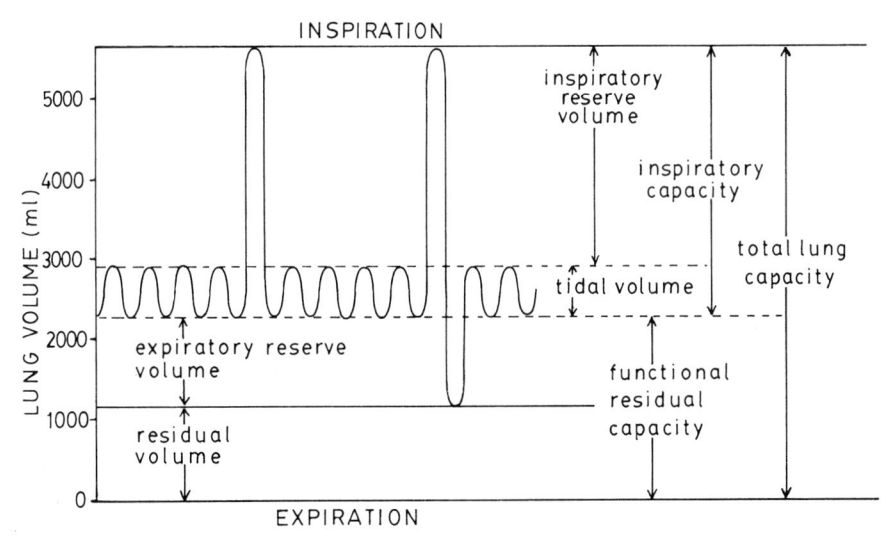

Figure 17.1. Lung volumes and capacities.

Static volumes

The volumes illustrated in Figure 17.1 are defined as follows:

Tidal volume (TV): The air moved in and out of the lungs during each successive breathing cycle.

Functional residual capacity (FRC): The air left in the lung following completion of a normal expiration.

Expiratory reserve volume (ERV): The air that can be expired by a forced maximum expiration after completion of a normal expiration.

Residual volume (RV): The air remaining in the lung after a forced maximal expiration.

Inspiratory capacity (IC): The air that can be inspired above the level of the functional residual capacity (that is, after completion of a normal expiration).

Vital capacity (VC): The maximum volume that can be expelled after a full inspiration.

Total lung capacity (TLC): The sum of the functional residual capacity and the inspiratory capacity.

Dynamic Volumes

Dynamic volumes are ventilatory volumes per unit time. There are two that are of significance in determining the extent of obstructive lung disease. These are the forced expiratory volume and the maximum voluntary ventilation.

Forced expiratory volume (FEV): The forced expiratory volume is the maximum volume that can be exhaled over a period of one second. It is expressed as a percentage of vital capacity. In the normal 20 year old man it is about 80 per cent of the VC. In obstructive disease of the lung expiration is interfered with more than inspiration. The flow is limited by the efficiency of the expiratory muscles and the extent of increase in airway resistance. Sometimes durations other than one second are used.

Maximum voluntary ventilation (MVV): The MVV is sometimes referred to as the Maximum Breathing Capacity (MBC). It defines the maximum volume of air that can be breathed in a 15 second period. The normal level for a 25 year old male is about 100–180 litres and for a 25 year old female about 70–130 litres.

Standardization of air volumes

The volume of air measured by, or calculated from, a spirometer is influenced by the ambient temperature, pressure, and humidity. It is thus desirable to bring all volumes to a standard level. This is accomplished by referring all temperatures to degrees Absolute (−273°), and referring water vapour pressures to a standard 760 mm Hg, as follows:

$$V_{BTPS} = V \times \frac{273 + t_B}{273 + t_S} \times \frac{P_B - P_{H_2O}}{P_B - 47}$$

where V_{BTPS} = volume at body temperature and ambient pressure, saturated (ml)
V = measured volume (ml)
273 = absolute temperature, (°C)
t_B = body temperature (37°C)
t_S = spirometer temperature (°C)
P_B = barometric pressure (mm Hg)
P_{H_2O} = ambient vapour pressure (mm Hg)
47 = lung vapour pressure (mm Hg) at body temperature (37°C)

Thus, assuming a measured FRC of 2800 ml, with a spirometer temperature of 12°C, an ambient vapour pressure of 40 mm Hg, and a respiratory vapour pressure of 47 mm Hg, then:

$$FRC_{BTPS} = V_{FRC} \times \frac{273 + 37}{273 + 12} \times \frac{760 - 40}{760 - 47}$$
$$= 2800 \times 1.09 \times 1.01$$
$$= 3082 \text{ ml (BTPS)}$$

Regulation of respiration

Respiration is regulated in the central nervous system via the respiratory centre at the base of the brain, and by way of chemical sensors, or chemoreceptors.

Respiratory centre

The respiratory centre comprises a network of neurons and interlinked fibres at the base of the brain. Two distinct components can be defined, namely an *inspiratory centre* and an *expiratory centre*. By way of their intrinsic rhythms these two centres set up a pattern of inspiration and expiration. Superimposed upon that pattern is stimulation from another centre in the base of the brain, the *apneustic centre*. The action of the apneustic centre is to impose on the inspiratory centre an unremitting demand for inspiration. Inspiration would then continue were it not inhibited by the action of still another centre, the *pneumotaxic centre*, and also by the Hering-Breuer reflex. The latter reflex is the term given to the neural signal which arises from stimulation of stretch receptors in the lung tissue itself which signal when the lung is being excessively expanded.

Chemoreceptors

In addition to direct neural control, respiration is responsive to excessive change in blood gas tension and hydrogen ion concentration. It would appear that there is a set of chemical receptors in the central nervous system, probably in close physical relationship to the respiratory centre. This detects changes in hydrogen ion concentration and partial pressure of carbon dioxide, and advises the respiratory centre accordingly; for example, to increase pulmonary ventilation and thereby remove excess carbon dioxide. The significance of partial pressures of gases in the lung is discussed in the next section.

Peripheral sensors are also found near the aortic and carotid arteries. They are known as the aortic and carotid bodies, and are responsible for detecting the partial pressures of oxygen and carbon dioxide, and advising the respiratory centre accordingly.

Increase in the partial pressure of carbon dioxide, and in the pH of the blood, and decrease in the partial pressure of oxygen, give rise to stimulation of respiration. Decrease in partial pressure of carbon dioxide and pH of the blood produce the opposite effect. The respiratory centre, however, is very much more sensitive to partial pressure of carbon dioxide than oxygen. Indeed it is the level of carbon dioxide in the blood and not the availability of oxygen that is the most significant factor in controlling respiration.

Gas exchange and partial pressure

As previously noted, gas exchange occurs between air in the alveoli and gas in solution in the blood. Oxygen enters capillaries from the atmosphere; carbon dioxide, which is derived from tissue metabolism, leaves the capillaries and enters the alveoli from where it is expired.

The extent and direction of gas exchange depends on the relative pressure gradients of oxygen and carbon dioxide across the wall of the pulmonary membrane, which is the name given to the aggregate of the alveolar wall, the capillary wall, and the intervening tissue fluid. These pressure gradients depend on the *partial pressures* exerted by the various gases that comprise the atmosphere.

The eighteenth-century chemist/physicist/mathematician John Dalton propounded these partial pressure relationships in what is termed Dalton's Law of Partial Pressures, which states that in a mixture of ideal gases the pressure exerted by the mixture is the sum of the pressures exerted by the components of the mixture, that is:

$$P_B = P_{O_2} + P_{N_2} + P_{XY}$$

where P = total barometric pressure
P_{O_2} = partial pressure of oxygen
P_{N_2} = partial pressure of nitrogen
P_{XY} = partial pressure of other gases

Conversely, the partial pressure of a component of a mixture, for example oxygen in the atmosphere, is proportional to the fraction of the mixture that it occupies, or its fractional concentration in the mixture, thus:

$$P_{O_2} = F_{O_2} \times P_B$$
where F_{O_2} = fractional concentration of oxygen

When a gas such as oxygen or carbon dioxide is passed into the blood stream and is dissolved in blood or tissue fluids, or carried by the transportation system, that gas exerts a type of pressure on its surrounds. This pressure is referred to as gas *tension*. Henry's Law outlines similar considerations for the tension of a gas dissolved in a liquid as does Dalton's Law for a mixture of gases in the atmosphere.

Water vapour

As noted earlier, inspired air becomes immediately saturated with water vapour. But, while water vapour is from other points of view a gas, its behaviour does not conform to the stipulations of Dalton's Law. The partial pressure of water vapour in a mixture of gases is not dependent on its fractional concentration in that mixture but is dependent only on the temperature in which it is found. As a result, at the nominal body temperature of 37 degrees C the water vapour maintains a partial pressure of 47 mm Hg regardless of its fractional concentration in the mixture.

Partial pressure of inspired oxygen

Bearing in mind that sea level barometric pressure is 760 mm Hg, and that oxygen comprises 21 per cent of the atmosphere, then in the trachea, which contains saturated air, the available oxygen partial pressure, using the parlance of the respiratory physiologist, will be:

$$P_{I_{O_2}} = (760 - 47) \times 0.21$$
$$= 149 \text{ mm Hg}$$
where $P_{I_{O_2}}$ = partial pressure of *inspired* oxygen

PARTIAL PRESSURE OF AVEOLAR OXYGEN

When atmospheric air, saturated with water, reaches the alveoli the air becomes diluted with carbon dioxide released from the venous blood. The volume and concentration of this carbon dioxide is dependent on the rate of tissue metabolism.

While it is extremely difficult to measure this volume it has a direct relationship to the volume of inspired oxygen. This relationship is known as the *respiratory exchange ratio* (R), thus:

$$R = \dot{V}_{CO_2} / \dot{V}_{O_2}$$
Where $\dot{V}$ = volume per unit time

Where tissue metabolism is based exclusively on the metabolism of carbohydrate that ratio would be unity. Commonly, however, there is also some metabolism of protein which changes the ratio in favour of oxygen. Thus, nominally, in the healthy fit person, R equals 0.8.

Recognizing these relationships, then, it is possible to calculate the effect of dilution of inspired air with carbon dioxide and to determine thereby the partial pressure of oxygen in the alveoli. The great Danish physiologist, Niels Bohr (1885–1962), derived an equation to define the alveolar partial pressure of oxygen. It is known as the Alveolar Equation, and is as follows:

$$P_{AO_2} = P_{IO_2} - P_{ACO_2} \left[F_{IO_2} + \frac{(1 - F_{IO_2})}{R} \right]$$

where P_{AO_2} = partial pressure of oxygen in the alveoli
P_{ACO_2} = partial pressure of carbon dioxide in the alveoli
F_{IO_2} = fraction of inspired oxygen in the atmosphere

Nominally, at rest, the partial pressure of carbon dioxide in the alveoli is 40 mm Hg, which of course may be modified by the value of R. When R equals unity no correction of course is necessary since;

$$F_{IO_2} + (1 - F_{IO_2}/R) = 1$$

When R equals 0.8, then:

$$P = 149 - 40 \left(0.21 + \frac{(1 - 0.21)}{0.8} \right)$$
$$= 149 - 40(1.2)$$
$$= 101 \text{ mm Hg}$$

For general purposes this number is rounded off such that at sea level, breathing air, the mean alveolar partial pressure of oxygen, which in turn is the pressure available for gas exchange, equals 100 mm Hg. Thus although the partial pressure of oxygen in the atmosphere at sea level is approximately 150 mm Hg, by the time the air reaches the alveoli, the oxygen pressure available for gas exchange is only 100 mm Hg. This pressure is then used to drive the oxygen across the pulmonary membrane into the blood, and ultimately into the blood cells.

Transport of oxygen in the blood

Once into the blood, a small quantity of oxygen, amounting to 0.3 ml per 100 ml of blood volume, is actually dissolved in the liquid portion of the blood, namely the blood plasma. This quantity, even when distributed throughout the entire blood volume, is inadequate to supply the body needs. Hence another mechanism of oxygen carriage is necessary. This mechanism involves the use of the chemical *haemoglobin* (Hb) in the red blood cells.

There are two basic types of cells in the blood, the red cells which carry oxygen, carbon dioxide, and occasionally other chemicals, and the white cells which are responsible for implementing the body's immune system, and defence against invasion by bacteria, viruses, and other agents.

Haemoglobin is the active principle and main constituent of the red cells. It is a complex, reddish coloured pigment which has a unique capacity to capture molecules of oxygen and carbon dioxide and hold them so long as the tension of the gas in the surrounding plasma remains sufficiently high. When the ambient gas tension drops, the haemoglobin will release its content until the tensions are equilibrated. Similarly, when the red blood cell is exposed to a higher oxygen or carbon dioxide tension the haemoglobin will acquire oxygen or carbon dioxide until once again equilibration is reached.

Thus, in the lung, where the haemoglobin of the returning venous blood is low in oxygen and high in carbon dioxide, the partial pressure of oxygen in the alveoli will load the haemoglobin with oxygen. Also, since venous blood has a higher concentration of carbon dioxide than is found in the alveoli, the carbon dioxide will pass from the red cells to the alveoli. Thus a full gas exchange will take place and oxygen-rich blood will proceed into the systemic circulation.

Oxygen-rich haemoglobin is called *oxyhaemoglobin* (HbO). Although the haemoglobin carries oxygen it is not itself oxidized in the sense that iron can be oxidized to produce iron oxide. It will be noted in the discussion later of industrial toxicology that haemoglobin can in fact be oxidized by certain toxic chemicals to form *methaemoglobin*, which renders the haemoglobin useless for its proper function.

The volume of oxygen that is actually carried by the haemoglobin is termed the *oxygen content*. The haemoglobin, however, is not necessarily saturated when one is breathing air, and carries only that volume of oxygen that is determined by the partial pressure of oxygen in the surrounding atmosphere. The maximum volume of oxygen that can be carried is referred to as the *oxygen capacity*; it amounts to 20 ml of oxygen per 100 ml of blood. The extent of saturation is determined by the ratio of content to capacity, and is termed the *oxygen saturation*, thus:

$$\text{oxygen saturation} = \frac{\text{oxygen content}}{\text{oxygen capacity}} \times 100$$

Normally, when one is breathing air at sea level, the arterial blood is saturated only to the extent of 97 per cent. When breathing oxygen at sea level, however, the arterial blood becomes fully saturated, that is to 100 per cent. During its passage through the body the blood loses oxygen to the tissues; mixed venous blood returning to the lungs is only 75 per cent saturated and is readily re-saturated to a level of 97 per cent in the presence of normal breathing air in the lungs.

310 Physical agents in the work environment

The process of release of oxygen to the tissues is known as *oxygen dissociation*. The extent and rapidity of this dissociation is a function of the partial pressure, or tension, of the gas concerned. The greater the pressure differential between the gas in the haemoglobin and the gas in the tissues the greater is the dissociation. The relationship is not rectilinear, however. It is expressed in what is known as the *oxygen dissociation curve* (Figure 17.2).

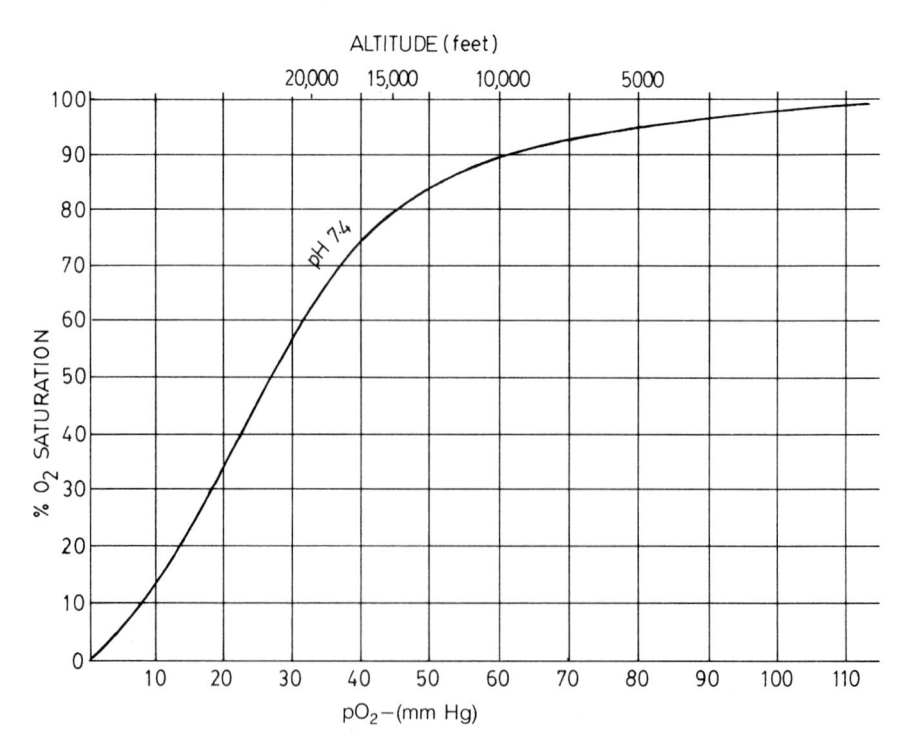

Oxygen dissociation curve

The oxygen dissociation curve is an S-shaped curve that relates oxygen saturation and partial pressure. It will be noted that as the partial pressure of oxygen decreases the oxygen saturation decreases.

Initially, however, from a saturation of 97 per cent at sea level, equivalent to a partial pressure of oxygen of 100 mm Hg, the saturation decreases relatively little, even to a partial pressure as low as 60 mm Hg. At that level, when the pH of the blood is maintained at 7.4, the saturation is about 88 per cent. The effect of pH will be discussed later. Homeostasis can in fact be maintained, and indeed very well in the acclimatized person, until the saturation drops below 88 per cent. It will be noted that this

Figure 17.2. Oxygen dissociation curve.

saturation level also occurs when breathing normal atmospheric air at 10 000 feet (approximately 3000 metres).

When the partial pressure of oxygen drops below 60 mm Hg there is a dramatic change in the slope of the curve such that relatively small changes in partial pressure will cause relatively large changes in saturation. The steepest change occurs between 40 mm and 20 mm Hg. This steep change now becomes significant with respect to internal, or tissue, respiration, since normally the oxygen tension in active tissues is between 20 and 30 mm Hg. Thus haemoglobin which has been saturated to 97 per cent in the lungs finds itself more or less suddenly among tissues where the tension requires a saturation of only 35 to 50 per cent. The haemoglobin then cannot retain its oxygen and consequently dissociates it to make it available to the tissues for metabolic purposes. A similar process in reverse operates for carbon dioxide.

Bohr effect

The Bohr effect is the term given to the change in slope of the dissociation curve that occurs with change in blood acidity. Change towards the acid occurs during normal metabolism. As can be seen from the figure, it has the effect of moving the dissociation curve to the right, which in turn allows the release of still more oxygen to the tissues where it is most needed. A similar effect occurs from expiration of excess carbon dioxide in overbreathing, either as voluntary (or sometimes emotionally induced) hyperventilation, or as one of the physiological compensations for breathing at low oxygen levels such as high altitude.

Chapter 18
The effects of barometric pressure

Small changes in barometric pressure occur as part of the natural phenomena associated with changing weather systems. These changes are not significantly sensed by humans nor do they have any observable effect on human wellbeing. Gross changes can occur, however, with major and potentially hazardous effect, on ascent to high altitude or descent underwater. These changes can also be simulated in specially designed pressure chambers.

Two basic, and fundamentally different, types of effect can be observed, namely those arising from changes in the partial pressure of the component gases, and those deriving from the effects of total pressure change.

Partial pressure effects

The partial pressure effects can arise from reduction in partial pressure, or increase in partial pressure, as described in Chapter 17. The major effects of reduction in partial pressure arise from the reduced availability of oxygen known as *hypoxia* or sometimes, erroneously, as anoxia. In this regard hypoxia means reduced oxygen while anoxia means no oxygen. Anoxia is rare. Increase in partial pressure of atmospheric gases can of course affect any constituent but is specifically found in relation to nitrogen, where it gives rise to *nitrogen narcosis*. Each of these will be considered in turn.

Hypoxia

Hypoxia can occur in various industrial situations where for one reason or another the concentration of oxygen in the atmosphere has been reduced. It should not, however, be confused with *asphyxia*, or suffocation. Asphyxia occurs where the air supply has been cut off, or where, for example, oxygen has been replaced by a gas such as carbon dioxide which is neither toxic nor irritant but at the same time cannot support life.

The most dramatic and potentially hazardous occurrence of hypoxia arises during ascent to high altitude, and indeed historically the study of hypoxia is the study of the human response to high altitude exposure.

Oxygen was discovered in 1774 by an English chemist by the name of Priestley. A similar discovery was also made by Schelle in Germany. At that time the gas was considered pretty much of a curiosity and its mandatory use for life support was not recognized. A further development took place about 10 years later, however, in 1782, when the Montgolfier brothers in France raised the first hot air balloon. The following year the first flight was made by a Frenchman, Count Pilatre de Rozier. For 75 years thereafter ballooning was an exciting diversion for the more daring at popular fairs until about the 1860's when manned balloons began to be used for scientific research. During this period several reports began to appear about the potential hazards, although even then the cause was not fully recognized.

A particularly dramatic report was written by the Frenchman Tissandier in 1875. He, along with two colleagues, Sivel and Croce-Spinelli, undertook a high altitude flight in order to make some measurements of physical conditions at the higher altitudes. On the basis of current work being done by the French physiologist, Paul Bert, they took with them a small amount of oxygen in the hope that it might be of use to them. However, having reached 7000 metres, disaster overtook them. Tissandier (quoted by Armstrong, 1939) writes:

> But soon I was keeping absolutely motionless, without suspecting that I had already lost use of my movements. Towards 7500 meters the numbness one experiences is extraordinary. The body and the mind weaken little by little, gradually, unconsciously, without one's knowledge. One does not suffer at all; on the contrary one experiences inner joy, as if it were an effect of the inundating flood of light. One becomes indifferent; one no longer thinks of the perilous situation or of the danger; one rises and is happy to rise. Vertigo of the lofty regions is not a vain word. But as far as I can judge by my personal impressions, this vertigo appears at the last moment; it immediately precedes annihilation, sudden, unexpected, irresitible ...

Tissandier goes on to describe progressive weakness leading to unconsciousness, in spite of recourse to the minimal oxygen supplies available. When he recovered, he found the balloon drifting slowly downward and his two companions dead. With great perspicacity, however, he recognized the relationship between oxygen and altitude, and went on to write:

> I am convinced that Croce-Spinelli and Sivel would still be living in spite of their prolonged sojourn in the higher strata, if they had

been able to breathe oxygen. Like me, they must have suddenly lost power of movement. The tubes conducting the vital air must have slipped from their paralysed hands! But these noble victims have opened new horizons to scientific investigaiton; these soldiers of science in death have pointed out the dangers of the way, so that their successors may know how to foresee and avoid them

And indeed their successors did foresee and avoid them such that today it is possible not only to fly in comfort at any altitude but even to proceed into the emptiness of space.

Acute hypoxia

What then are the effects of hypoxia? Hypoxia can be of relatively sudden onset, as in ballooning or flying, or it can occur much more slowly as in the arduous climbing of high mountains. The effects are somewhat different; the former, acute hypoxia, will be considered first. As noted in Chapter 17 a person can adapt relatively easily to an altitude of 10 000 feet. Exposure to that altitude, however, whether sudden or gradual, will give rise initially to a general feeling of malaise accompanied by a slight shortness of breath and early onset of fatigue. A curious, and measurable, accompaniment, is some loss of night vision. The visual organ of the eye, namely the retina, is very demanding of oxygen. Oxygen supply which would be adequate for all other physiological requirements can be inadequate for vision, and in conditions of even mildly reduced oxygen the vision suffers.

As the altitude rises towards 14 000 to 15 000 feet one of the striking effects of oxygen lack becomes manifest, namely a progressive reduction in the higher mental faculties such as reason, judgment, and memory. The effect is one of euphoria, as described by Tissandier, but the significance of this effect is that the subject may be completely unaware of the fact that he is impaired; or even if he is aware he is not concerned. In this respect the effects are somewhat similar to the early effects of alcohol. Accompanying the cerebral effects are various physiological responses which attempt to compensate for the reduced oxygen. These include, for example, an increased depth and rate of breathing, an increased heart rate and blood pressure. Since the blood is no longer being adequately oxygenated the normal red colour of the arterial blood is lost and the blood at the surface of the skin and mucous membranes becomes bluish, in the condition known as *cyanosis*. This phenomenon is most evident in the lips and fingernails, but since the subject is commonly euphoric by this time he may not observe it.

Towards 20 000 to 25 000 feet a relative loss of consciousness may ensue, depending on the individual susceptibility of the subject. Initially the loss of consciousness is not full, as in fainting, but partial. It is referred to as loss of *useful* consciousness and describes a condition in which the subject is conscious but totally unaware of his surroundings or what is demanded of him for survival. With continued exposure over several minutes at 25 000 feet the subject will become completely unconscious. The unconsciousness will deepen into coma and eventual death.

At about 40 000 feet unconsciousness with convulsions will occur within seconds to minutes, depending on individual susceptibility, and again will lead to coma and death.

Time of useful consciousness

The various effects described above are normally found either in experimental situations in high altitude chambers, or in the relatively unusual conditions of rapid decompression of the pressurized cabin of an aircraft flying, for example, above 25 000 feet, which is a common altitude for commercial aircraft. Under these circumstances, the passengers and perhaps even more importantly the pilot, have a limited time to don oxygen equipment and take whatever other action may be required. The time of useful consciousness at different altitudes is given in Table 18.1 below:

Table 18.1 Time of useful consciousness at altitude

10 000 feet:	indefinite
20 000 feet:	several hours (unadapted)
25 000 feet:	2-5 minutes
30 000 feet:	1 minute or less
40 000 feet:	20-30 seconds

Chronic hypoxia

Chronic hypoxia can be distinguished from acute hypoxia by the time taken to attain altitude and the duration of exposure. The term chronic in this context refers to exposure to relatively mild levels of hypoxia for periods of hours to days or longer. It is sometimes referred to as *mountain sickness*, or altitude sickness, and as the name implies it tends to occur among unacclimatized persons exposed to mountain conditions over 10 000 feet.

The condition was first described by Father Joseph Acosta in 1604 during a passage over the Andes mountains with the Spanish *conquistadores*. He wrote:

> ... I was surprized with so mortall and strange a pang that I was ready to fall from my beast to the ground; and although we were many in company, yet every one made haste (without any tarrying for his companion) to free himselfe speedily from this ill passage ...

> To conclude, if this had continued I should undoubtedly have died: but this lasted not above three or four houres that we were come into a more convenient and naturall temperature, where all our companions, being fourteene or fifteene, were much wearied. Some in the passage demaunded confession, thinking verily to die ...
> I therefore perswade myself, that the element of the aire is there so subtile and delicate, as it is not proportionable with the breathing of man, which requires a more grosse and temperate aire ...

While there is considerable individual variation in response, most persons will develop symptoms such as shortness of breath, severe headache, insomnia, impaired ability to concentrate or perform complex tasks, and in many instances severe nausea and vomiting, after exposure to altitudes of 11 000 to 12 000 feet for periods of eight hours or more. In most instances these symptoms decline in frequency and severity over a period of two to five days, although the ability to perform muscular work is moderately to severely impaired and persists for prolonged periods (Billings, 1973).

Permanent adaptation is possible up to an altitude of 15 000 to 16 000 feet. Native populations are known to live in the Andes mountains at 16 000 feet, and indeed to work during the day at 18 000 feet, although they have to return to their homes at the end of the day. This adaptation is accompanied by considerable physiological changes, including increase in the number of red blood cells and increased blood viscosity. These adapted persons also have difficulty in living at sea level. High mountain climbers, such as those engaged in climbing Mount Everest at a height of about 25 000 feet spend weeks adapting themselves to altitudes at various levels up to 16 000 feet or even higher. When adapted, two persons will be selected to make a dash to the summit without oxygen under the most arduous of circumstances. Successful attempts have also been made using oxygen equipment.

Protection from hypoxia

It will be apparent that protection from hypoxia requires the addition of increasing proportions of oxygen to the breathing air as the altitude rises above 10 000 feet. The objective, of course, is to maintain the partial pressure of oxygen in the alveoli within an acceptable range close to 100 mm Hg.

The addition of oxygen to breathing air, however, is acceptable only to a certain limit. Let us consider that appropriately calculated, proportionally increasing, amounts of oxygen are being added to the breathing air from 10 000 feet up to 30 000 feet or more. At 34 000 feet the ambient barometric pressure is 187 mm Hg. At this pressure, even when breathing pure oxygen, the alveolar oxygen pressure will be as follows:

$$P_{AO_2} = 187 - (P_{H_2O} + P_{CO_2})$$
$$= 187 - (47 + 40)$$
$$= 100 \text{ mm Hg}$$

It will be recognized from Chapter 17 that 100 mm Hg is the alveolar partial pressure that is required to maintain an arterial oxygen saturation of 97 per cent, or the equivalent partial pressure of alveolar oxygen found at sea level. Thus, breathing 100 per cent oxygen at 34 000 feet is the same as breathing atmospheric air at sea level. Any rise in altitude above 34 000 feet will produce increasing hypoxia even when breathing 100 per cent oxygen.

At 38 000 feet the ambient barometric pressure is 154 mm Hg. Thus, breathing 100 per cent oxygen, the alveolar oxygen pressure will be as follows:

$$P_{AO_2} = 154 - (47 + 40)$$
$$= 67 \text{ mm Hg}$$

From the previously discussed oxygen dissociation curve it will be seen that an oxygen partial pressure of 67 mm Hg is equivalent to that found at 10 000 feet breathing air and will produce an arterial oxygen saturation of 87 per cent, generally regarded as being the border of life support. Any rise in altitude above 34 000 feet, even breathing 100 per cent oxygen, will inevitably give rise to progressive non-compensable hypoxia, unless the oxygen can now be added under increasing pressure.

The mechanisms for the addition of oxygen has evolved considerably since the days of Tissandier and his colleagues. By the end of World War I aircraft were flying sufficiently high to need oxygen which at that time was provided from tanks by way of pipe stems. By the middle of World War II complex delivery systems with close-fitting face masks had been developed. After World War II the increasing altitudes led to the need to develop pressurized mechanisms for the delivery of oxygen at heights above 38 000 feet, and the need to provide pressurized cabins for passengers in commercial aircraft, and of specialized pressure suits for extra-vehicular astronauts and cosmonauts. The details of these requirements are not relevant here. It might be noted, however, that the altitude equivalent of the interior of a commercial aircraft is normally allowed to rise to 8000 feet, at which it is maintained as the aircraft itself climbs higher. Only under extreme circumstances does the cabin altitude rise further, and never above 10 000 feet except in emergency conditions or in accidental decompression. In the latter case, of course, special emergency oxygen is provided for passengers by way of temporary oxygen masks.

Hyperoxia

It is perhaps surprising to recognize that oxygen, without which we cannot survive, can become highly toxic when breathed under sufficient pressure. Hyperoxia, or oxygen poisoning, is uncommon, but can occur among workers in hyperbaric (or high pressure) chambers, or in users of self-contained underwater breathing apparatus (SCUBA). Indeed, oxygen poisoning can occur at sea level if pure oxygen is breathed for a period of 24 hours. At a pressure of 3 atmospheres (20 metres of sea water) convulsions will occur in most persons within about 3 hours.

Mechanism of occurrence

As noted earlier, oxygen is normally carried in the haemoglobin of the blood to the extent of 20 ml per 100 ml of blood. At sea level oxygen is also carried dissolved in the blood plasma to the extent of 0.3 ml per 100 ml of blood. At 2 atmospheres, or 10 metres of sea water, the dissolved oxygen in the blood plasma will amount to:

$$[1520 - (40 + 47)] = 4.3 \text{ ml oxygen}/100 \text{ ml blood}$$

Under normal circumstances regular oxygen consumption is about 250 ml per minute, and cardiac output, that is, the blood flow from the heart, amounts to 5 litres per minute. At 2 atmospheres, breathing oxygen, the amount of oxygen dissolved in 100 ml of blood will comprise:

$$\frac{5000}{100} \times 4.3 = 215 \text{ ml}$$

A rate of flow of 215 ml per minute is fairly close to the demanded rate of 250 ml per minute for normal body purposes. Consequently, at a little over 2 atmospheres, breathing pure oxygen, the dissolved oxygen meets most if not all of the body requirements. Meanwhile, of course, oxygen is still being carried by the haemoglobin, but this oxygen is now in gross excess of demand and acts to disrupt the enzyme systems that are involved in all metabolic activities within cells.

As already noted 10 metres (33 feet) of sea water is equivalent to 2 atmospheres. At that pressure the oxygen tension in body fluids is approximately 200 mm Hg. A depth of 40 metres (132 feet) comprises 4 atmospheres of water plus 1 atmosphere of air, to a total of 5 atmospheres. Since oxygen comprises one-fifth of the air, then at 5 atmospheres, breathing air, the tension of the oxygen is again equivalent to 1 atmosphere of oxygen. Thus breathing air at 40 metres depth is equivalent to breathing oxygen at 10 metres. Similarly, breathing air at 90 metres, or 10 atmospheres, is equivalent to breathing oxygen at 20 metres, at which level it becomes toxic. Thus air itself becomes toxic at about 90 metres (300 feet) of sea water.

Manifestations of hyperoxia

Respiratory symptoms, comprising initially coughing and irritation behind the sternum, begin to become apparent in susceptible persons at oxygen tensions of about 200 mm Hg, although they are uncommon below 250 mm Hg (Billings, 1973). At levels around 250 mm Hg and above, damage to red blood cells has been reported (Berry and Catterson, 1967). While these effects are noted only after prolonged exposure (several days), lung irritation has been reported after 12 to 72 hours of exposure to tensions up to 750 mm Hg (Billings, 1973). Fifty years ago it was common practice to place premature babies in oxygen tents under pressures at those levels. Many of these babies suffered damage to the retina of the eye from hyperoxia in a condition known, but not understood at the time, as retrolental fibroplasia, which commonly led to blindness.

At considerably higher tensions, in the order of 2000–5000 mm Hg, potentially hazardous cerebral symptoms begin to occur. Initial warning symptoms are nausea, dizziness, twitching, tinnitus, and disturbances of vision. There may be an associated anxiety and irritability. If these symptoms are ignored, the subject will begin to become confused. Convulsions will follow, leading to unconsciousness and death.

The duration to convulsions when breathing oxygen is widely variable from person to person, lasting for minutes to hours. The U.S. Navy Diving Manual notes that a diver in 50 feet (15 metres) of water may suffer convulsions in 30 minutes. Advantage is taken of this variability to select non-susceptible persons by exposing them to an oxygen tolerance test. Acceptable oxygen tolerance is found in a subject who can breathe oxygen at one atmosphere for a duration of 2 hours and 10 minutes during moderate exercise (U.S. Navy, 1970). Most persons, however, can tolerate 3 atmospheres for about 90 minutes if at rest and recumbent in a dry chamber breathing oxygen by mask (Kindwall, 1975). The U.S. Navy limits underwater diving operations breathing oxygen to 1 hour at 25 feet (7.5 metres), and 10 minutes at 50 feet (15 metres), with proportional increases in duration between 25 and 40 feet (12 metres).

Nitrogen narcosis

Nitrogen comprises four-fifths of the atmosphere. We breathe it in and out of our lungs with no ill effects every day of our lives. It is absorbed across the pulmonary membrane, just as is oxygen; it is dissolved in the blood plasma, in the interstitial fluid, and passes across cell membranes into the cells themselves. Indeed, it is commonly regarded as an inert gas, and even in the chemical laboratory it is difficult to combine nitrogen with other chemicals. When breathed under pressure, however, it ceases to be a physiologically inert gas and becomes a narcotic. The term narcotic in this sense refers to the ability to produce narcosis, or depression of cerebral and neural activity.

One of the earliest effects of narcosis (and hence the use of narcotics as recreational drugs) is to reduce the inhibitory action of certain cerebral centres on some of the higher functions. The result, as is familiar to many from the use of alcohol, is to produce an initial euphoria, loss of judgment, and a higher tolerance of unwise or antisocial behaviour. The great French scientist and underwater explorer, Jacques Cousteau, coined the term 'rapture of the deeps' to describe the effects of breathing nitrogen in deep water, and while perhaps more poetic than scientific it does depict the loss of reality associated with early nitrogen narcosis.

The onset of effects generally occurs at depths greater than 30 metres (100 feet), breathing air, although there is a considerable individual variation in susceptibility. Significant effects are not generally observed under 60 metres (200 feet), and motor skills and coordination remain good until about 90 metres (300 feet). Thereafter, intellectual processes deteriorate rapidly; some persons become unconscious at 325–350 feet (Kindwall, 1975). Some workers, indeed, perhaps more experienced in the effects of more familiar toxic substances, have described what they refer to as Martini's Law, namely that every 15 metres (50 feet) under water is equivalent to one gin Martini. The U.S. Navy considers 190 feet (58 metres) to be the limit of air diving.

Prevention of nitrogen narcosis

Since the offending substance is atmospheric nitrogen the obvious control measure is to remove it. As already noted, however, replacement with oxygen is unacceptable. Attention therefore has to be directed to the use of other constituents of the atmosphere to replace it as a breathing gas. The capacity of a gas to produce narcosis is related to its affinity for lipids, or body fats. Of the atmospheric gases, helium has the least affinity, and is the least narcotic, followed by neon, hydrogen, nitrogen, argon, krypton, and xenon, in that order. With any of those gases, however, the intensity of the narcosis increases with pressure.

Because it is the least narcotic, and the trace gas in greatest supply in the atmosphere, helium has come to be the gas of choice as a replacement in most deep diving. Neon is some 25 per cent less narcotic than nitrogen and could be used as a replacement gas. It is, however, denser than helium and in deep dives greatly increases the work of breathing occasioned by the need to move air through the respiratory passages. Hydrogen is, of course, the least dense substance, and although it is only half as narcotic as nitrogen it has the grave disadvantage of being explosive when mixed with oxygen in concentrations greater than 4 per cent. Nevertheless it is used in extremely deep dives, sometimes in conjunction with helium, such that an oxygen/helium mixture is used in the early part of the dive while hydrogen is added later to ultimately replace the helium. As the depth increases, of course, the actual pressure of gas introduced into the lungs, whether it be oxygen, helium, or hydrogen, has to be progressively reduced to avoid toxicity.

Although helium reduces much of the narcotic problem, its use in deep dives unfortunately leads to the development of a new problem, namely the high pressure nervous syndrome (HPNS). HPNS, which is manifest as nausea, vomiting, shaking, and loss of coordination, occurs after rapid compression to depths of greater than about 150 metres (500 feet). It can be minimized by employing extremely slow compression, lasting over several days, to reach maximum depth, and that depth would appear to be about 915 metres (3000 feet) (Bennett et al., 1969).

In addition to the physiological problems associated with its use, helium can present other difficulties. These arise since helium is significantly less dense than the nitrogen it replaces. In particular, the human vocal cords have evolved to function via the passage of air with a density sufficient to cause vibration and subsequent resonance in the pharynx and sinuses, at what have come to be recognized as the speech frequencies. This density, of course, is largely determined by the presence of nitrogen. One of the striking effects, then, of breathing helium is a dramatic change in the timbre of the voice, which becomes squeaky and difficult to interpret. While this, indeed, may be laughable to the listener, it can render important or emergency communication difficult. While the interpretive skill can be learned, it is not always feasible to do so. Various electronic deciphering systems have been developed for this purpose and are used in professional diving situations.

Another problem, also related to density, arises from the fact that in a helium atmosphere there is increased heat conduction. The effect on a worker, say in a chamber or suit using a helium atmosphere, is one of continued subjective coldness and eventual disabling hypothermia. Still a third problem, this time with respect to hardware, is that helium has the capacity to penetrate the glass walls of electronic equipment, for example, television or cathode ray tubes, rendering the display fuzzy and indecipherable. While this equipment can be replaced, a problem is nevertheless created, if only one of cost.

Total pressure effects

Apart from the problems arising from changes in partial pressure of various atmospheric gases, barometric pressure can exert physical effects on the body in two ways, namely by the effects of gas trapped in body pockets, or trapped gas effects, and the effects of gas evolved from absorbed atmospheric gases, or evolved gas effects.

Increased pressure by itself has no mechanical effects on body tissues since the tissues are mostly water and essentially incompressible. However, should the body be, for example, in a diving suit, and should there be a failure in equalization of pressure inside and outside the suit, the whole body will be subjected to a crushing squeeze even at depths as low as 3–4 metres. Fortunately this is uncommon. More common effects occur because the body contains various pockets of air or other gases in such places as the ears, the sinuses, the chest, and the abdomen. Boyle's Law has already been noted, to the effect that with changes in pressure there are consequent changes in volume, since the product of the two remains constant. Thus, when pressure changes as in ascent or descent, the air volume must also change to compensate. When this occurs in a closed or semi-closed space within the body, discomfort, pain, and even injury can develop. These conditions are known by the divers as various forms of squeeze.

Trapped gas squeeze

Sinus squeeze

The effects of barometric pressure on sinuses and ears are well known to the air traveller and the snorkel diver. Air in sinuses is normally in direct communication with air in the nose. Should the sinus openings be blocked, however, by a cold for example, then when the pressure is decreased as in ascent to altitude, the air volume within will expand and give rise to sinus pain which will not be relieved until the sinuses are re-opened, by, for example, a nasal decongestant or on return to starting pressure, Similarly, increase in pressure during descent under water when the sinuses are bocked may cause rupture of the capillaries in the lining mucous membrane with severe pain and bleeding.

Middle ear squeeze

Middle ear squeeze is somewhat more complex in its pathogenesis. The middle ear is a bony box closed at one end by the elastic tympanic membrane or ear drum. Leading from the middle ear is the Eustachian tube which connects the ear cavity to the back of the throat. Under normal circumstances the tube allows easy equilibration of pressure across the eardrum. Insertion of the Eustachian tube into the throat, however, is such that there is a fold of mucous membrane akin to a flap valve which allows ready egress from the ear to the throat but impedes passage the other way. Obstruction to the air flow is grossly increased in either direction in the presence of inflammation, and/or excess mucus from a cold or sore throat.

Middle ear problems can arise, however, even without added inflammation because of the above noted valve. Thus, for example during ascent to altitude, air volume in the middle ear is increased as pressure falls. That volume is relieved through the valve of the Eustachian tube and comfort is maintained as long as that altitude is maintained or increased. On descent, however, the pressure is reduced. Air volume now has to be restored to the middle ear to compensate for the pressure change.

Because of the flap valve it is difficult to push air back through the Eustachian tube, and, as noted, this difficulty is grossly exaggerated in the presence of inflammation or mucus. A similar mechanism operates in reverse on descent under water. The ear drum can rupture at a differential pressure of 5–10 psi, across the drum. Healing normally takes place over 10–14 days, commonly with no residual damage.

To restore the air pressure it is often necessary to perform a Valsalva manoeuvre, which is the term given to the action of holding the nose firmly closed with the fingers, keeping the mouth shut, and forcibly attempting to expire air against the closed mouth and nose. This action has the effect of opening the Eustachian tube, if such is possible under the circumstances, and pushing a bubble of air into the middle ear. Repeated actions will relieve pressure change and restore the equilibrium.

Chest squeeze

Chest squeeze is another manifestation of the same principle. It will be recalled that the total lung capacity is 5000 ml, while the vital capacity is 3500 ml, and the residual capacity is 1500 ml. On descent the volume of the air in the lungs contracts unless it is replaced by continued breathing. At 30 metres, for example, which is 4 atmospheres, the total lung capacity is reduced by a factor of 4, or in other words, the lung capacity of 5000 ml is reduced to less than 1500 ml, which, as noted above, is the value of the residual volume. Consequently, any further increase in depth without breathing will cause squeezing of the lung tissues and injury.

Lung rupture

Chest squeeze should not be confused with lung rupture which can also occur as a result of pressure change. The mechanism, however, is somewhat different. Lung rupture fortunately is rare. It can occur, however, on rapid return to the surface from diving or during a rapid decompression in high altitude flight. In either case it is dangerous and potentially fatal. An essential requirement is that the subject must be holding his breath during the decompression. The breath holding may be very short, a matter of a few seonds, but this allows a transient overpressure to build up across the lung. It was shown many years ago that the mammalian lung can rupture at 80 mm Hg, although the values for the human lung vary widely and have not been clearly established for obvious reasons. Assuming a value of 100 mm Hg, however, and noting that the density of mercury is 13.6, then, theoretically at least, a more or less instantaneous ascent through 1.36 metres, or about 4½ feet, of water during breathholding could give rise to rupture of the lung. Fortunately it is not easy to hold the breath against the forceful outrush of air that accompanies the decompression, but it can happen, and indeed has happened (personal communication) when a diver was thrown violently upwards by a wave while in a shallow dive. Other incidents have occurred in experimental decompression chambers.

When a rupture of the lung occurs, the escaping air can break into the pleural space between the lung and the chest wall and cause collapse of the lung in the condition known as pneumothorax. It may break into the central tissues of the mediastinum between the lungs, or into the pericardium surrounding the heart, or upwards into the neck tissues. Each of these conditions is dangerous, and constitutes a serious medical emergency. Even more potentially dangerous, however, is the passage of air bubbles into the pulmonary capillaries or veins, where they constitute air emboli which can be carried through the heart to any part of the body and notably the brain. As will be discussed later, however, the effects of air emboli must be distinguished from those derived from air bubbles generated by the inert components of breathing gas in what is known as the evolved gas effects.

Evolved gas effects

As noted in the discussion of nitrogen narcosis, when nitrogen is breathed into the body it subsequently permeates every body tissue in concentrations proportional to its partial pressure in the air. As the pressure is increased, for example by underwater diving, more and more nitrogen is absorbed.

When the diver returns to the surface, then according to the previously noted Dalton's law of tensions of gases in fluids, the change in pressure has to be equilibrated by the release of the excess nitrogen to the air. Thus, nitrogen from the cells has to pass to the interstitial fluid, to the blood, and ultimately to the lungs, where it is exhaled. If the gas from the tissues is evolved at a higher rate than it is carried away by the blood then actual bubbles of gas will form in the tissues. A similar situation can occur, although not so readily or dramatically, on ascent from sea level to high altitude or during a rapid decompression in flight or in a decompression chamber.

Although the practical manifestations of decompression arise most commonly in connection with diving procedures and work under high barometric pressure, the condition was first recorded experimentally in the seventeenth century by the English physicist, Robert Boyle, of Boyle's Law fame, who had recently discovered the principle of the vacuum pump and was putting it to various experimental uses. He writes:

> The little Bubbles generated upon the absence of Air in the Bloud juyces, and soft parts of the body, may by their Vast number, and their conspiring distention, variously streighten in some places, and stretch in others, the Vessels, especially the small ones, that convey

the Bloud and Nourishment; and so by choaking up some passages, and vitiating the figure of others, disturbe or hinder the due circulation of the Bloud? ... to shew how this production of Bubbles reaches even to very minute parts of the Body, I shall add on this occasion (hoping I have not prevented myself on any other), what may seem somewhat strange, what I once observed in a VIPER, furiously tortured in our Exhausted Receiver, namely that it had manifestly a conspicuous Bubble moving to and fro in the waterish humour of one of its Eyes.

The presence of the bubbles causes a painful, disabling, and sometimes fatal disorder that goes by a variety of names. Properly it is called dysbarism or decompression sickness. Because it was first brought to public attention during the era of railroad building where underwater caissons were required on which to mount bridges, it was long known as caisson disease, while the divers themselves gave it the popular name by which it is commonly known to-day, namely the *bends*, perhaps from the fact that the early stages of pain and discomfort are found in the 'bends' of the elbow, shoulder, knee, and so on.

Development of gas bubbles

The rate of evolution of the bubbles is dependent on the tissues from which they are being evolved. Fat has five times as much affinity for nitrogen as other tissues and consequently absorbs more of the available nitrogen. It also releases it more slowly than do the other tissues. Bubble formation, however, is not simply a result of reduced pressure. Experimentally, for example, if an open container of soda water is rapidly decompressed in a decompression chamber while at complete rest and undisturbed, bubbles will not form spontaneously without a pressure change of 100–1000 atmospheres. Clearly other contributing factors are required over and above the change in pressure. Various other factors have been proposed. Since the condition tends to occur around joints it has been suggested, for example, that cavitation may take place in body fluids during the joint movement, producing close to infinite changes in pressure at the centre of the cavity. Alternatively, vortices may occur as blood flows round a sharp bend in an artery. Other suggestions postulate the presence of gas or solid nuclei in the blood or body fluids on which a bubble can form. No finally acceptable explanation has been established.

Once the bubble is formed, however, its future development depends on two contributory factors, namely the relative solubility of the gas in body tissues as compared with its solubility in blood, and the capacity of the gas to diffuse through body tissues. Roth (1967) defined the *permeation coefficient* as an index of the effectiveness of a gas as a bubble former. The permeation coefflcient is the product of the relative solubility and the diffusibility.

He determined that, in terms of their permeation coefficients, nitrogen and helium are somewhat similar in their propensity towards bubble formation. Helium would be less likely to produce bubbles in fatty tissues and more in non-fatty. It would appear that there would be some advantage in the use of neon. It has been noted that argon, krypton, and xenon would be inappropriate since they do not remain inert under pressure. Tests with neon as a diluent breathing gas in a pressure chamber, however, have shown that the advantage is slight as compared with helium, while the cost is greater. Thus while helium and other gases are used as atmospheric dilutents the objective in so doing is not primarily to reduce bubble formation but to reduce narcosis.

Occurrence of dysbarism

Dysbarism tends to occur when the ambient pressure is reduced by half at a rate suffiiently rapid to induce the formation of bubbles; thus, for example, it will tend to occur when there is a rapid decompression from sea level to at least 18 000 feet, or from 132 feet underwater (4 atmospheres) to 66 feet (2 atmospheres). While there has been one fatality from exposure at 18 000 feet (Billings, 1973), and other isolated incidents, decompression sickness is rare in flying below 30 000 feet. Indeed, because of pressurized cabins, decompression sickness is rare in aviation to-day, although it was a hazard among bomber crews in World War II and later. It may occur occasionally among drivers or pressure workers who return from normally tolerable pressures and proceed to fly at normally acceptable altitudes of, say, 10–15 000 feet. It is wise to avoid such combinations for 24 hours. At this time, decompression sickness, when it occurs, is a feature of underwater diving, or working in pressurized atmospheres such as are sometimes used in subway construction.

Influencing factors

Just as the physical formation of bubbles in the experimental environment requires the co-existence of other factors, there are various factors which influence the onset and severity of decompression sickness in the person. These include the following:

Rapidity of decompression: The more rapid the rate of decompression the more likely is the onset and severity of the condition.

Duration of exposure: The longer the duration of the exposure until saturation is reached the more severe will be the resulting effects.

Exercise during exposure: Other than the physical factors of pressure and time, the addition of exercise to exposure is one of the greatest influencing factors. In altitude exposure in particular exercise produces a much earlier response.

Ambient cold: Symptoms will occur more rapidly and severely when the exposure is accompanied by cold.

Age and experience: The condition tends to occur more commonly in persons above the age of thirty, and also among persons who have previously experienced an attack. Indeed bends will tend to occur in the same location on subsequent occasions.

Fitness and injury: As with most exposure to adverse environments the more physically fit a person is, the greater resistance he or she will have to sickness and injury. Bends, however, will tend to occur at the site of a previous bone or joint injury, particularly if that injury has been compound, that is, if the bone has penetrated the skin. Perhaps in these situations minute pockets of gas are trapped within the tissues and act as a nidus for bubble formation.

Manifestations of dysbarism

As previously noted dysbarism gives rise to a painful, disabling, and sometimes fatal disorder. The old divers gave colourful names to the different manifestations. The term 'bends', already noted, describes severe, intractable pain usually around the larger joints such as the elbow, shoulder, and knee, caused by bubbles in these regions. 'Creeps' is the name given to an itching, crawling, sensation under the skin, commonly on the chest or back, and often followed by a spotty rash. It is caused by bubbles in the skin itself. These two sets of symptoms are in the nature of general warnings, and, while not disabling in themselves, are indicative of severe potential hazard.

The term 'chokes' is given to the occurrence of coughing and choking, with difficulty in breathing and varying degrees of disability. The condition is caused by bubbles in the tissues of the respiratory system and can be serious if not immediately dealt with.

The foregoing are conditions of progressive seriousness which although causing various degrees of incapacitation are not fatal in themselves or leave lasting disability. The other manifestations, however, can be both permanently disabling, and/or ultimately or rapidly fatal. These can be defined as cardiovascular effects and central nervous system effects.

Cardiovascular effects, of course, involve the heart and circulation, and are caused by bubbles forming in the heart itself. These occurrences are rare, but when they do appear they manifest themselves as disturbances of heart action, specifically arrythmia. Arrythmia is the term given to the development of irregularity in the heart beat. Depending where the bubbles occur the irregularity may appear as runs of supraventricular extrasystoles (palpitations), experienced as uncomfortable flutterings in the chest, or in the worst situation as ventricular fibrillation, or total failure of heart rhythm, which can be rapidly fatal. Where the condition is not sufficiently severe as to be fatal it may still induce fainting because of resulting inadequate blood supply to the brain. Urgent action is needed in these situations.

The second potentially fatal situation occurs when bubbles develop in the brain or in the blood vessels supplying the brain. Again, this occurrence is rare, but can be potentially disabling and/or fatal when it does occur. The nature of the resulting effects, of course, varies according to where the bubbles are found. Commonly there are initial sensory disturbances, such as tingling and numbness in the limbs, bright lights, flashes, and other disturbances of vision. There may be violent headaches, and twitchings of fingers and limbs leading to convulsions and unconsciousness. The end result, if not fatal, may include loss of vision, blindness, loss of sphincter control, and temporary or permanent paralysis or muscle weakness.

As already noted, these signs and symptoms should be differentiated from those of cerebral air embolism. The latter is much more sudden, occurring within a minute or so of surfacing; it may be accompanied by the signs of respiratory distress such as pain, coughing, bloody sputum, and increased rate of breathing, or with signs arising from the presence of air in the central chest or neck tissues, such as swelling in the neck, change in voice quality from pressure on the laryngeal nerves, nausea, and general malaise. The urgent need for treatment of air embolism has already been stressed.

Effects of cold in underwater activity

Because of the relatively high thermal conductivity of water, cold can be a major problem among divers. Comfort in water can be maintained with a skin temperature of 34 degrees C. When skin temperature falls below 31 degrees C there is significant discomfort, with shivering induced at 30 degrees C. At 15 degrees C pain is intolerable (U.S. Navy, 1970).

The U.S. Navy, however, considers that a well trained, well motivated diver can tolerate, without protection, about 3 hours of exposure at 18–21 degrees C, about 50 minutes between 13 and 15 degrees C, and about 30 minutes between 10 and 13 degrees C. Some kind of protection, however, such as a 'wet suit' is required outside these ranges. The wet suit is a close-fitting garment of rubberized material which traps a layer of water between the suit and the skin and encourages the retention of body heat. It provides decreasing protection down to a water temperature of about 1–2 degrees C, below which some form of heated suit becomes mandatory. Heating is generally provided via a system of tubing in the suit through which hot water is pumped. At very great depths, however, even the inspired air has also to be heated.

Need for medical supervision

Trained medical personnel should be on site during compressed air, caisson, or tunnelling work, as well as during supervized diving. Arrangements should be made for access to a physician knowledgable in the management of pressure disorders, and to a hospital with recompression facilities.

The physician, industrial hygienist, or delegated person trained for the purpose, should also be responsible for sampling, testing, and ensuring the quality of breathing air in pressure operations, and for ensuring that workers and their supervisors are knowledgable and properly trained both in work procedures under pressure and in routine decompression procedures.

Many jurisdictions define in detail the requirements for training of technical and professional medical personnel, and the on-site need for equipment, including in some cases a portable recompression chamber. These regulations and guidelines should be consulted for appropriate information.

Prevention of decompression sickness among divers and pressure workers

Stage decompression

As a result of public outcry over the incidence of death and disability in divers and caisson workers, the British government in 1908 commissioned the British scientist, J. B. S. Haldane, to investigate the cause and prevention of decompression sickness. He and his colleagues not only defined the mechanisms involved in its causation but also proposed the relatively simple solution of controlled decompression to prevent the onset of symptoms. They devised, in fact, a set of empirical tables defining stops at discrete depths, for various durations during ascent, which would so control the rate of decompression that bubbles would not form. The depth of each stop was based on the consideration that bubbles would not form if the pressure were reduced by no more than half at each stage. The concept was referred to as *Stage Decompression*, and is illustrated in Figure 18.1.

These Haldane tables were very effective and became the basis for numerous other sets of tables developed later. Many of these, such as the U.S. Navy Standard Decompression Tables, the British Blackpool Tables, and U.S. Bureau of Labor Standard Tables, have already become widely accepted. Each set of tables, which are basically similar, defines

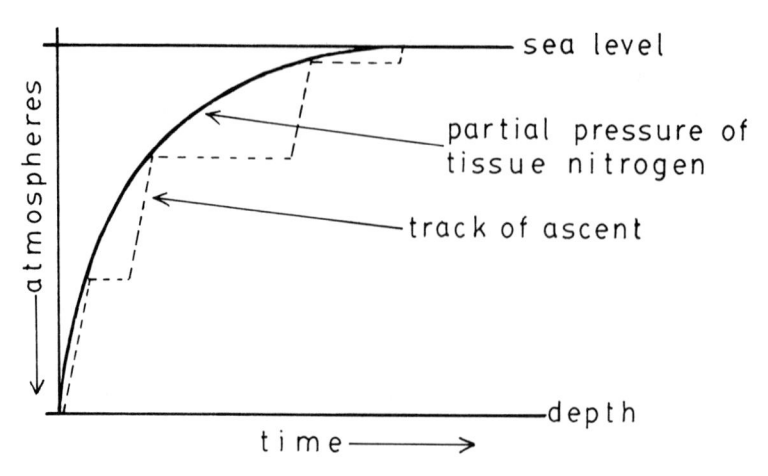

Figure 18.1. Principles of stage compression.

for a given depth, or a given duration of bottom time (that is, time at the bottom of a dive), the rate of ascent to the first and subsequent stops, and the duration of stay at each stop until the surface is reached.

Other tables have been devised to define the amount of time that may be spent at various depths *without* any requirement for decompression, for example, no limit at 10 metres, 60 minutes at 18 metres (60 feet), 15 minutes at 37 metres (120 feet), 5 minutes at 46 metres (150 feet), and so on. The various tables of different jurisdictions should be consulted for details.

It will be apparent that diving without subsequent decompression may allow very little bottom time, often insufficient to complete a job. On the other hand, prolonging the duration of bottom time may require extremely long decompression time, using standard tables. For example, working at 200 feet (61 metres) for two hours will require a total decompression time of 8 hours. This is not only tedious, uncomfortable, and potentially hazardous for the diver, it is also not cost-effective in terms of productive work.

These problems and others led to two approaches towards solution, one being the development of saturation diving techniques, and the other being the use of a miniature portable analogue computer carried by the diver to provide a more accurate determination of the required rate of ascent, with a consequently shorter duration of ascent. These are considered below.

Saturation diving

Since gas at depth is only absorbed until saturation of the tissues is reached, and since no problem can arise until that gas is evolved in ascent, then it would appear reasonable, if feasible, to maintain a diver at depth until the work is completed and then return him/her to the surface. The requirement for controlled decompression will remain the same regardless of the duration of exposure at pressure.

Recognition of this concept led to the construction of underwater habitats which have been used for both scientific and commercial activities. A controlled pressure is maintained within the underwater habitat, with an appropriate gas mix equivalent to the underwater pressure where the habitat is located. Locks allow ingress and egress. The internal environment requires a sophisticated control system for the provision of oxygen and the removal of carbon dioxide and other contaminants, along with maintenance of the proper diluent mixtures at the appropriate pressure. There is a need also for careful control of temperature (26–36 degrees C), and for the provision of suitable living and hygiene facilities, as well as for the requirements of scientific and/or commercial work.

On completion of the work, the habitat may be returned to the surface if it is portable, or the divers may be returned via a portable pressurized system. Thereafter, the divers, after transfer to a fully equipped recompression chamber, must undergo at the surface the prolonged decompression ritual demanded by their exposure.

Analogue computer mediated ascent

The analogue computer that is used in determining the optimum rate of ascent was developed initially by Stubbs and his colleagues at the Defence and Civil Institute in Toronto. It utilizes the principle that gas is evolved from various tissues at different rates depending on the extent of saturation of these tissues with gas at the operating depth at any given time. Thus, on ascent from depth, first one tissue, for example muscle, will be predominant in the evolution of gas, followed for example by fat, and so on. Rather than following an overall mean rate of gas evolution, as is done in the standard tables, the computer method allows the diver to ascend at a faster rate, just less than the rate which would permit the evolution of bubbles from the tissue which is the predominant bubble producer at that depth.

The analogue computer, which is carried by the diver and exposed to the same pressure conditions, simulates three such representative tissues or tissue compartments, and monitors the evolution of gas from each. The resulting information is processed and presented to the diver on a small visual display which indicates by pointer and dial the changing actual depth of the diver, and the depth where he could be according to the rate of evolution of gas in the leading compartment. By maintaining his actual depth at or below his permissible depth he can avoid the generation of bubbles. Use of the device has greatly reduced the duration of required decompression.

Emergency recompression

The primary approach to the management of decompression sickness and air embolism is recompression in a properly equipped recompression chamber. Recompression chambers are also used in the treatment of, for example, carbon monoxide poisoning where the ability of the haemoglobin to carry oxygen is destroyed. Other usage is found in the treatment of such medical conditions as tetanus and gas gangrene where there is infection by anaerobic organisms, that is, by bacteria which require a mandatory oxygen-free environment. Exposing the infected patient to a high pressure oxygen environment assists in the destruction of the organisms.

When decompression sickness or air embolism is diagnosed or suspected, the subject should be placed in a recompression chamber at the very earliest opportunity, even if he or she requires transportation to the chamber over a period of an hour or more. If all possibilities of chamber recompression are exhausted, the subject could be returned under water or to wherever the problem originally developed. It might even be necessary to fly the patient to a recompression chamber; if so the flying should preferably be in a helicopter at a very low level.

The depth to which the subject should be recompressed is still a matter of some controversy. Some suggest recompression to the original working pressure, others the depth of relief plus one atmosphere, but normally the depth should not be greater than 50 metres (5 atmospheres). Many jurisdictions define specific standards which should be consulted. The subject should then be retained at the appropriate pressure until all symptoms are completely relieved and then be slowly decompressed according to the standard table schedule, or more slowly if necessary. The management should be supervized by a knowledgable physician with all necessary medical support, including drugs, intravenous fluids, and other medications, as required.

Oxygen recompression

Since recompression may expose the subject to pressures of 4–5 atmospheres or higher, and commonly to greater pressures than the original exposure, there is not only a possibility of creating an even greater potential bubble formation, but there is also the possibility of producing nitrogen narcosis. Accordingly, the idea of using pressurized 100 per cent oxygen was developed by Goodman and Workman (1965). The oxygen is introduced at 3 atmospheres absolute (approximately 20 metres, or 66 feet, equivalent) and maintained, interspersed with initially brief, but increasingly prolonged, periods of air breathing, as the subject returns to the surface. The schedule of oxygen and air breathing is defined in standard oxygen decompression tables which should be consulted for details. The procedure is also used in treatment of air embolism and indeed has become the treatment of choice in decompression sickness.

Aseptic necrosis of bone

A serious and delayed effect of compression and decompression is aseptic necrosis of bone, which is the term given to a progressive destruction of the bone matrix, that is, the internal substance of bone. The condition would appear to be caused by blockage of the rather meagre capillary blood supply by minute bubbles in the bone substance which are not completely resolved during decompression, or perhaps by the sludging of blood components such as platelets and cells within these capillaries (Kindwall, 1975). As a result the active bone cells which are responsible for continued maintenance of the integrity of bone are destroyed, and the dependent areas of bone become dead and calcified. Where the damage occurs at joint surfaces, there is destruction of the joint with grave disability. The joints most commonly affected are the shoulder and hip. When the condition occurs in the shaft, however, there is no pain or disability.

The time of onset of the condition may be significantly delayed, from 3 or 4 months minimum to as much as 5 years, while the minimum exposure pressure would appear to be from 17–35 psi, or approximately 40–80 feet (12–24 metres) of sea water (quoted by Kindwall (1975), from the British Central Decompression Registry).

The earliest, and sometimes the only, manifestations are found by X-ray. Many jurisdictions and commercial companies are now demanding X-ray of the shoulders, hips, and knees of their divers, both for protection of the divers, and to protect against financial loss through insurance claims and workmen's compensation requirements.

Treatment of the condition is a medical responsibility. There is no concern unless the articular surfaces of the joints are involved. Even in that situation little can be done other than prolonged avoidance, over many months, of weight or load bearing until a degree of natural healing can occur.

Part V
Chemical agents and aerosols in the work environment

Chapter 19
A review of organic chemical terminology

The occurrence of undesirable chemical agents and aerosols in the work environment gives rise to the study of industrial toxicology. The chemistry of industrial toxicology is very largely organic chemistry, that is, chemistry mostly based on study of the relationships of carbon. It is the intent of this chapter to provide an overview of the nature of some of these relationships and the terminology involved for those who have no background in organic chemistry. Of necessity, this review is superficial. For a more detailed study the reader should refer to a full chemical textbook.

Carbon

The carbon atom has the capacity for attachment to four, and only four, other atoms, by way of linking mechanisms known as *valency bonds*. thus:

$$-\overset{|}{\underset{|}{C}}-$$

Each valency bond is capable of being attached to another atom to form a *molecule*, for example:

$$H-\overset{\overset{\displaystyle H}{|}}{\underset{\underset{\displaystyle H}{|}}{C}}-H \qquad \textit{Methane}$$

For the molecule to be stable all valency bonds must be occupied at all times. In fact, however, only certain types of linkage can occur. Carbon has the capacity to link itself in simple or complex chains and loops, or cyclic variants of these loops. It must be recognized, however, that the molecular formats illustrated by these diagrams are symbolic only. The molecule does not actually exist in the format shown. It exists in a whirling and constantly changing atomic complex. When one of the hydrogen atoms of methane is replaced by another molecular fragment, for example, $-CH_3$, a new compound is formed. The $-CH_3$ fragment is referred to as a *methyl* group. A fragment with two carbon atoms and associated hydrogen atoms (H_3C-CH_2-) is an *ethyl* group. A fragment with three carbon atoms is a *propyl* group, and so on with different names as the number of carbon atoms is increased. Some of the combinations are shown below.

Aliphatic hydrocarbons

The term aliphatic refers to a simple straight chain of two or more carbons linked with hydrogen to form various different hydrocarbons, thus:

$$H-\underset{\underset{H}{|}}{\overset{\overset{H}{|}}{C}}-\underset{\underset{H}{|}}{\overset{\overset{H}{|}}{C}}-H \qquad \textit{Ethane}$$

The foregoing symbol can be written as CH_3-CH_3

When a third carbon atom is added the formula becomes:

$$H-\underset{\underset{H}{|}}{\overset{\overset{H}{|}}{C}}-\underset{\underset{H}{|}}{\overset{\overset{H}{|}}{C}}-\underset{\underset{H}{|}}{\overset{\overset{H}{|}}{C}}-H \qquad \begin{array}{c}\textit{Propane}\\(CH_3\ CH_2\ CH_3)\end{array}$$

Further additions in the same pattern create butane (4 carbon atoms), pentane (5), hexane (6), and so on, with the prefix 'pent-', 'hex-', and so on referring in Greek numbers to the number of carbon atoms in the chain.

Halogenated aliphatic hydrocarbons

An aliphatic hydrocarbon can have one or more of its hydrogen atoms replaced with one or more members of the *halogen* family of atoms, namely chlorine, bromine, iodine, and so on. Some of these that are of particular interest as industrial solvents are:

H_3C-Cl	methyl chloride
H_2C-Cl_2	methylene chloride
$HC-Cl_3$	chloroform
$C-Cl_4$	carbon tetrachloride

Alcohols

Alcohols are hydrocarbons in which a hydrogen atom is replaced with a hydroxyl group. A hydroxyl group ($-OH$) is a combination of hydrogen and oxygen with a free bond. Two common alcohols are as follows:

H_3C-OH methyl alcohol
H_3C-CH_2OH ethyl alcohol (common alcohol)

Aldehydes

Aldehydes are hydrocarbons in which the terminal carbon is replaced by a special molecular fragment, or suffix, known as an aldehyde group. The aldehyde group is expressed as follows:

$$-\overset{\overset{\displaystyle O}{\|}}{C}-H$$

In this grouping it will be observed that the oxygen is joined to the carbon by a double bond. Oxygen always has a two-bond valency. In this manner carbon also retains its four bonds.

Two common industrial aldehydes, often occurring as unwanted end-products, or used in the manufacture of other more complex compounds, are:

$$H-\overset{\overset{\displaystyle O}{\|}}{C}-H \quad \text{formaldehyde (HCHO)}$$

$$CH_3-\overset{\overset{\displaystyle O}{\|}}{C}-H \quad \text{acetylaldehyde } (CH_3\ CHO)$$

Ketones

Ketones are hydrocarbons containing the *carbonyl* grouping, which is represented as:

$$-\overset{|}{C}=O$$

Two common solvents widely used in industry occur as ketones, namely:

$$CH_3-\overset{\overset{\displaystyle O}{\|}}{C}-CH_2 \quad \text{acetone } (CH_3\ COCH_2)$$

$$CH_3-C-CH_2-CH_3$$
$$\|$$
$$O$$

methylethyl ketone (CH CO CH CH)

Acids

An organic acid is a chemical which has the suffix:

$$\begin{array}{c} O \\ \| \\ -C-O-H \end{array}$$

This suffix is referred to as a *carboxyl* group, and is commonly written as:

$$-COOH$$

Acids are very common in industry. Two of the simpler are:

$$\begin{array}{c} O \\ \| \\ H-C-OH \end{array}$$ formic acid

$$\begin{array}{c} H_3C-C-OH \\ \| \\ O \end{array}$$ acetic acid

Other functional groups

It will be noted that there are several standard molecular groupings which serve to characterize various organic chemicals. Already presented are the hydroxyl group and groupings for acids, aldehydes and ketones. Some other common groupings are as follows:

amino	$-NH_2$
isocyanate	$-NCO$
nitro	$-NO$
sulfhydryl	$=SH$
carbonyl	$\begin{array}{c} -C- \\ \| \\ O \end{array}$

There are numerous others, but the foregoing are some of the more common.

Alkenes, halogenated alkenes

An alkene is an aliphatic hydrocarbon in which one pair of carbon atoms is linked by a double bond, for example:

$$H_2C = CH_2$$

A review of organic chemical terminology 341

These compounds are sometimes called olefins. As with ordinary aliphatic hydrocarbons these molecules can be halogenated. One potentially hazardous example in industry is the vinyl chloride monomer. A monomer is a molecular unit which can be polymerized into a long chain of similar units which have characteristics that are different from the monomer. The use of monomers is common in the plastics industry. Vinyl chloride, which is hazardous as a monomer but not as a polymer, is represented as follows:

$$H_2C = CHCl$$

Benzene ring, aromatic hydrocarbons

The benzene ring (note the spelling, with two 'e's) is the term given to a hexagonal cyclic form of hydrocarbon in which six carbon atoms, each attached to a hydrogen atom, are linked together in the form of a hexagon, so:

```
        CH
      /    \\
    CH      CH
    ||       |
    CH      CH
      \    //
        CH
```

It will be noted that the bond linkage between the carbon atoms in this molecule alternates between single and double. Variations on the basic ring give rise to a group of potentially hazardous compounds, widely used in industry, and known from their spicy odours as the aromatic hydrocarbons. In recording these chemicals it is common practice, on the grounds of simplicity, to omit the $-CH$ groups in the symbolic structure and merely use the hexagon shape with alternating single and double linkages. Some examples are:

Toluene: This compound has one methyl group attached to one carbon atom, so:

```
        CH
      /    \\
    CH      CH−CH_2
    ||       |
    CH      CH
      \    //
        CH
```

Xylene: Xylene has two methyl groups which can be placed in different positions on the hexagon. These positions are referred to as ortho-, meta-, and para- (o-, a-, and p-) respectively, and can be expressed as follows:

o-xylene:

$$\begin{array}{c} CH \\ CH CH-CH_2 \\ \| | \\ CH CH-CH_2 \\ CH \end{array}$$

m-xylene:

$$\begin{array}{c} CH_2 \\ CH \\ CH CH \\ \| | \\ CH CH-CH_2 \\ CH \end{array}$$

p-xylene:

$$\begin{array}{c} CH_2 \\ | \\ CH \\ CH CH \\ \| | \\ CH CH \\ CH \\ | \\ CH_2 \end{array}$$

Phenol

Phenol, which is wrongly but commonly categorized as carbolic acid, is another compound which makes use of the benzene ring, this time with a hydroxyl suffix. It is represented as follows:

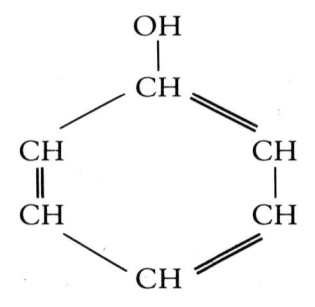

$$\begin{array}{c} OH \\ | \\ CH \\ CH CH \\ \| | \\ CH CH \\ CH \end{array}$$

There are many other modifications of the benzene ring, and indeed many other types of ring structure and other complexities. The foregoing are merely intended to be illustrative of organic chemical terminology, with particular emphasis on some compounds which are both common and hazardous. The characteristics and industrial control of various hazardous and potentially hazardous materials will be considered later.

Chapter 20
A review of the principles of industrial toxicology

Industrial toxicology can be considered as the study of those chemicals used in the work place which have a toxic, poisonous, or deleterious effect on the body. The definition can be enlarged to include those dusts and particulates which, while perhaps not toxic within a strict definition, can also be harmful. Industrial toxicology is a complex and broad based subject which cannot be studied in depth here. The objective of this and the following chapters is to define some of the problems and explore some of the approaches to their solution. The reader who requires a more comprehensive study is referred to specific reference texts such as the classic three-volume edition of *Industrial Hygiene and Toxicology,* by F. A. Patty (Wiley interscience), or *Toxicology* by Casarett and Doull (MacMillan).

Before proceeding, however, some further definitions should be considered.

Definitions

Aerosol

An aerosol is a suspension in a gas, such as air, of fine liquid or solid particles of size fine enough to remain in suspension for an observable period of time. The term applies to all such particles, and includes the following:

* dust
* fume
* smoke
* gas
* vapour
* mist

Each of these will be considered in turn.

Dust

Dust comprises solid particles derived from such activities as handling, crushing and grinding of solid materials. An individual particle may vary in diameter from 1–25 microns.

Fume

Fumes comprise solid particles which have been condensed from the gaseous state of a metal, such as, for example, lead oxide in smelting, iron oxide in welding, zinc and magnesium in grinding. The term is frequently misused to describe an aerosol which would more properly be described as a gas or smoke.

Smoke

Smoke comprises particles of carbon or sooty products arising out of incomplete combustion of a material. It may be based on a watery droplet to which a fume might adhere. Smoke particles vary in diameter from 0.1–0.25 microns. Tobacco smoke lies towards the end of the range.

Gas

A gas is defined as a formless fluid which completely occupies an enclosed space at 25°C and 760 mm Hg. It may be changed to a liquid or solid state depending on the ambient temperature and pressure.

Vapour

A vapour is the gaseous form of a substance which is normally found in the solid or liquid state, but has been changed because of change in temperature or pressure.

Mist

A mist is a finely divided liquid suspended and dispersed in the air. It is generated by condensation from gas to a liquid state, or by break up of a dispersed liquid. In industry it is commonly found as an oil mist from the use of cutting and grinding oils.

Industrial toxicology

Industrial toxicology, then, is concerned with the biological action of noxious materials, that is with their effect on living tissues, organs, and cells. Tissues, of course, are made of a complex of cells; muscle tissue is made of muscle cells, liver tissue of liver cells, and so on. The cell is the basic independent biological unit. It can be considered as a microsystem, that is, an aggregate of interactive components operating together to perform a function. Like any system it has an input of raw materials such as carbohydrates and oxygen, some form of active processing such as protein generation or nerve transmission and an output depending on the nature of the process within. It functions within a very critical range of environmental variables such as temperature, acidity, and so on. Toxic, or otherwise noxious materials disturb that system, and interfere with homeostasis.

Effects on the body

Perhaps a little surprisingly, when reduced to the simplest of terms, toxic materials have only two basic effects on cells, namely interference with function, and destruction of substance. Thus the effects can be considered as physiological or pathological.

Physiological effects

Physiological effects, in general, are implemented by interference with, or modification of, enzyme action in the cells, resulting in some change or changes in the function and activity of the cell. Enzymes, of course, are the chemical catalysts which modulate biochemical activity in the cell. For every chemical activity there is an enzyme or a system of enzymes that control it.

Again reducing to the simplest of terms, there are only two basic changes in function that can occur in a cell, namely stimulation of activity or depression of it. In passing, it might be noted that when a virus enters a cell it takes over the nuclear control system of the cell and changes the nature of its processes, to the point of death. Stimulation of a cell causes over-function, depression causes reduced function. Thus, the gas acetylene, or for that matter various anaesthetic gases, act as depressants of the cell, and reduce cerebral activity to the point of loss of sensation and unconsciousness. On the other hand, a chemical such as mercury, while having in addition severe pathological effects, acts as a stimulant to cell activity, and in particular gives rise to classical cerebral symptoms of irritability, neural excitement and even euphoria.

Pathological effects

The effect of certain materials such as acids, corrosives, and other irritants, is more pathological than functional. It is manifested initially by irritation and inflammation, but goes on if continued, or if in sufficient concentration, to cause actual destruction of tissue. Depending on the site and the nature of the chemicals concerned the effects may vary from simple skin irritation from external application, to destruction of lung tissue from inhalation, or damage to organs such as the kidney, liver, and brain. Even minor degrees of destruction may be followed by repair with non-functioning scar tissue and the result may lead to secondary loss of function.

Entrance to the body

Three modes of entrance to the body can be defined for toxic chemicals, namely, inhalation, ingestion, and transcutaneous absorption. It is not uncommon for some combination of these to occur. Each of them is examined below.

Inhalation

By its nature, inhalation, which is entrance via the respiratory system, applies mainly to gases, mists, smokes, vapours, and small particulate dusts. The problems of smokes and dusts will be examined separately in a later chapter.

With respect to the more volatile materials, consideration has to be given to their relative solubility in water and water vapour. As previously noted, when air is inhaled it is rapidly brought to full saturation in the upper respiratory passages, particularly the sinuses. Where the toxic material under consideration is highly soluble in water then the main irritation tends to be found in the upper respiratory passages. On the other hand, if its solubility in water is low then the material tends to be carried to the inner reaches of the lung where it can give rise to such clinical entities as pulmonary irritation, with characteristic discomfort, pain, coughing and spitting of mucus, or, if sufficiently intense, to the much more critical condition of pulmonary oedema and even death.

Ingestion

Ingestion is the term given to gaining entrance to the body by the mouth and digestive tract. By its nature it applies most to particulates and liquids. Entrance can take place directly when something is placed into the mouth or some material gains accidental access into the mouth. Not uncommonly in industry, however, access to the mouth can take place from contact with contaminated hands, either in touching the mouth or in the act of cigarette smoking. This form of access is of particular significance among lead workers where exposure to minute amounts of lead over many years will ultimately result in the appearance of lead poisoning. The same situation occurs with some of the other heavy metals such as mercury and arsenic.

Skin contact

The skin is a major body organ. In addition to its participation in heat exchange which has already been considered, it acts of course to protect the underlying structures. Although a number of subdivisions of its structure can be defined, it is composed essentially of two layers, an outer, horny, protective layer, or epidermis, and an inner layer containing blood vessels, nerve endings, and so on, called the derma. The fat component of the skin cells is such that the skin is resistant to the passage of watery solutions of chemicals, which then tend to do their damage to the outer surface of the skin. Fat solvents, however, which are often used in cleaning to dissolve oily materials, can pass through the fat barrier.

Toxic or irritant materials can affect the skin in three basic ways, namely, by surface irritation, by penetration with subsequent sensitization, and by penetration and systemic action.

Surface irritation: Application of irritant chemicals to the skin, and commonly repeated application, will produce an adaptive response of the skin cells in an attempt to protect themselves from further damage. This is manifest by dryness, and scaliness where the skin becomes hard and even horny. The dry scaly skin may ultimately crack to produce fissures. With continued exposure there will be skin inflammation with pain, swelling, heat and redness; the skin may become infected by pathogenic bacteria with destruction of the surface and moist weeping areas.

Penetration and sensitization: Sensitization, with its associated skin reactions, is an abnormal response on the part of the body's immune processes. Immunity to infections, and resistance to the attack of foreign materials, is a phenomenon that is partly genetic and partly acquired at a very early age. Two fundamental and related types of immunity occur in the body. One of these, humoral immunity, involves the presence and activity of circulating antibodies which are globulin molecules with the ability to attack and destroy an invading agent. The other derives from the activities of certain specially developed white cells in the blood known as lymphocytes, which, when suitably sensitized by the presence of a foreign agent, have the capacity to attach themselves to that agent and destroy it.

Allergy, or sensitization, occurs after repeated exposure to a sensitizing agent which causes the production of excess numbers of some of these specialized lymphocytes. These in turn diffuse into the skin and set up an over-reactive defensive response. Part of that response involves the release of chemicals from the lymphocytes, which, in addition to attacking the invading agent, may cause extensive skin inflammation and serious tissue damage. Common sensitizing agents, such as plasticizers are found in the chemical and plastics industry.

Rarely, in the process of *anaphylaxis,* when a specific allergy-inducing substance is passed directly into the blood stream, the reaction is sudden, very violent, and may be sufficient to cause circulatory shock and death.

Penetration with systemic action: As already noted, certain chemicals, notably solvents and tetraethyl lead, have the capacity to penetrate the skin. While skin irritation may be a problem with these chemicals a more serious situation arises when they are absorbed in sufficient concentration to affect internal body organs. Carbon tetrachloride, for example, which is a very popular and hazardous solvent, may pass through the skin and ultimately affect the liver, kidney, and brain.

Action within the body

Immediate action and elimination

Once toxic materials are inside the body, several courses of action are possible. With some chemicals, such as carbon monoxide or some of the long-chain hydrocarbons, there is an immediate action on the part of the chemical.

For example, carbon monoxide when it is inhaled will be immediately absorbed into the blood stream where it will become attached to the haemoglobin of the red cells and block the carriage of oxygen. It will continue to exert its action until the red cell is ultimately destroyed in the natural life cycle of red cells. If the concentration is high enough its toxic effects will become manifest. Carbon tetrachloride, once into the body will be carried by the blood to all areas including the liver and the brain where it will act to destroy the liver and brain cells. It will however ultimately be broken down and eliminated. The amount of damage again will depend on the concentration.

Accumulation to threshold

With some chemicals, and particularly lead, the material is taken into the body and accumulated until it reaches a threshold. It is then released into the body and diffused into body cells where it exerts its toxic action.

For example, when lead is ingested, if the concentration is sufficiently high it will exert an immediate toxic effect before it is sequestered and stored in the bones, particularly the long bones. It will accumulate in the bones until they can store no more. This process may take as long as 5–20 years. When the bones are saturated lead will begin to circulate in the bloodstream again and once more exert increasingly toxic effects as the lead accumulates from the bones.

Detoxification

A third possibility also exists to which is given the name detoxification. This is the term given to the process whereby a toxic chemical is transformed into a less toxic form by way of enzyme action. Although the affect may be desirable, the process is a random one. In other words, many different enzymes exist, a large number in the liver where many metabolic processes take place. Each of these enzymes is capable of effecting some particular chemical reaction. If a chemical and an enzyme are brought together and the chemical change governed by that enzyme is capable of occurring it will, and the chemical will now be changed into another chemical. If that second chemical is now less toxic than the first detoxification is said to have taken place. It is emphasized again, however, that it is a random and not purposeful process. If the mechanism is available it will happen.

A number of different detoxification processes can be defined and are briefly outlined below.

Oxidation: Oxidation is the name given to the addition of oxygen to a molecule, or the removal of hydrogen. Commonly acetic acid is oxidized to carbon dioxide and water in a detoxification process, thus:

$$CH_3COOH \longrightarrow CO_2 + H_2O$$

Reduction: Reduction involves the addition of hydrogen to a molecule, or the removal of oxygen. In rats, at least, carbon tetrachloride is reduced to carbon dioxide and chloroform by this process. The chloroform then undergoes further processing.

$$CCl_4 \longrightarrow CO_2 + CHCL_3$$

Synthesis: The most common form of synthesis as a detoxification procedure is *hydrolysis,* which is the term given to the addition of the components of the water molecule. The solvent ethyl acetate goes through a complex group of processes, including hydrolysis, to its ultimate breakdown to carbon dioxide and water, so:

$$C_2H_5OOCCH_3 \longrightarrow C_2H_5OH \text{ (alcohol)} + CH_3COOH$$
$$\text{(acetic acid)} \longrightarrow CO_2 + H_2O$$

A second form of synthesis is *conjugation* in which the chemical of interest is combined by enzyme action with various other organic molecules, for example:

* methylation (CH_3-)
* acetylation (CH_3CO-)
* hydroxylation ($-OH$)
* amination ($-NH_2$)

* sulfhydrylation (—SH)
* acidification (—COOH)

as well as with various organic amino acids (the constituents of protein) such as glutamine, cystine, and glycine, and a variety of other chemicals.

Elimination

Ultimately it is the objective of the body defence mechanisms to eliminate all toxic chemicals from the body, either in their original state, or broken down through various stages of detoxification. The process of elimination takes place through the usual channels, chiefly the kidney, but to some extent also by the bowel. Normally the bowel eliminates largely unwanted material which has been ingested but not absorbed, as well as various secretions from the digestive tract and the liver. Most chemicals which have been absorbed are carried, after whatever processing they undergo, to the kidney where they are excreted in the urine. The kidney, however, may be damaged in the process. A classic example of this damage occurs in the elimination of mercury during which the kidney tubules may be seriously damaged and destroyed.

Some chemicals, notably the more volatile, are excreted in part during exhalation, as anyone will know who has smelt the breath of a person who has partaken of alcohol; some, and particularly some of the heavy metals, are indeed excreted in the sweat, and even cut off in the hair.

Lethal dosage

Although on the basis of experiment and clinical experience it is relatively easy to determine that certain chemicals are toxic it is desirable for predictive and comparative purposes to know on a quantitative basis how toxic a given chemical might be. This of course is very difficult to determine. Some of the necessary information can be derived from skilled observation and some indeed from human experimentation with volunteer subjects. Human experimentation, however, to the extent required for definitive results, is neither acceptable nor practicable, thus much of our knowledge must be derived from controlled animal experimentation. It can be misleading, however, to extrapolate from animal findings to human predictions. There are situations where animals react differently from man when exposed to the same material in the same relative concentration. And of course, even differences in size can lead to difficulties in interpretation. Thus one has to be very cautious in applying the results of animal experimentation to the human condition.

Nevertheless, through careful application of the findings derived from skilled observation of human response both in the work place and in the laboratory, amplified by the results of experimentation on animals, much of value can still be learned.

Animal experimentation, in particular, has led to the concept of what is termed the LD_{50}, or in other words the dose of a substance that is lethal to 50 per cent of the animals exposed. Thus a certain substance might be shown to have an LD_{50} of 100 mg for rats. This does not mean that it also has the same LD_{50} for humans, but it at least gives some indication of its relative toxicity in comparison with other materials.

Measurement of concentration

Toxic materials are commonly found as aerosols within the breathing air. It therefore becomes necessary to consider ways of recording their concentration in that atmosphere.

It is convenient to divide these materials into two physical categories. One category includes the gases and vapours, and the other comprises the liquids and solids. Gases and vapours, of course can be distinguished physically, but there is nothing to be gained by doing so in this instance. Each is measured in parts per million (ppm), that is units of volume of gas or vapour per million volumes of air. Usually the measure is referred to as so many ppm, but if it is necessary to define the volumes concerned, for example 1 ml per 1000 litres, then that volume has to be corrected to standard atmospheric conditions as indicated in Chapter 17, namely at 25°C and 760 mm Hg atmospheric pressure.

The unit, ppm, is a volume/volume unit. It is of course possible to convert a volume/volume unit into a weight/volume unit such as mg/litre. To do so one must recognize that 1 gm molecule, or mole, of an ideal gas, which is the weight in grams of a molecule numerically equivalent to its molecular weight, assumes the volume of 24.45 litres. Hence:

$$1 \text{ ppm} = \frac{\text{molecular weight}}{1000 \times 24.45} \text{ mg/l}$$

and,

$$1 \text{ mg/l} = \frac{24.45 \times 1000}{\text{molecular weight}} \text{ ppm}$$

In contrast with gases and vapours, liquids and solids are normally measured in weight/volume units, namely mg of substance per cubic metre of atmosphere, that is mg/m^3.

Bearing this in mind, it is of interest to recognize that during moderately heavy work over an eight hour day a worker will inhale 10 litres of air per minute, or 4800 litres per day, which in turn will occupy the space of 4.8 cubic meters. Thus knowing the concentration of a material in mg/l one can in fact calculate how much of that material is inhaled during a day's work. Thus, for example, if the concentration of lead in the breathing air were 0.15 mg/l, which is the normal acceptable maximum, then in the course of a normal work day doing moderately heavy work, a worker would inhale 720 mg of lead.

From the foregoing, then, it can be seen that we can establish lethal doses, at least with respect to animals, and also we can establish concentrations of toxic material in the atmosphere. One further element remains to be considered in this brief introduction to toxicology and that is the definition of toxic thresholds, or in other words permissible limits of toxic exposure in the work place.

Permissible exposure limits

The most widely used concepts pertaining to thresholds of toxic exposure have been developed by a body known as the American Conference of Government Industrial Hygienists (ACGIH) who have defined the terminology and the recommended limits of exposure to a large variety of toxic chemicals. These recommendations are not only widely accepted but have been adopted to a greater or lesser extent into legislation in many administrations. A proprietary booklet entitled *Threshold Limit Values for Chemical Substances and Physical Agents in the Working Environment* (copyright American Conference of Government Industrial Hygienists) is published on an annual or biennial basis by ACGIH, listing recommended limits. All reference to such limits herein, and all use of the term TLV, and its derivatives, should be considered proprietary. These limits are defined in a standard terminology, given below:

Threshold limit value (TLV)

A Threshold Limit Value refers to the concentration of an airborne substance, and defines conditions under which it is believed that nearly all workers may be repeatedly exposed day after day without adverse effect.

There are three categories of TLV, namely:

* time weighted average (TWA)
* short term exposure limit (STEL), and,
* ceiling (C)

Each of these is considered in turn below.

TLV-TWA

This category is a time-weighted concentration averaged over an 8-hour day and a 40-hour week, during which no adverse effect will occur on repeated exposure. When averaged over that period the concentration can indeed exceed the recommended TLV for periods of time, provided that these periods are compensated and the average level is not exceeded when measured over an 8-hour day. For example, the TLV-TWA for the solvent methyl ethyl ketone (MEK) is listed at 200 ppm.

TLV-STEL

This category refers to the maximal exposure permissible for a 15-minute period which does not give rise to irritation, chronic or irreversible tissue damage, or narcosis (i.e. sleepiness, stupor, unconsciousness). Each exposure must be separated by at least 60 minutes and the daily TLV-TWA may not be exceeded. That is, a high concentration exposure must be compensated by a comparable low concentration exposure. The STEL is not a separate independent exposure limit, rather it supplements the time-weighted average limit where there are recognized acute effects from a substance whose toxic effects are primarily of a chronic nature. STELs are recommended only where toxic effects have been reported from high short-term exposures in either humans or animals. For example, the TLV-STEL for methyl ethyl ketone is listed at 300 ppm.

TLV-C

The ceiling TLV is the maximum concentration that is permitted even instantaneously. For some substances, for example, irritant gases, only one category, the TLV-Ceiling, may be relevant. For other substances, either two or three categories may be relevant, depending upon their physiological action. It should be noted that a TLV-C limit, unlike a TLV-TWA limit, sets a definite boundary which may not be exceeded even for a short time regardless of compensation. It is also important to observe that if any one of these three TLVs is exceeded, a potential hazard from that substance is presumed to exist.

'Skin' notation

Certain chemicals have a special predilection for absorption through or damage to the skin, mucous membranes, and eye. They are awarded a 'skin' notation in the listings of recommended values, that is the word 'Skin' is added after the name of the chemical, for example, methyl alcohol.

Maximum allowable concentration (MAC)

The maximum allowable concentration is an obsolete term originally used to define exposure limits. The MAC defined a ceiling above which exposure was not permitted, similar to the TLV-Ceiling, applicable to all listed chemicals. The concept proved to be unworkable in practice and was replaced with the TLV approach. The term however is still found in the older literature.

Threshold limit value for mixtures

It is not uncommon to find that toxic agents exist in the atmosphere in mixtures rather than singly. When two or more hazardous substances are present their combined effect should be considered. Normally, and unless there is reason to believe to the contrary, the effects of the hazard, and consequently the concentration of an agent in relation to its TLV should be considered additive. Thus in a mixture where independent activity is not considered significant the threshold limit should be developed in accordance with the following relationship:

$$\frac{C_1}{T_1} + \frac{C_2}{T_2} + \cdots + \frac{C_n}{T_n} = 1$$

where, C_1 = observed concentration of chemical No. 1
T_1 = TLV of chemical No. 1
C_2 = observed concentration of chemical No. 2
T_2 = TLV of chemical No. 2
C_n = observed concentration of other chemicals
T_n = TLV of other chemicals

In other words, the sum of the ratios of the concentrations of chemicals to their Threshold Limiting Values should not exceed unity. For example, consider a mixture of organic solvents with concentrations and TLVs which might be as follows:

Solvent	Concentration	TLV
Solvent A	5 ppm	10
Solvent B	20 ppm	50
Solvent C	10 ppm	25

Then:

$$\frac{5}{10} + \frac{10}{50} + \frac{10}{25} = \frac{65}{50} = 1.3$$

Under these circumstances the permissible threshold would have been exceeded.

On the other hand, where the effects of the components of a mixture are considered to be independent of each other then the threshold of the mixture has to be determined in terms of each component. In other words:

$$\frac{C_1}{T_1} = 1; \text{ and } \frac{C_2}{T_2} = 1$$

As an example let us consider a mixture comprising a metal particulate and some acid mist with which it does not react. The concentrations and TLVs might be as follows:

Substance	Concentration	TLV
Metal	0.15 mg/cu.m.	0.20 mg/cu.m.
Acid	0.70 mg/cu.m.	1.00 mg/cu.m.

Then:
$$\frac{0.15}{0.25} = 0.75; \text{ and } \frac{0.7}{1.0} = 0.7$$

The threshold limit, then, in this situation has not been exceeded.

The foregoing are examples of simple calculations for mixtures in air, and are given as examples. Other more complex calculations can be undertaken. The reader is referred to the previously noted annual brochures of the ACGIH for further information.

Chapter 21
Toxicity in the work environment: Some selected chemicals

Every year thousands of new chemicals are added to the working environment, some of them hazardous, some of them not, some of them which have been studied at length, some of them about which little is known. In many administrations today manufacturers of chemicals are required by law to provide, for the use of management and labour, information data sheets for the chemicals they manufacture and distribute. These sheets may include such information as the proprietary and generic names of the chemicals, their natural occurrence and origins, their usage in industry and elsewhere, relevant physical and chemical data pertaining to them, their effects on humans and animals at different concentrations, both short and long term, their accepted permissible concentrations in the work place, their measurement, control and management, including protective measures and prevention of harmful effects and the necessary first aid or other treatment that should be used in the event of exposure.

Several definitive texts have been written about toxic materials in the work place, including the already noted Patty's *Industrial Hygiene and Toxicology*, as well as others, such as the popular *Fundamentals of Industrial Hygiene* published by the United States National Safety Council, and *Chemical Hazards in the Workplace* by N. H. Proctor and J. P. Hughes. Others include the *Toxicity of Industrial Metals* by E. Browning, *The Basic Science of Poisons* by L. J. Casarett and J. Doull, as well as the NIOSH publication, *Occupational Diseases* edited by M. M. Key, and the World Health Organization publication *Early Detection of Occupational Diseases*, while numerous groups, including the American Industrial Hygiene Association, have published guidelines which give detailed outlines of many toxic chemicals. The material in this chapter, although modified, is derived from these and other sources.

It is not the intent in this chapter to provide an exhaustive review of toxic materials in the work place. Instead it is intended to highlight certain substances which by reason of common usage, special toxicity, or other interest, might be of special concern.

It would be convenient if one could identify a given chemical or material and state that since it belongs to a certain class of chemicals it will therefore have a certain effect on an exposed person. Unfortunately this is not the case, although certain broad groups producing similar effects can be defined. Some of these chemicals will be examined here. The TLV's quoted are proprietary and are derived from the 1886–87 listing of the American Conference of Government Industrial Hygienists.

Respiratory irritants and asphyxiants

While many materials, liquid and solid, act as irritants to the skin, eyes, and mucous membranes, the most serious effects are commonly found among the respiratory irritants. Some of the most irritant of these are the following:

Table 21.1. *List of selected irritant gases and vapours*

Substance	Symbol
ammonia	NH_3
chlorine, hydrochloric acid	Cl_2
formaldehyde	$HCHO$
sulphur dioxide, sulphuric acid	SO_2, H_2SO_4
hydrogen sulphide	H_2S
oxides of nitrogen	NO, NO_2, N_2O_4
ozone	O_3
phosgene	$COCl_2$
phosphine	PH_3

These materials, with the exception of phosgene, present their major effect as irritation to the upper respiratory tract, and in particular to the nose, throat, and trachea (windpipe). If the concentration is sufficiently high the effect is immediate, giving rise to coughing, gasping, and spluttering, with watering eyes and nose, and much production of mucus. Since the effect is so immediate and recognizable they are readily avoided provided one can escape from their presence. Thus, for example one can run out of an open room, although one might be caught in a closed container.

Ammonia (TLV 25 ppm)

Ammonia is a colourless gas used in refrigeration, petroleum refining, fertilizer manufacture, and in the manufacture of plastics and other chemicals. It is a severe irritant to the eyes, respiratory tract, and skin. Mild irritation can be felt around 30–50 ppm, increasing at 75 ppm, and becoming severe to most persons over 125 ppm. Tolerance can be acquired by exposure.

Chlorine (TLV 1 ppm)

Chlorine is a yellowish gas used in metal fluxing, water sterilization, bleaching and chemical processing. Hydrochloric acid, which is the aqueous form of the gas hydrogen chloride, is a liquid used in steel picking, and as an intermediate in chemical processing.

Each is highly irritant to the eyes, mucous membranes, skin, and respiratory tract. Mild irritation with chlorine will be observed at as low as 1 ppm, and with hydrogen chloride at 5–10 ppm. Exposure to higher concentrations (for example, chlorine 50–100 ppm) will produce intense respiratory irritation with pulmonary oedema leading to pneumonitis.

At even higher concentrations it rapidly causes respiratory paralysis. Inhalation of 1000 ppm can cause coma after only one breath, while prolonged exposure at lower levels such as 250 ppm can cause pulmonary oedema.

Formaldehyde

Of recent years formaldehyde has given rise to potential problems. It has a wide usage in the plastics industry in the making of formic esters and resins. It is also used in leather manufacturing and in rubber making as an additive, while it is a significant constituent of fibre board, particle board and in furniture made with these materials. Urea formaldehyde plastic foam is used in the home to provide thermal insulation within the walls of a house. It is formed *in situ* using pressurized hoses containing urea and formaldehyde which unite in a thermoplastic reaction when they meet to produce a foamy material which ultimately hardens into an effective insulating material. Sometimes, when the foam is improperly formed, an excess of formaldehyde develops and seeps through the walls in concentrations which can occasionally be above the recommended level. Much distress has been caused to the owners of these houses when problems have arisen, as well as to the owners of houses insulated in this manner who have not had problems but fear that the value of their property might drop because of the presence of foam. Its use as an insulating material in walls has been banned in some jurisdictions.

Formaldehyde, of course, is a respiratory irritant. At 10 ppm it is tolerable for only about 10 minutes. Concentrations of 4–5 ppm are tolerable for about 30 minutes, while at concentrations of as low as 2–3 ppm there is manifest irritation in the eyes, the mucous membranes, the upper respiratory tract and the skin. Hypersensitivity has also been observed in some persons who develop asthmatic attacks in the presence of even very low concentrations of formaldehyde.

In high dosages, with levels of 15 ppm for prolonged periods of weeks or months, it has been shown that rats may develop cancer. There is no evidence of carcinogenesis in humans.

Hydrogen sulphide (TLV 10 ppm)

Hydrogen sulphide is a colourless gas found as a by product in sulphuretted chemical processes, as well as an effluent around oil wells and petroleum processing. At low concentrations (up to 50 ppm) it acts as an irritant to the eyes, mucous membranes, and respiratory tract. In higher concentrations (over 250 ppm) it can cause pulmonary oedema, and with increasing concentration, respiratory paralysis and asphyxia.

It has a very offensive odour even at very low concentrations which can provide an effective alarm. Unfortunately the sense of smell is rapidly overwhelmed and ceases to act as a warning.

Phosgene (TLV 0.1 ppm)

Phosgene is different from the other irritants. It may not in fact be immediately irritant even in hazardous concentrations. When inhaled it has a slightly unpleasant odour of rotting grass. However, over a period of several hours while still in the lung it is converted by body enzymes to chlorine in increasing quantities deep in the alveoli. Thus there is a considerable latent period between inhalation and effect. Since also the chlorine is formed deep in the lung, the resulting outpouring of watery mucus can fill the interstices of the alveoli and produce the dangerous and potentially fatal condition of pulmonary oedema, such that, if the oedema is sufficiently great the subject can, in effect, drown in his own secretions. Even if the oedema is less than fatal it can still act as an area for infection with a resulting potentially fatal pneumonitis.

Phosgene is generated by oxidation of chlorinated hydrocarbons and can occur in welding either by direct heating of metals dipped, for example, in a chlorinated hydrocarbon solvent such as carbon tetrachloride, or in the case of arc welding, by the action of ultraviolet light acting in line of sight on a chlorinated hydrocarbon, such as vapour or drips from a solvent tank.

As noted it is a respiratory irritant. In concentrations of 50 ppm a brief exposure can be rapidly fatal, although in lower concentrations, as previously mentioned, the onset of symptoms may be delayed up to even 72 hours after exposure.

With moderate exposures (5–10 ppm), there may be dryness and burning in the throat, vomiting, chest pain and difficulty in breathing. In concentrations of 4–5 ppm there is coughing, respiratory irritation, and irritation of the eyes. Coughing may occur at 3 ppm, while the smell is apparent about 1 ppm. Smell, however, is an unreliable sense, and with continued exposure the smell is lost.

Phosphine (TLV 0.3 ppm)

Phosphine is a gas which also can be generated under certain conditions of welding. It can occur during the welding of steel which has been treated with some form of phosphate rust inhibitor. While this is the most common way in which it occurs it can also be found as an impurity in acetylene gas. It has a fishy or garlic odour at 2 ppm.

Like the others in this group it is a respiratory irritant. Indeed it can be a severe pulmonary irritant at high concentrations (greater than 300 ppm). At lower levels (10–30 ppm) it can produce nausea, vomiting, diarrhoea, chest tightness, coughing, headache, and dizziness. Fatality can occur with exposures of 400–600 ppm for half an hour.

Oxides of nitrogen

The oxides of nitrogen are commonly found in welding, and occur as nitric oxide (NO), nitrogen dioxide (NO_2), and nitrogen tetroxide (N_2O_4) as a result of oxidation of atmospheric nitrogen in the welding process. Thus they tend to occur in arc welding, brazing and braze welding. They also occur in gas welding as a result of oxygenation catalyzed by ultraviolet light. Nitrogen oxides are also a serious hazard among farm workers as a result of biological generation in farm silos. Another form of nitrogen oxide, with anaesthetic qualities, is nitrous oxide (N_2O), or 'laughing gas'.

The three gases nitric oxide, nitrogen dioxide, and nitrogen tetroxide exist in equilibrium one with another, such that, in the presence of oxygen, nitric oxide rapidly oxides to nitrogen dioxide while nitrogen tetroxide and nitrogen dioxide change from one to the other in a spontaneous fashion. Thus all three tend to be found in the same conditions.

The gases, commonly known improperly as 'nitrous fumes' are respiratory irritants which can, in high concentration, give rise to pulmonary oedema. On exposure the following effects may be demonstrated:

100 ppm:	pulmonary oedema and death
50 ppm:	pulmonary oedema with temporary or permanent lung damage
25 ppm:	respiratory irritation, chest pain, coughing, gasping
5–10 ppm:	respiratory irritation

Ozone (TLV 0.1 ppm)

Ozone, which comprises three atoms of oxygen, can also be generated in most conditions of welding, and in particular during inert gas welding. It is produced by the action of ultraviolet light on oxygen.

Despite popular belief to the contrary, ozone is a highly irritant and toxic gas with a characteristic and not unpleasant smell redolent of seaweed and the ocean. It is, however, a classical respiratory irritant. With exposure in a concentration of 0.6 to 0.8 ppm for two hours there is manifest impairment of lung function, accompanied by coughing, pain, and/or interference with breathing.

At levels between 0.1 and 0.5 ppm there may be considerable respiratory irritation along with some changes in visual acuity, while even in the range of 0.5 to 0.01 ppm there may be irritation and dryness of the throat, upper respiratory passages and eyes.

Other irritants of special interest

As noted earlier there are many other irritant materials which are not gases. Two of particular interest are the isocyanates and the polychlorinated biphenyls.

Isocyanates

There are two industrial isocyanates, namely toluene-2, 4-diisocyanate (TDI), and methylene bisphenyl diisocyanate (MDI). While TLVs may be listed there is a move towards no permissible exposure. Isocyanates occur as colourless liquids which can be inhaled as aerosols. They are extensively used in the manufacture of polyurethane foam plastics.

The acute effects are those of an irritant to the eyes, mucous membranes and skin occurring at low concentrations in the order of 0.5 ppm or less, with respiratory irritation occurring at slightly higher levels.

A much more significant effect, however, is respiratory sensitization which takes the form of bronchospasm and asthma which may progress to pulmonary oedema. This may occur with minimal exposure to the lowest feasible concentrations. A pattern of repeated attacks may develop, with an exposure to even the lowest concentration of isocyanates, which will lead to total disability. The condition may develop suddenly, with no previous warning, in a worker who may have been working with isocyanates for a prolonged period.

Polychlorinated biphenyl (PCB)

There are a number of chlorinated biphenyls of slightly different chemical structure. The TLV depends on the percentage of chlorine in the molecule. For 42 per cent chlorine the TLV is 1 mg/m^3, while for 54 per cent it is 0.5 mg/m^3. They occur as oily liquids, and although their usage has been largely discontinued because of suspicion of carcinogenesis they have in the past been used as dielectrics in capacitors and transformers in the electrical power industry, as well as heat exchange fluids and hydraulic fluids. Exposure is by inhalation, and skin contact. PCBs give rise to a special form of dermatitis known as chloracne which resembles the common acne of adolescence which affects chiefly the face, back and chest. The skin may be very dry, with extreme itching and cyst formation. Liver damage has occurred in animals.

Asphyxiant gases

Asphyxia is the term given to physical blockage of respiration. It is not a toxic effect, nor does it occur as a result of physiological change. Certain gases encountered in industry exert their lethal action simply by choking out the available oxygen supply. They include the following:

Table 21.2. List of selected asphyxiant gases

Name	Symbol
carbon dioxide	CO_2
methane	CH_4
helium	He
nitrogen	N_2

Carbon dioxide

Carbon dioxide is, of course, a natural physiologically occurring gas produced by the body in the normal course of metabolism. It is used in industry for various purposes, including its 'asphyxiant' action as a fire extinguisher. When found in a high concentration, however, which could occur, for example, when breathing air in a sealed tank, it will block the available oxygen simply by taking up room that would be otherwise occupied by air containing an adequate oxygen supply. The TLV for safe breathing is listed at 5000 ppm.

Methane

Methane, (marsh gas, fire damp) is again a non-physiological agent which can block the air supply. It has another hazard in that it is potentially explosive. A specific TLV for methane is not listed, but it is considered to be breathable provided that the oxygen proportion in the atmosphere be maintained at not less than 18 per cent (normal 21 per cent).

Helium and nitrogen, as noted before in consideration of the effects of changed barometric pressure, are normal constituents of the atmosphere. Helium is, of course, virtually inert, but nitrogen under pressure can produce nitrogen narcosis, as already discussed. In either case it is considered that the gases will not be asphyxiant so long as the oxygen in the atmosphere is maintained at a level of 18 per cent.

Solvents

Solvents are used for degreasing of metallic objects prior to further processing as well as for cleaning purposes and solution of other chemicals.

There are many different varieties, some of which are more toxic than others. It is unfortunate that one of the more toxic, carbon tetrachloride, is also one of the most common. While various differences can in some instances be identified from solvent to solvent, many, if not most, tend to have certain characteristics in common with respect to their effect on humans.

Effects of solvents

Long term, low concentration

In low concentration over a long period the effects of solvents are largely the result of damage to the fat layer of the skin. This gives rise to a form of dermatitis characterized by irritation of the skin with dryness, scaling, and fissures. There is also irritation of the lining mucous membranes of the nose and throat and also the eyes..

Ultimately the solvents will penetrate the skin and be carried to the internal organs, with subsequently damage to the liver, where they concentrate, to the kidney, where they are excreted, and to the brain, where slow damage to brain cells may occur.

Short term, high concentration

In high concentrations, generally by inhalation over a short period, the effects are chiefly manifest in the brain, with fairly rapid onset of dizziness and incoordination, sometimes with euphoria. These lead to anaesthesia, and unconsciousness, sometimes with convulsions. The unconsciousness may deepen into coma and eventually death by respiratory failure. Should there be recovery there may be permanent residual damage to the liver, kidney, and brain.

Methyl butyl ketone (MBK), (TLV 5 ppm)

MBK is an exception to the general rule of solvents in its effects. While at high concentrations (several hundred ppm) it is irritating to the respiratory system and the mucous membranes, a very significant effect, namely peripheral neuropathy, is found with concentrations of less than 100 ppm, with exposure over a few months.

Peripheral neuropathy manifests itself as slowly developing loss of motor activity such as weakness of the hands, difficulty in maintaining a pincer-type grasp, weakness of the ankles, and so on. There may be, in addition, tingling and numbness (paraesthesia) in the hands or feet. Even when the subject is removed from the exposure the condition may continue to progress, with a prolonged, and sometimes incomplete, recovery.

Welding and other fumes

A curious sensitization-type condition may arise in workers exposed to hot fumes of certain metals or their oxides, as for example during metal turning on a lathe. The materials include, but are not restricted to:

- beryllium
- cadmium oxide
- cobalt
- copper oxide
- manganese oxide
- magnesium oxide
- tin
- zinc oxide

The illness is referred to as metal fume fever. In those affected it tends to occur a few hours after exposure over a working day. There is a sudden onset of an acute, feverish illness, with chills, and fever, often with nausea and vomiting. The condition will last for several hours, with a complete recovery over a period of 24 hours. It may recur on subsequent exposure, but ultimately disappear.

Other effects of exposure to various metal fumes are shown in Table 21.3. Some of these materials are examined in further detail below.

Table 21.3. *Effects of Exposure to Metal Fumes*

Fume	Source	Chief effect
beryllium	metal alloys (base and filler)	lung irritation, lung fibrosis, ? cancer, metal fume fever
cadmium oxide	silver soldering brazing	pulmonary oedema, fibrosis, kidney damage metal fume fever
chromium trioxide	plating	lung irritation, bronchospasm, bronchitis
cobalt	metal alloy	metal fume fever
copper oxide	metal alloy, brazing	metal fume fever
iron	metal	siderosis
manganese oxide	steel alloy	metal fume fever, weakness incordination
magnesium oxide	metal alloy	metal fume fever
molybdenum	steel alloy	lung irritation, damage to liver and kidneys
tin	metal solder	metal fume fever
titanium oxide	steel alloy	respiratory irritation
vanadium pentoxide	filler metals	respiratory and conjunctival irritant
zinc oxide	galvanized, or painted metals	metal fume fever
zinc chloride	soldering fume	respiratory and conjunct-irritation

Beryllium (TLV 0.002 ppm)

Beryllium is a highly toxic material which occurs in metal alloys, and as a hardening agent. It is used as a filler with copper, magnesium, and aluminium, and in the manufacture of special steels.

Its effects depend on the concentration and duration of exposure. When the fumes are inhaled in high concentration for a short time it gives rise to a severe inflammation of the lung (pneumonitis) which, if not fatal, may be grossly debilitating over a period of six months or more.

In low concentrations for more prolonged periods it will cause a classic metal fume fever, along with eye irritation and eventually damage to the liver. It is also irritant to the eyes and skin where it may cause skin ulcers, or non-cancerous skin tumours.

An unusual feature, in low concentrations over prolonged periods, is the development of non-cancerous lung tumours, called granulomas (or granulomatas), which in turn produce a slow progressive disease with weakness, shortness of breath, and coughing. The illness may last many years, and full recovery even after removal from exposure is uncommon.

Cadmium oxide (TLV 0.05 mg/m^3)

In welding type operations, cadmium oxide is found in brazing, braze welding and soldering. It occurs in filler materials, such as copper-silver-zinc with cadmium, and as a cadmium plated basemetal.

Its effects again depend on the concentration and the exposure. In high concentration, up to 10 hours, there is a sudden onset of gasping, coughing, and frothing typical of sudden pulmonary oedema, which may be rapidly fatal if sufficiently severe, or may provoke a non-fatal pneumonitis if less so.

In low concentration, a typical metal fume fever may occur. On continued low level exposure there may be development of a progressive lung irritation with emphysema, where there is destruction of lung tissue and loss of lung function. Other effects are found in the kidney, the blood and the nervous system.

Fluorides (TLV 2.5 mg/m^3)

Fluorides occur as components of the fluxes used in brazing and braze welding. As well as occurring as fumes, they also occur as gases and dusts. They are also found in various other metallurgical processes, as well as in glass and ceramic manufacture.

High concentrations of fluorides are very irritating to the eyes and the respiratory tract, and cause cough, bronchospasm, and reduced lung function with wheezing and shortness of breath. Still higher concentrations produce pulmonary oedema. Although now very rare, repeated exposure to high but tolerable concentrations over a period of years used to give rise to a crippling condition of overgrowth of bone, called fluorosis.

Heavy metals and their compounds

The heavy metals are used in many different industries in a wide range of occupations. They are all toxic to a greater or less extent, and have the same broad general effects, except that, for example, while both lead and mercury affect the digestive and nervous systems the effect of lead is relatively greater on the digestive system while the effect of mercury is relatively greater on the nervous system.

Lead (TLV 0.15 mg/m^3)

Lead is one of the most common industrial toxic materials, and can exert its toxic action either in its metallic form, or in its chemical combinations, including the organic chemical tetraethyl lead, which, although much reduced in use, is still a common additive to motor vehicle fuels.

Lead is used in a wide variety of activities including:

- smelting
- battery making
- plumbing
- metal paint manufacture
- lead heating baths
- foundries
- vehicle body work
- soldering
- alloy making
- lead fuel additives

When absorbed into the body, lead acts as a general body poison, affecting to a greater or lesser extent all body cells and tissues. Entrance to the body can be by ingestion (most common), inhalation of fume or fine particle, or absorption through the skin (particularly tetraethyl lead).

Once in the body it accumulates in the bones until they reach their capacity, a process which may take from 5–20 years. When no more lead can be stored it is released into the bloodstream, whence it is distributed throughout the body with continual turnover between bones and blood. Thus it can cause damage during the initial circulation and again after release from the bones.

While lead can gain entrance to the body in all possible ways, the greatest danger lies with the dust generated in handling. Lead is often used in lead pots into which objects may be hand dipped. Although this technique would appear particularly dangerous because of potential exposure to lead fume, in fact the melting temperature of lead at which the pots are normally kept (about 500°C) is below the fuming temperature (900°C).

Effects of lead toxicity

Lead acts to disturb the function of all body organs and systems. Unfortunately the early effects are vague, generalized, and easily missed unless the observer is alert to the possibilities. Although commonly of slow and gradual onset, they may indeed occur suddenly after leaving the exposure. They tend to be manifest as a generalized weakness, with lassitude, some loss of weight, and insomnia. Since each and all of these are common occurrences even without exposure to lead they may be easily overlooked or attributed to emotional problems, stress, mild infection, and so on. A not uncommon early symptom, however, namely impotence, may quickly be brought to the attention of medical or other personnel.

In addition to its generalized toxic effect lead has a special predilection for three body systems, specifically the digestive system, the blood or haemopoietic system, and the nervous system.

> *Effects on the digestive system:* Again the effects begin by being somewhat vague, with some loss of appetite, which eventually becomes marked, and generalized abdominal discomfort. Constipation is a feature. It may be very marked and difficult to manage. As the condition continues to develop, colic, or abdominal griping, can be a focus of attention. The colic can be extremely severe and may lead to hospitalization of the sufferer. In the presence of poor dental hygiene a characteristic blue-black line may appear along the gum line from the presence of lead sulphide. This at one time was thought to be diagnostic of lead poisoning. It can occur in other conditions, however, and may be late in occurrence in lead poisoning if it occurs at all.
>
> *Effects on the haemopoietic system:* In the blood, lead is absorbed into the red blood cells where it will ultimately lead to their destruction. Initially the effects can be observed only microscopically, but as more and more red blood cells are destroyed the effects become manifest as anaemia, with increasing loss of the capacity of the blood cells to carry oxygen. The anaemia contributes to the weakness, lassitude and weight loss, and is shown as pallor of the face and the eye grounds. There may be abdominal discomfort and palpable enlargement of the spleen.
>
> *Effects on the nervous system:* One of the earliest effects on the nervous system is found in the peripheral nervous system, that is the nerve supply to, for example, the hand and the foot, Muscle weakness at the wrist and ankle may indeed be one of the first symptoms that draws the patient's attention to his condition. It shows as a dragging of the foot ('foot drop'), or an inability to maintain the wrist in a raised or horizontal position ('wrist drop'). Hand tremor, or shaking, and tremor in other parts of the body may become evident, and

as the situation becomes full, paralysis will develop.

The most serious effect on the nervous system occurs when lead is absorbed into the brain and central nervous system. This is uncommon in adults from inorganic lead, but may occur in children ingesting lead paint. In adults it is a result of exposure, usually through the skin, to the organic tetraethyl lead used as a fuel additive. Lead encephalopathy, as it is known, is a very serious and not uncommonly fatal condition which shows itself very differently from that of exposure to inorganic lead. The onset of symptoms may be delayed for as much as eight days after exposure and shows initially as irritation, insomnia, wild dreams, anxiety, along with tremor, sudden spasmodic contractions, and convulsive seizures. Recovery, if it takes place, may be accompanied by gross disability.

Lead, organic or inorganic, can be measured in the urine of the blood. Blood concentrations give a more accurate picture of the extent of the absorption. The upper limits of normal are generally considered to be 40 micrograms per 100 ml of blood, while the lower limit of excessive exposure is 150 micrograms per 100 ml blood.

Mercury (TLV 0.05 mg/m^3)

Mercury is a fairly common industrial metal. It is found in a number of very different situations including the following:

- raw materials
- laboratories
- instruments
- photography
- plating
- electrical equipment
- amalgams
- pharmaceutical

Like lead, mercury is a general body poison. While in acute exposure at high concentrations it is a respiratory and digestive irritant, followed by marked kidney damage, chronic exposure, or chronic mercurialism, is more common.

Again, like lead the onset is insidious, but unlike lead the central nervous system is a primary target. Tremor of the hands, eyelids, lips and tongue is a common early symptom, while jerky movements and incoordination develop a little later. Emotional disturbances, varying from euphoria to irrigation and insomnia, and inability to make decisions are common. Alice in Wonderland's Mad Hatter, who in his trade a that time would have rubbed mercury by hand into the head bands of felt hats, provides a not unsatisfactory description of the psychic effect of mercury.

The cerebral effects, however, are not the only effects. Digestive disturbances like those of lead can occur, while mouth irritation and excessive salivation are common. Again like lead, mercury circulates in the blood and is excreted in the urine, but unlike lead there are no critical levels which might indicate potential intoxication.

Antimony (TLV 0.5 mg/m)

Antimony is still another widely used metal. It is found in, among other instances:

- mining
- refining
- typesetting
- pigment making and using
- glass and pottery making
- pharmaceuticals
- smelting
- abrasives
- alloys
- plasticizers, catalysts
- lacquers
- matches, explosives

In direct contact it can have local effects on the skin and mucous membranes, or, in the form of dust or fume, it can act as an irritant to the eyes, throat and upper respiratory tract.

It can also have general, or systemic, effects. In acute exposure to high concentrations of fumes or ingestion of excess quantities, it will give rise to severe digestive effects, with nausea, vomiting, and bloody diarrhoea, followed, if sufficiently severe, by circulatory or respiratory failure.

In low concentrations over years of exposure it may give rise to vague symptoms of dry throat, nausea, headaches, sleeplessness, loss of appetite, and dizziness, eventually leading to liver and kidney damage.

Metal carbonyls

A metal carbonyl is a compound of a metal with carbon monoxide. They are relatively uncommon and mostly found in metallurgical operations. Most carbonyls are solid and cause little problem; some, namely iron, nickel, osmium, and ruthenium are liquid, volatile and produce respiratory irritation leading to pneumonitis. Nickel carbonyl, however, is a special case.

Nickel carbonyl (TLV 0.05 ppm)

Nickel carbonyl is formed and then decomposed during the Mond process for refining nickel. It occurs in that process as a vapour.

After exposure there may be an asymptomatic period of 12–36 hours followed by chest pain, coughing, hyperventilation (overbreathing), and cyanosis (blueness of the skin from lack of oxygen). The condition will worsen to bronchopneumonia or pneumonitis, which may lead to delirium, convulsions, and death in as little as 3 days. Recovery and convalescence for survivors is prolonged over several months.

Epidemiological studies have shown a higher incidence of cancer in the sinuses and lungs of workers in nickel refineries. Although nickel carbonyl causes cancer in rats it is not clear whether the causative agent in man is nickel carbonyl or some other substance such as nickel oxide. Suspicion rests on the carbonyl.

Agents acting on blood and bone marrow

Many chemicals, in addition to lead, have a specific action on the constituents of the blood and the bone marrow where the blood cells are formed. Some of these effects involve actual destruction of blood elements, as with lead; some, such as carbon monoxide, block the oxygen-carrying capacity of the haemoglobin in the red blood cells, while some, such as aniline, chemically alter the haemoglobin. Benzene, on the other hand, acts directly on the bone marrow. Some of the more common, and/or more significant will be examined here.

Carbon monoxide

Carbon monoxide is an odourless, colourless, tasteless, gas. Unlike carbon dioxide, which is a natural physiological product containing two oxygen atoms in its molecule, carbon monoxide is a toxic chemical which contains only one oxygen atom and is the product of incomplete combustion.

Carbon monoxide can occur in any situation where there is low level combustion. Accordingly it can be found in association with arc or gas welding, or with unventilated internal combustion engines, in foundries, in closed rooms with fuelled stoves or heaters, and so on.

It gains entrance to the body through inhalation. After inhalation it is absorbed into the blood stream where it immediately combines with the haemoglobin of the blood to form carboxyhaemoglobin (COHb). Oxygen is carried as atoms in special sites on the haemoglobin. It does not combine with the haemoglobin and, as already discussed, can be acquired or released according to the partial pressure in the breathing air, or the tension in the tissues. Carbon monoxide occupies these same sites, but does so irreversibly such that the molecule is held until the haemoglobin breaks up in the course of normal attrition. Unfortunately the haemoglobin has a greater preference for carbon monoxide by a factor of 200–300 and consequently will accept carbon monoxide in favour of oxygen, resulting in a special type of hypoxia. The effects depend on concentration of carbon monoxide in the breathing air.

At 3000–4000 ppm, which is 0.3–0.4 per cent, inhalation leads rapidly to unconsciousness with or without convulsions, coma, and death from respiratory failure. At much smaller concentrations, such as 50–100 ppm, the effects are much less dramatic and more insidious. They include, headache, which may be severe and prolonged, nausea, dizziness, weakness, and mental confusion along with hallucinations. At the lower end of the range, the effects are of course less severe and more insidious. As with the hypoxia of high altitude, the victim indeed may be unaware of his danger. There is no shortage of breath, nor cyanosis. Indeed, the skin colour is bright. Since carboxyhaemoglobin is cherry red in colour, the skin, and particularly the cheeks, are high-coloured and the lips are bright red. The unusual colouring can be a diagnostic aid.

For obvious reasons the condition is aggravated by high altitude. It is also, however, aggravated by smoking. A confirmed smoker carries anything from 2–10 per cent of COHb in his blood stream without any other exposure to carbon monoxide, as opposed to the 1 per cent carried by non-smoking town-dwellers.

The reaction to a given blood level is very variable. Levels of COHb over 60 per cent tend to be fatal while levels of 40 per cent generally lead to unconsciousness. Levels below 15 per cent rarely produce symptoms. Even if the exposure is not fatal there may be residual damage on recovery varying from gross neuropsychiatric problems to memory impairment.

Benzene (TLV 10 ppm), *and related chemicals*

The aromatic cyclic hydrocarbons include benzene with its characteristic structure, and its derivatives including dinitro-benzene, toluene, and dinitrotoluene. Benzene is a colourless and aromatic smelling liquid readily inhaled as a vapour. It can also be absorbed through the skin. It is used as a solvent, and as an intermediate in the manufacture of many other chemicals such as phenol, detergents, paint removers and so on.

Exposure to benzene has both local and systemic effects, of which the latter are much the more serious. Locally it is a primary irritant to the skin, eyes and upper respiratory tract. Systemically it affects two major body systems, namely the central nervous system and the blood forming system. It acts as a depressant to the central nervous system. In moderate concentrations of about 200 ppm it causes headache, drowsiness, dizziness, and nausea. Higher concentrations, in the region of 300 ppm, will produce euphoria and incoordination, and will lead to unconsciousness and coma.

Of greater significance, however, since it is more likely in lower concentrations, is the effect on the bone marrow. The bone marrow is the body structure responsible for the growth and development of red and white blood cells which mature there and are then passed into the blood stream as needed. Initially benzene acts as a stimulant to red cell production, but stimulus gives way to depression with progressive loss of red cells, until finally the condition of *aplasia* is reached where no more red cells are produced. Loss of red cells is associated with anaemia which becomes more profound as the red cells decrease. Consequently, if not already present from the central nervous system depression, there is headache, dizziness, nausea, abdominal discomfort, with weakness and even difficulty in breathing on exertion if the anaemia is sufficiently severe. The skin and mucous membranes are pale, and since there is an associated destruction of the platelets, which are responsible for blood clotting, there is also evidence of easy bruising, bleeding from the gums, minor haemorrhages, and so on.

Epidemiological evidence suggests that benzene also has an action on the white blood cells. These are involved in the body immune response. In benzene workers it has been shown that there is a significantly higher incidence of leukaemia, which is an abnormal, and commonly fatal, overgrowth of white cells akin to cancer.

Dinitrobenzene (TLV 0.15 ppm)

Dinitrobenzene is a derivative of benzene used in the manufacture of dyes. It can gain entry by inhalation and skin absorption. Like benzene, it also attacks the blood system, but it is different from benzene since it is a methaemoglobin former, that is, it reacts with haemoglobin to produce methaemoglobin which no longer has the capacity to carry oxygen. The onset of symptoms is often delayed, perhaps up to 4 hours after exposure, and is insidious in its development. The first indication may be cyanosis of the lips and face, noticed by other workers. This can occur when the concentration is around 15 per cent. More signficant features of weakness, dizziness, nausea, headache and drowsiness may occcur at levels up to 70 per cent, to be followed by unconsciousness and coma as the level rises higher.

Toluene (TLV 100 ppm)

Toluene is a colourless liquid used in the manufacture of benzene and as a solvent. The vapour can be inhaled and the liquid can be absorbed through the skin. Although closely related chemically to benzene it has no effect on the blood or bone marrow, but acts as a central nervous system depressant at levels of around 200 ppm or more.

Dinitrotoluene (TLV 1.5 mg/m^3)

Dinitrotoluene occurs as yellow crystals. It is used as an intermediate in chemical manufacture and in dye-making. It gains access to the body by inhalation and skin absorption. Although related chemically to toluene it is quite different in its effect, and indeed is similar to dinitrobenzene as a methaemoglobin former.

Aniline (TLV 2 ppm)

Aniline is a colourless or light yellow liquid which is used in chemical synthesis and, in particular, in the manufacture of dyes, pharmaceuticals, shoe polish, and plastics. It can be inhaled as a vapour and absorbed through the skin. Like dinitrobenzene and dinitrotoluene it is a methaemglobin former. As with other methaemoglobin formers the effects of aniline tend to be insidious in onset, with delay for up to four hours. Its effects are similar to those of the others.

Carcinogens

Over 50 substances in industrial use are recognized to be carcinogenic, or suspected on the grounds of epidemiological or experimental evidence.

Substances so recognized by the ACGIH include a group for which TLVs have been assigned, namely:

- acrylonitrile
- asbestos
- bis (chloromethyl) ether
- certain chromium derivatives
- coal tar pitch volatiles
- nickel sulphide roasting, fume, and dust
- vinyl chloride

Certain others, including 4-aminodiphenyl, benzidine, beta naphthylamine, and 4-nitrodiphenyl are recognized carcinogens for which no TLV has been established, and a large group is suspected on the basis of less substantial evidence.

Vinyl chloride (TLV 5 ppm)

As noted above vinyl chloride is a carcinogen. It has, however, certain other significant effects. It is a colourless gas, normally handled as a liquid under pressure. It is the monomer from which polyvinyl chloride (PVC) plastic is derived. PVC is non-toxic. Vinyl chloride can be inhaled, and consequently it is normally handled in a closed system.

Although carcinogenesis is the most significant hazard, vinyl chloride can also produce the condition of acroosteolysis, which is a degeneration of the finger bones. This begins with interference with the blood circulation to the fingers in the condition known as the Raynaud Phenomenon, already considered in discussion of the effects of segmental vibration. As in white finger disease the condition develops to destruction of bone.

Chapter 22
Dusts and other solid particulates

Except in a few special instances, normally covered by legislation, the presence of dust in the industrial environment has not uncommonly come to be regarded as a normal occurrence, something to be considered as a nuisance and cleared up once in a while. This view, of course, is not only erroneous but, if persisted in, may give rise to conditions that encourage the onset of a wide variety of occupational diseases. Indeed, it is not normally the dust that one can see that gives rise to the problem but more commonly the dust that one cannot. The dust that one can see can cause dermatitis and irritation, but the dust that one cannot see can be inhaled and absorbed into the blood stream, or it can accumulate in the lung. The former can give rise to a number of systemic diseases, from lead poisoning to metal fume fever, while the latter gives rise to a group of lung dust diseases collectively known as the pneumoconioses. This chapter will focus on the latter, although mention will be made of some other problems. The material for the chapter is derived in part from the sources noted in the previous chapter. In addition, valuable information is available from the National Safety Council Data Sheet 1-532-Rev.80, *Dust, Fumes, and Mists in Industry* (1980), as well as from *Threshold Limit Values and Biological Exposure Indices* (1986-1987) published by the American Conference of Government Industrial Hygienists.

In will be recalled that dust comprises solid particles generated by abrasion as in handling, crushing, grinding, and so on. In size, a particle may range from 0.1 to 25 microns in cross section. Because of its size and mass dust does not diffuse but settles under gravity.

Factors determining dissemination and retention of dusts

Dusts and other particulates are disseminated in the air and ultimately inhaled or not depending on a number of factors. Inhaled particles also may or may not be retained in the respiratory passages and the lung. The factors involved include the following:

Concentration and size

The source of dust in the atmosphere does not require to be great to create a large volume of dust, or a concentration large enough to be significant. One cubic centimetre of quartz when crushed into fine particles of 1 cubic micron will generate one trillion particles with a surface area of 6 square metres, and a volume, because of interspersed air, of nearly 600 cubic metres (NSC, 1971). Looking at these numbers another way one can say that dispersion in finely divided form of as small an amount of lead as 0.0015 ounces in a space of 10 000 cubic feet will give a concentration of lead in the atmosphere equivalent to the TLV of lead which is 0.15 mg/m^3.

Dust particles of 50 microns in diameter can be detected readily by the normal eye. Even smaller particles can be detected when reflecting light. Dust of respirable size, however, (that is, below 10 microns) cannot be seen without a microscope. To enter the inner recesses of the lung, however, the dust must be less than 5 microns in diameter.

Settling and dispersion

Whether or not dust is inhaled depends to some extent on the extent of its dispersion in air, which in turn depends on the rate of settling. Industrial dust normally comprises particles that vary in size and mass, although small particles tend to outnumber the large. Although the dust is dispersed in air it is, of course, responsive to gravity, but because of differences in density and mass of the components, the settling rates of these components can be very different. Settling rates for different sizes of particle are shown in the table below (NSC, 1971).

Table 22.1. Settling rates for different sizes of particle (after NSC, 1971)

Size of particle (microns)	Time to fall 1 foot (min)
0.25	590
0.50	187
1.00	54
2.00	14.5
5.00	2

Since the setting rates for different components of dust are different, the composition of the dust at the source may be different from the composition of the dust inhaled. Thus, for example, foundry moulding sand comprises clays and free silica. The silica, which is denser than the fine clays, tends to settle more quickly, hence the inhaled dust contains relatively more clays than the source.

The dispersion of dust is also dependent on the kinetic energy of the ejected particles. Where the mass of the particle and the applied velocity are high the particle will travel a significant linear distance before the air resistance and gravitational forces overcome the kinetic energy. This phenomenon tends to apply to the large non-respirable particles (for example, >50 microns) which give rise to skin and mucous membrane irritation, dermatitis, and eye damage, rather than to the respirable particles. Indeed, the impact of the particles on the skin and mucous membranes plays a part in any subsequent damage.

Where the mass is too low for the kinetic energy to overcome air resistance, and where the effects of gravity are similarly reduced, particles will remain in suspension for prolonged periods and be subject to inhalation.

Structure of lungs and respiratory passages

The structure of the lungs and the respiratory passages has already been considered in discussion of the effects of barometric pressure on the production of high altitude hypoxia and underwater problems. Some of the anatomical features, however, are of significance in the development of dust diseases.

For example, the entrance to the nose, and the mucous membrane within the nose, and for that matter, all the upper respiratory passages are lined with hairs known as cilia. The cilia wave with a slow whip-like motion, forward and back. The forward stroke, towards the exterior, is more powerful and consequently the concerted motion of the cilia tends to propel to the exterior particles entrapped in the hairs. The mucous membrane also secretes sticky mucus which assists in collecting particles.

Dust that penetrates the respiratory passages, may, depending on size, ultimately accumulate in the alveoli. These accumulations may lead to the production in the lung of fibrous non-functional tissue and scarring, in the condition previously referred to as pneumoconiosis. When this occurs the alveolar walls may be broken down and the alveoli opened up in the condition called emphysema, where the surface area of the alveoli is now much less than normal, and is insufficient to meet the needs of gas exchange. The lung is supported by a fine structural network known as interstitial tissue. It too can become infiltrated by dust and replaced by fibrous scar tissue.

Blood vessels, to and from the heart, also pass throughout the lung tissue, while in addition to the system of blood vessels there is also a system of lymphatic vessels carrying the fluid called lymph which assists in carrying away debris from the lungs. The lymph vessels are interrupted frequently by small nodules called lymph glands which act as a lodging for the white blood cells involved in the process of immunity and scavenging. The latter are known as phagocytes and are found outside the lymph glands as well as inside. They have independent motion and are capable of ingesting and carrying away foreign bodies with which they come in contact. The lymph system is important in dust disease since not only does it assist in protection against the development of disease, but it too may become clogged and initiate the occurrence of fibrous nodules and other agglomerations.

Fate of dust in the body

Although knowledge of the structure of the respiratory system is necessary to understand the development of dust diseases, not all dusts exert their main effect through inhalation, nor are all inhaled dusts retained. The larger particles, from about 1 to nearly 100 microns, tend to be filtered out in the airways. Those that are not expelled through the media of mucus entrapment and ciliary rejection may give rise to irritation of the upper airways, or may be swallowed and exert their effects on the digestive system.

At the other end of the scale, below 0.5 microns, inhaled particles may be too small to be retained and are promptly exhaled again. In the range between, namely from 0.5 to 5 microns, the particles may be deposited in the alveoli. From there they may be removed by phagocytes. If the phagocyte is not itself killed, in which case further phagocytes take its place, then the particle will be transported via the lymph and blood systems for eventual excretion by the kidney, or it may be trapped by a lymph node to initiate activity there.

If, however, particles are not removed by phagocytes or trapped in lymph nodes, they will, according to the extent of their solubility, be retained in the lung. If the particles are not toxic they will exert a mechanically obstructive effect in the lung as they accumulate. If, on the other hand, they are intrinsically toxic, they will evoke a more or less violent acute reaction with perhaps inflammation of the lung and the formation of nodules and agglomerations, or they may generate a more diffuse, chronic reation involving the interstitial tissue and leading to obstruction and obliteration of functional lung tissue.

Classification of dusts by effect

From the point of view of their effect on the body, dusts and other particulates can be classified in three broad categories, namely, the toxic dusts, and inert or nuisance dusts, and the proliferative dusts. Proliferative dusts generate a growing fibrosis in the lung. It is emphasized that the classification is broad and that some overlap is inevitable. Thus wood dust, commonly considered to be a nuisance, can produce irritation and allergy. Beryllium, at an acute exposure, can be very toxic in sufficient concentration, but in low level long term exposure it produces a special form of proliferative lung scarring of fibrosis. Free silica, normally considered to produce a proliferative fibrosis of the lung, may in fact be toxic to the very phagocytes entrusted with removing it. Each of the above noted forms is examined below.

Toxic dusts

A toxic dust, for example inorganic lead, or zinc, is a dust which on absorption by inhalation, ingestion, or through the skin, gives rise to adverse physiological or pathological change in body cells. For example, lead on ingestion will act on the digestive system and ultimately on all body cells and organs; zinc fume will give rise to the acute systemic condition of metal fume fever, and so on. Toxic dusts have been considered in the previous chapter and will not be examined further here except incidentally.

Nuisance (inert) dusts

Nuisance dusts have little adverse effects on the lung. Although sometimes referred to as inert dusts they are not in fact inert. All dusts cause some reaction in the cells and tissues with which they come in contact. The reaction of the lung to nuisance dusts, however, is significantly different from the reaction to the proliferative dusts. With exposure to nuisance dusts the form and structure of the lung and the lung spaces remain intact; fibrosis or scarring of lung tissue does not occur to any significant extent, and irritant reactions are potentially reversible. The exposure may be physically unpleasant with deposit of dust in the nose, mouth, ears, and eyes; there may be irritation of the respiratory passages, skin, and mucous membranes, but significant damage will tend to be mechanical and will only occur on prolonged exposure to massive concentrations. There are no specific TLVs for individual nuisance dusts.

The ACGIH recommends a threshold limit of 10 mg/m^3, or 30 million particles of particulates per cubic foot (mppcf), or, 5 mg/m^3 respirable dust for substances in these categories for which no specific threshold limits have been assigned. The limit for a normal work day does not apply to brief exposures at higher concentrations, nor does it apply to those substances which may cause physiologic impairment at lower concentrations but for which a threshold limit has not been adopted. A large variety of dusts can be found. The following are of special interest.

Carbon (smoke, soot)

Accumulation of carbon in the lung is known as anthracosis. It occurs from deposits of coal dust, smoke and so on, and is normally benign.

Excessive deposits over many years may utlimately give rise to some destruction of lung tissue, with varying degrees of emphysema.

Calcium

Exposure to calcium occurs in workers in limestone quarries, and other limestone workers, as well as in workers with cement, marble, and gypsum. Some accumulation will occur over prolonged exposure, but it is normally symptomless. Because of the low density of calcium accumulations are not visible on X-ray.

Iron

Exposure to iron dust occurs in many different conditions involving welding, grinding, steel cutting, and metal burning. The exposure is commonly found in the form of iron oxide fume. Exposure over some 6–10 years will give rise to a benign pneumoconiosis with pigmentation which is visible on X-ray although symptoms and other signs are commonly nil to negligible.

Abrasives

Abrasives are used for grinding and smoothing metals and other materials. the traditional grinding wheel was made of sandstone containing silica which could be inhaled into the lung and give rise to the dangerous form of pneumoconiosis known as silicosis. Sandstone has been largely replaced with artificial abrasives such as carborundum, emery and aluminum oxide, or some combination of abrasive materials.

No problems have been found with carborundum or emery. Aluminum dust has been studied extensively. While in some studies, particularly in Europe, aluminum has been shown to produce a pneumoconiotic type of fibrosis, with coughing, shortness of breath and weakness, these findings have not been replicated in the United States where pulmonary fibrosis has not been observed in aluminum workers. Indeed, in a misguided attempt to prevent the onset of silicosis, workers in Canada some 40–50 years ago were encouraged to inhale aluminum dust without any adverse effect.

Barium and tin oxide

Barium and tin oxides occur as powders which can be inhaled into the lung. In both cases a benign pneumoconiosis can develop after prolonged inhalation at high concentrations with some respiratory irritation but no proliferation.

Mica

Mica is a non-fibrous silicate which occurs in plate form in a variety of different chemical structures. It is used in some electrical equipment, and in rubber and wallpaper manufacturing.

When inhaled over prolonged periods it is known to give rise to pneumoconiosis, with chronic cough and dyspnoea (difficulty in breathing). The deposits are visible on X-ray and changes in lung function tests can be measured.

Kaolin

Kaolin, which is a form of china clay, is found in grinding and handling of the clay. It is normally considered to be benign in its effects although occasional problems have been observed with massive exposures over prolonged periods.

Biological dusts

Exposure may occur in industry to various biological dusts, including, but not limited to, wood, grain, flour, wool, starch, and sugar. These dusts are usually relatively harmless although each can give rise to irritation of the skin and/or the upper respiratory tract. In some susceptible persons disabling allergies can arise, while some tropical woods, fortunately relatively uncommon, are intrinsically toxic. Certain hardwoods, such as beech and oak, with a listed TLV of 1 mg/m^3, are considered to be more irritant than soft-woods.

Cotton

Cotton is distinguished from other biological dusts in its capacity to produce, in susceptible persons, a respiratory condition called byssinosis. The condition is characterized by recurrent 'asthma-like' reactions such as coughing, wheezing, and tightness of the chest, which progress ultimately to obstructive lung disease. The condition would appear to be caused by a combination of exposure to an undefined chemical in the cotton along with mechanical obstruction of the respiratory passages by cotton dust. In non-susceptible persons large accumulations of cotton dust can occur over many years with relatively little effect, although eventually mechanical obstruction will occur. The condition is aggravated by tobacco smoking.

Proliferative dusts

Proliferative dusts can be considered as those which give rise to pneuomoconiosis, or dust disease of the lung with fibrous scarring of lung substance and destruction of lung tissue. The diseases are characterized by coughing, shortness of breath, and progressive impairment of lung function. In the case of silicosis the disease may be complicated by tuberculosis, which is a serious, prolonged, and sometimes fatal bacterial infection, as well as emphysema, and progressive heart failure. The dusts that are involved include the following:

Silica

Accumulation of silica in the form of free silica, silicon dioxide, quartz, flint, and so on, gives rise to the condition of silicosis which shows itself initially in the lung tissue and lymph nodes as small fibrous particles containing silica. The nodules develop to form conglomerations which increase in size and number until a large volume of lung is involved. The condition is accompanied by progressive impairment of lung function, demonstrated by the tests such as forced expiratory volume and maximum voluntary ventilation discussed in Chapter 11. The onset tends to occur after years of exposure, and, if sufficiently developed, will continue after exposure ceases.

Silica is found in a large number of different forms, including chalcedony, which is a chemically inert, decorative material, chert, which is used in abrasives, flint, which is used in abrasives and in ceramics, jasper, used for decorative purposes, quartz, a constituent of sand and sandstone, tripoli, used in scouring powders, polishers, and fillers, cristobalite, used in high temperature casting and speciality ceramics, diatomaceous earth, used in filtration processes and as a filler, and finally, silica gel, used in dehydrating and drying. Silicates, the salts or esters of silicic acid, do not cause pneumoconiosis.

The TLV for silica as determined by the ACGIH is given as follows (Table 22.2). The respirable mass is considered to be 50 per cent of the total mass:

Table 22.2. Threshold Limit Values for Silica Compounds (ACGIH, 1986–87)

Type	Resp. mass (mg/m^3)	Total mass (mg/m^3)
amorphous	3	6
cristobalite	0.05	0.15
fused silica	0.1	0.3
quartz	0.1	0.3
tridymite	0.05	0.15

Asbestos

Asbestos occurs as a hydrated mineral silica fibre, in various physical forms known respectively as chrysotile, crocidolite, tremolite, anthophyllite, and a complex of varieties known as amosite. It is found in scattered areas throughout the world, notably in Canada, where the most common form is chrysotile.

Accumulation of asbestos in the lungs gives rise to a form of pneumoconiosis called asbestosis, which is characterized by a diffuse fibrosis of lung tissue, sometimes involving the pleura, or chest cavity lining. It has also been found to cause cancer of the lung and digestive tract, as well as a rare form of cancer called mesothelioma. This latter cancer may involve the pleura and the peritoneum which lines the abdominal cavity. Asbestosis gives rise to progressively increasing cough, difficulty in breathing, and impaired lung function. It may develop fully after seven to nine years of exposure, with death as early as 13 years after the first exposure. Once established the condition will progress even after exposure ceases.

Exposure to asbestos arises in mining and milling, manufacture and use of sprayed asbestos for thermal and electrical insulation in public and office buildings, residences, and ships; it also occurs in fire smothering blankets and safety clothing; as a filler in plastics and cement; in asbestos pipes; and formerly in asbestos tile.

The TLV for asbestos has not been established by the ACGIH at this time. A representative guideline might be considered from the regulations developed by the province of Ontario, Canada (Anon, Ontario, 1978), which provide for limits as follows:

> The time-weighted average exposure to airborne asbestos is reduced to the lowest practical level and in any case shall not exceed,
>
> (a) in the case of amosite, 0.5 fibres/cm^3 of air
> (b) in the case of crocidolite, 0.2 fibres cm^3 of air
> (c) in the case of other asbestos, 1 fibre/cm^3 air.

Talc

Talc is used in industry for clarifying liquids by filtration, for lubricating moulds and machinery, and for electrical and heat insulation.

Talc can occur in a non-fibrous form as a crystalline magnesium sulphate, or in a fibrous form akin to asbestos. It may also contain free silica. The non-fibrous form has not been shown to cause anything other than nuisance problems. The fibrous form can give rise to a form of chronic obstructive lung disease similar to asbestosis.

Chapter 23
Control of chemical agents and aerosols in the work environment

There are three very obvious reasons why there should be control of hazardous chemicals in the work environment, namely (a) to maintain the health and safety of the workers, (b) to ensure compliance with the law, and (c) to maintain or enhance productivity. As noted in Chapter 1, maintenance of health and safety has not always been a concern of management, nor even the workers themselves. Some of the former have been prepared to sacrifice the health of their workers in favour of the demands of the work, and even some of the workers have been prepared to take what should have been unacceptable risks for increased pay. Fortunately, much of that attitude has disappeared or is disappearing.

Some of this change has come about from a developing humanitarian understanding of responsibilities among both management and labour, but certainly much of that understanding has been developed because of the requirements of legislation originated through the efforts of far-sighted individuals and institutions.

Not the least of the reasons, however, although often difficult to prove in terms of cost-effectiveness, is the fact that maintenance of a healthy and safety environment not only encourages high quality productive work, but also reduces the inevitable costs of lost work time and increased compensation or insurance payments.

Control of hazardous materials in the workplace involves three elements, namely, hazard recognition and identification, hazard evaluation, and hazard control itself.

Hazard recognition and identification

Before control can be initiated it is necessary to recognize a problem or potential problem and to define its extent. Although this may be a simple process in principle, it is not necessarily so in practice. Nor is it strictly the function of management. Labour, regulatory, and social pressures are demanding that not only employers but also employees take an in-depth look at the workplace. Labour unions require assurance of health at work, and are not necessarily satisfied that management alone can undertake this task. They demand that the worker on the shop floor, whose health is ultimately at stake, also be involved, a demand which is often backed up by legislation. Legislation, indeed, calls for regular formal assessments to determine to what extent chemical and other hazards exist.

Walk-through survey

The first and most important step in that formal assessment is the walk-through survey. An outline of survey technique has been presented by Fraser (1985). This survey is not just a caasual amble through the plant to see that everyone is performing properly or wearing the appropriate respirator. Instead, it is a systematic examination of all relevant plant activities to determine, define, and record the presence and extent of current hazards, and the probability of potential hazards arising from these activities. The general process involved in the walk-through survey is shown in Figure 23.1. (Fraser, 1985).

Large corporations usually have occupational hygienists on staff trained to conduct the walk-through survey. Unfortunately, the majority of workplaces do not have that luxury. Responsibility for the survey, then, falls upon such persons as the owner, the plant manager, the production manager, the health and safety officer, or the occupational nurse. In fact, the walk-through survey can be conducted by any responsible person as long as that person understands what is involved and has the ear of top management to obtain action when required.

Preparation and preliminary analysis

Although the walk-through survey is not a sophisticated task it is one that requires preparation and preliminary analysis. Complete familiarity with all operations is essential before the walk-through can begin. Therefore, the extent of the preparation will depend on who is doing the survey and how familiar he/she is with plant operations.

Perhaps the easiest approach to analysis of plant activities is from the end-product backwards. The first step is to identify the end-product and by-products that are produced by the plant operations. The end-product can be a chemical or combination of chemicals; it can be a component of yet another product, or it can be a finished piece in its own right. By-products could include effluent gases, fumes, dusts, or other chemicals.

Identifying the end-product and by-products serves a two-fold purpose: it helps the investigator pinpoint any hazards presented by the end-product and by-products, and also defines the raw materials,

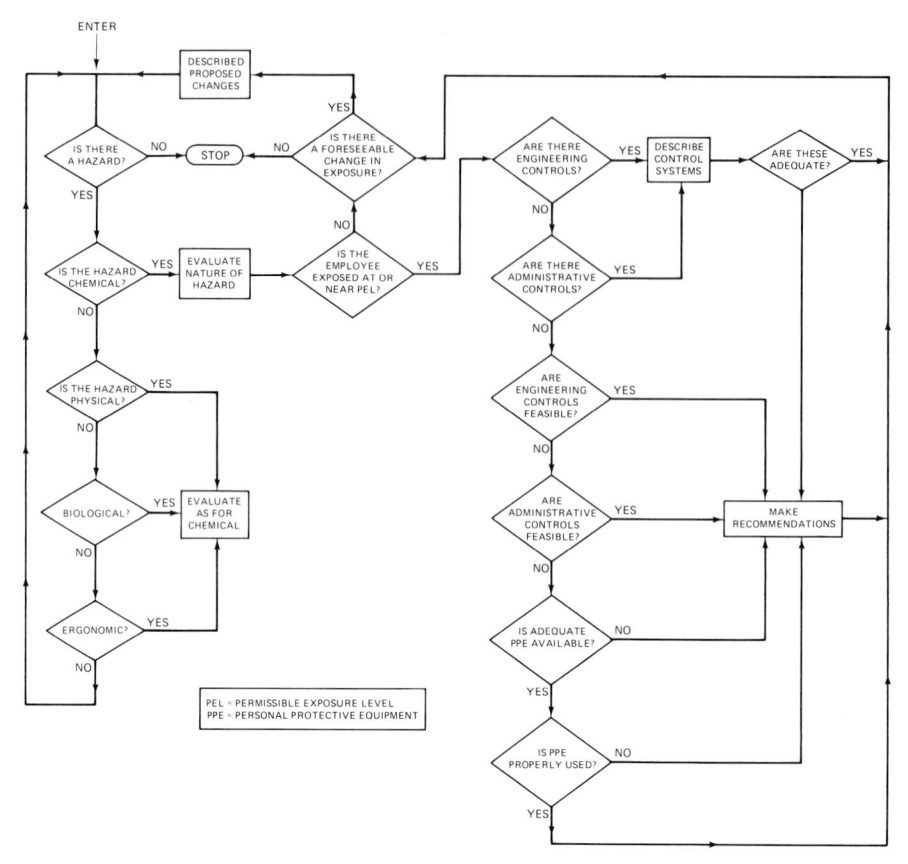

Figure 23.1. Decision flow chart for chemical hazard evaluation (after Fraser, 1985).

processes, tools, equipment, and material flows that make up the total activity.

Table 23.1 will aid the investigator in defining the plant activities, and the hazards and health problems connected with each activity.

To answer the questions posed, some consultation and reseach will probably be required. Existing knowledge and expertise within the workplace should be tapped, by contact with, for example, the plant manager, production manager, floor supervisors, safety and health officer, health and safety committee members and other health personnel. Specialized information about hazards and health effects may be obtained from manufacturers' health and safety data sheets. Specialist consultation may also be needed.

After answering the questions, a simple flow chart can be drawn that depicts the total activity from entrance and storage of raw materials,

Table 23.1. Questions in a preliminary analysis

End-products
What is produced?
How are the end-products stored?

By-products
What are the by-products?
What are the hazards associated with the by-products?
How are the by-products disposed of?

Raw materials
What are the primary materials
How are they stored?

Process
What is the process? Open? Closed?
What materials are added during the process?
What are the relevant hazards?
What equipment is used?
Is hazardous material used within the equipment?
What is the operation cycle(s)?
What ventilation system is available?
What is the expected level of exposure to agents?
What are the appropriate regulations?

Employees
How many employees are exposed?
    How often?
    How long?
    Age? Sex? Duration of employment
What health systems are in the plant?
    Adequacy?
    Effectiveness?
What are the health patterns of employees?
    Sickness and injury?
    Absenteeism?
What is compensation or insurance history of plant?

---

through the various processes, to the final product. The chart can be used to indicate the nature of the activity, for example, storage, mixing, drill pressing, degreasing, blending, reacting, painting and so on, as well as the hazards and health problems to look for at each stage.

Preliminary inspection

The actual survey comprises a preliminary inspection and a detailed examination. The preliminary inspection can be by-passed if the investigator is completely familiar with the plant. Its purpose is to visually observe the conditions and processes that the investigator has analyzed and defined.

The preliminary inspection, which is a continuous observational walk from the intake of raw materials to the output of finished product, will confirm topographically, and in terms of equipment and activity, the previously determined relationships and process flows. During this period, the investigator will also take note of obvious points of potential hazard, and areas or processes for special consideration.

On completion of this preliminary inspection, which could take from 10 minutes to half-an-hour, the investigator should have a clear overview of the conditions and processes involved and may need to reconsider the previously prepared flow chart. The investigator can now define more clearly the various activities, processes, or functional units where a particular hazard may exist, or where excessive exposure may be found. Further discussion with experts, or information from reference sources, may be necessary to clarify points of uncertainty. At the end of this process the investigator can begin the detailed examination.

Detailed examination

A formal checklist, such as that shown as Appendix A, is a valuable tool in the conduct of the detailed examination.

The questionnaire shown on this table can be adapted or designed specially for the operations of a particular workplace. A questionnaire of this nature can be used in two ways. The investigator, questionnaire in hand, can go to each process in turn and check off the answer to each question. Although comprehensive, this approach is tedious and repetitive. A simpler approach is for the investigator to become thoroughly familiar with the contents of the questionnaire, then conduct the examination from memory. Of course, the investigator should check the questionnaire afterwards to ensure that nothing has been missed.

Armed with the guidelines from the questionnaire and background information on the hazardous substances in use, including their characteristics and health effects, the investigator should now be well prepared to apply this information during the detailed inspection.

The usefulness of the inspection hinges on the investigator's ability to use his five basic senses to recognize a hazard by looking, smelling, listening, touching and tasting, where appropriate, and to confirm his suspicions by talking with persons involved, including both management and labour, and conducting such tests that might be necessary or helpful. The requirements and value of instrumentation for this purpose will be considered later. Other matters of concern to the investigator in his inspection will also be considered later. These include the nature and effectiveness of ventilation, the insistence on, and effectiveness of, personal hygiene and plant housekeeping, as well as the use, storage, care and appropriateness of personal protective equipment.

Attention should also be paid to the unexpected, such as the possibility that a container may rupture and spill its contents, that toxic material extracted to a roof vent may be pulled back into circulation again by an air intake, or that a tank of chlorinated solvent, acted on by ultraviolet light emanating from a welding process, may release phosgene gas. Alert awareness is the key to the whole survey.

The required frequency of the survey depends on the frequency of changes to the work environment, including the introduction of new processes, new materials, or new equipment. A high incidence of accidents, illnesses, and/or absenteeism would also point to the need for a repeat survey.

Records

A survey is incomplete unless there is some permanent record of the findings for implementation of recommendations and future reference. During the survey, a running record of notes, illustrated by diagrams where useful, should be made in a field notebook, or on forms prepared for the purpose. A Polaroid camera, or equivalent, or even better a video camera, is an additional asset, although not a necessary one. On completion, the field record should be transferred to a properly organized formal report, with illustrations and recommendations as required, for retention or submission to a higher authority for implementation. The extent of the implementation should be checked at a later date and the appropriate action taken.

Air sampling

A walk-through survey will indicate that a hazard does, or might, exist. It is desirable to provide a more definitive quantitative, or even qualitative, measure of the extent of that hazard. Various sampling devices have been developed for this purpose. By way of sampling and subsequent analysis it is possible to derive the time-weighted average exposures and compare them against standards. Techniques of sampling and types of sampling devices are changing rapidly. The following are representative of those in use.

Detector tubes

Detector tubes are pencil-shaped glass tubes, pointed at each end, each of which contains a measured quantity of a chemical reagent which changes colour on exposure to a specific chemical. The length of reagent which changes colour in the tube is measurable against a scale on the glass and indicates the concentration of the material in the air. In some tubes a change in colour intensity gives a qualitative reading. In use, the pointed ends of the tube are broken off and the tube inserted into a small hand pump which pumps air through the tube. Tubes for over 200 materials are available. The technique is very useful for general purposes but the accuracy is often ±25 per cent at best.

Dosimeter badges

A dosimeter badge comprises a small plastic case about the size of a pocket watch which contains a chemical on to which ambient air is passively absorbed. The badge is worn on the person near the breathing zone during the course of the working day. The chemical, which is specific to the contaminant under investigation, is then removed with its absorbed content. The latter is analyzed to determine the exposure.

Cassettes and filters

Cassettes are clear plastic filter holders, in cylindrical form, about 1 inch in diameter and varying up to 1½ inch in height. They contain one or more filters in the form of treated paper, glass fibre, or other materials such as cellulose acetate. A tube at one end allows air to be drawn into the cassette, and through or on to the filter by a pump. Filters are chosen according to the properties of the substance or substances to be collected, which could be particulate or other.

The cassette is worn on the person near the breathing zone, for a time sufficient to make a collection adequate for later analysis, commonly 4-8 hours. The pump is normally worn on the waist belt and attached to the cassette by a tube. On completion of the sampling period the cassette is dismantled to remove the collecting medium, and the material on the medium is either, in the case of chemicals, extracted for further analysis, or, in the case of dusts, processed for weighing and microscopic examination.

Sampling pumps

Low volume sampling pumps are precision devices capable of consistent action at up to 2 litres per minute for up to eight hours without variation in performance. The minute volume, and the duration of performance are adjustable in advance. Some pumps can be programmed for varying volumes and durations, and some special purpose pumps are manufactured to work at a higher volume than 2 litres per minute. The pumps have a compensation mechanism to allow them to maintain constant action against varying resistance. The pumps are battery operated, and as noted, are worn on the person during sampling. Each pump must be calibrated for consistency of flow against an independent standard on a regular basis, and re-checked on each occasion before use. Pumps operating at a much higher volume per minute are also available.

Direct reading instruments

One or two direct reading instruments are available for chemicals and for particulates. These include, for example, portable infrared gas analyzers, such as the Miran 1A (Wilks Scientific Corp.), or a device such as the RDM-101 Respirable Dust Monitor (GCA/Technology Division) which collects dust on to a sampling disc through which is passed radioactivity from a carbon-14 source. The change in intensity of the beam occasioned by the presence of the dust is calibrated in terms of particulate concentration. Other devices are available for specific gases such as carbon monoxide, or oxides of nitrogen, and so on.

Hazard evaluation

In this connection, the term hazard indicates that there exists a state which could give rise to harm to the person or damage to property; the risk involved expresses the cost of that harm or damage should it occur, and the danger is an expression of the likelihood of occurrence. In industry, of course, the possibility of harm to the person is normally, or at least should be, of greater concern than the possibility of damage to property, but that decision, in turn, is part of the risk evaluation.

In evaluating chemical hazards, then, it becomes necessary to set priorities. Some hazards clearly present greater risks and greater danger than others. Each must be evaluated on its merits to define the extent of the risk and the imminency of danger. The development of safety standards, such as TLVs, and regulations, such as those in most jurisdictions governing the use of lead, are attempts to codify controls. However, where these do not exist, and often even where they do, it is still necessary to consider the factors rendering a material more or less hazardous. These factors include the characteristics of the material or materials under consideration, (for example, physical state, toxicity, corrosiveness), the source of the exposure, (for example, storage of raw materials, processing of materials, housekeeping), the type of the exposure (for example, inhalation, ingestion), the population at risk in terms of numbers and special considerations such as pregnancy, the frequency of exposure, the nature of the effects (for example, cancer, respiratory irritation, skin lesions), the immediacy of these effects, and their severity.

Each of these, and others as they present themselves, must be evaluated on whatever tems can be made available, and the relative hazards be given a formal or informal rating before a rational approach can be made towards control. There are, in fact, three types of approach towards control that can be made, either alone or in concert. Each of these will be considered in turn.

Hazard control

Hazard control is undertaken on the basis of the findings generated by the walk-through survey, or for the management of a specifically identified hazard or hazards.

As in the management of noise and other health and safety problems in the workplace, and as noted above, there are three approaches to control of chemical exposure in the plant, any or all of which may be used to meet a given set of circumstances. These are the administrative approach, the engineering approach and the use of personal protective equipment.

Administrative

There are a variety of methods by which management can ensure that the workplace is at the highest feasible level of air quality. These will be considered below, but before examining the techniques it is necessary to emphasize the importance of developing an appropriate climate in which these techniques are applied. This requires consideration of attitude.

Attitude

In the development of any health and safety control programme it is essential to have the approval and cooperative support of management, from the highest level down. Without that support, which then disseminates through all levels of management and supervision, the control measures are ineffectual. However, not only is it necessary to have the support of management it is also necessary to develop an attitude of positive health and safety awareness among the employees so that they recognize that maintenance of healthy and safe operations is their individual and personal responsibility as well as that of the management.

Substitution of materials

It is not uncommon that materials are used in a process for no reason that is unique to that process. In such situations, which can be identified in the survey, it may be possible to replace a hazardous or potentially hazardous substance with one that is less so. For example, carbon tetrachloride as a solvent might be replaced with methyl chloroform, or even with detergent and water, if suitable. Benzene might be replaced with toluene, and free silica parting compounds in foundries with aluminium oxide. The substitute materials are not necessarily non-hazardous, but at least they are less hazardous than those they are replacing. It should be the objective, wherever feasible, to seek and use less hazardous materials. There is, unfortunately, a limitation in this approach. Sometimes the substitute materials might be more expensive than the originals.

Much useful preventive work, indeed, can be done at the purchasing level by evaluating in advance the potential hazards of materials about to be purchased and alerting the appropriate authorities to consider alternatives.

Change of process

Change of process is commonly undertaken for economic reasons or because there is a change in the intended product. While it is not often feasible to change a process for health reasons alone, unless these are very demanding, advantage can be taken during a process change to reduce or eliminate any hazard which might have been associated with the previous process. For example, dust disseminated by excessive grinding might be reduced by precision filing; the mist and droplets of spray painting can be minimized by dipping; degreasing can be done by dipping rather than washing by hand, and so on.

Housekeeping

Much unnecessary exposure to hazardous materials arises from the accumulated dust, leaks and spills, as well as containers of solvent, grease, paint, and other chemicals which are allowed to accumulate around individual workplaces, or are stacked untidily in open areas. Maintenance of cleanliness and tidiness throughout the plant can play a key role in reducing hazardous exposures. A regular clean-up schedule should be instituted, with immediate removal of spills, covering of otherwise open containers, and establishment of tidy practices.

Employee rotation

There are some situations, particularly, to quote an extreme example, in uranium mining, where, since the exposure is to natural radiation and otherwise uncontrollable, the only method of reducing exposure is by rotating the employees through the environment for a few hours at a time. A similar situation might apply in cleaning a tank or container which had contained some toxic chemical, and where protective equipment was otherwise inadequate.

Training and education

It was noted earlier that development of a positive health-oriented attitude among both management and labour is of vital importance in ensuring a healthy and safe environment. This attitude is achieved in part by training and education. The training is not necessarily formal, although it should be supplemented wherever feasible with formal educational sessions, either developed by knowledgable persons within the plant or from outside sources. Much of the education, however, can be on the job, by precept and example on the part of knowledgable supervisory personnel, aided by brochures, signs, labels, and meetings, both management sponsored and labour sponsored. Again it is emphasized that in developing such programmes it is essential to have the support of management from the top down, as well as labour organizations, and to ensure that any educational process is applied not only to labour on the shop floor but also to supervisory and higher levels of management.

Engineering control

The use of engineering methods in the control of hazardous exposure should not be considered as an alternative to administrative control, but supplementary to it. The two act together in a joint approach. Again there are several techniques that can be applied. There is no doubt, for example, that provision of proper ventilation is the most important engineering measure than can be undertaken. Ventilation will be examined in the next chapter. In addition to ventilation, however, there are certain other engineering techniques that may be applied, as follows:

Isolation and enclosure

Where it is not feasible to control a chemically hazardous process by, for example, changing the materials used, or the actual process itself, consideration should be given to isolating the process by so locating it as to minimize the number of workers who might be exposed. This is feasible where there is plenty or room for locating activities, and where little attention is required to maintain the operation. Where this cannot reasonably be undertaken it may be useful to consider instead the possibility of enclosing the process by some physical barrier which would retain the process within it, as is done for example in foaming procedures used in the manufacture and moulding of polyurethane. Some processes, indeed, such as the manufacture of tetraethyl lead, are completely enclosed from start to finish. On a lesser scale, it is common practice to enclose spray painting operations, sometimes behind a water barrier rather than a solid barrier.

Wetting down

Dispersal of dust can be controlled by wetting down the source, as for example in rock drilling where pressurized water is applied through the drill bit, or in a less sophisticated manner by simply spraying dusty floors, sometimes with an added wetting agent, prior to sweeping.

Personal protective equipment

Except under unusual conditions, the use of personal protective respiratory equipment should be considered as a last resort, to be used only when other administrative or engineering control methods are unavailable or impracticable, or where the requirement is only occasional and for short periods. Other protective equipment, such as eye, ear, and body protection, may be used supplementary to administrative and engineering control, but, unless no other approach is feasible, it should not be used as primary protection.

Respiratory protective equipment

It should be recognized, of course, that protective equipment does not eliminate the hazard, and that should the device fail, perhaps unknown to the worker, he or she will be exposed perhaps in conditions where escape is not immediately possible. In addition the equipment may be uncomfortable and its use resisted by the worker.

As noted, therefore, the use of respiratory protective devices should normally be restricted to intermittent exposures, or those where it is impracticable to achieve control by any other methods, (for example, oxygen deficiency in a confined space).

Selection of devices: A number of different respiratory devices is available for use. Before selection of one of these devices it is necessary firstly to identify the substance or evaluate the hazard or hazards presented by each of those substances, and to determine the conditions of exposure under which the device will be worn, including the duration.

Secondly, it is necessary to determine the personal characteristics and capabilities required of the user of the equipment. Not all persons, for example, can wear with equanimity a breathing apparatus which is not only cumbersome, but restricts to some extent both their breathing and their vision, nor do all persons have the physical strength to manipulate, for example, lengths of air hose in closed and cluttered surroundings. Some people, indeed, are psychologically incapable of wearing restrictive masks and proceeding into closed containers such as holding tanks.

Thirdly, it is necessary to determine what facilities might be required for support and maintenance of the selected equipment—in storage, in preparation, and in use. Facilities, for example, are required for cleaning and storing breathing apparatus, ensuring it is in proper condition for use, in maintaining supplies of consumables, and in training personnel in its use. In another context, if an air hose and pump is in use, it is essential to ensure that the supply of air to the pump is uncontaminated, and in particular is not taken from the same source that requires the apparatus in the first place.

Having defined the requirements, selection can be made from a wide variety of devices serving different purposes. The specifications for respiratory protective equipment are provided by various national standards bodies, such as are found in *Practices for Respiratory Protection Z88.2 (ANSI)*, and by the regulations of various legislative bodies.

Air purifiers

Air purifiers have the capacity to remove contaminants from the air by filtration or chemical absorption. The term embraces both filter respirators and gas masks.

Filter respirators are made of paper or plastic material commonly mounted within a mask or frame which holds the filter in place. They cover the mouth and nose, and are held in place by a band around the head. The filter can be made of varying porosity to be selective for different sizes of dusts. It is essential to recognize, however, that particulate filters are of no value against solvent vapours, gases, or for that matter in the face of inadequate oxygen.

Gas masks, in place of the filter, use activated charcoal and sometimes added chemicals to absorb and react with toxic gases, mists and vapours. Masks are available for acid gas, such as hydrogen sulphide, sulphur dioxide (sulphuric acid), chlorine (hydrochloric acid), hydrocyanic acid, and for organic vapours such as aniline, benzene, ether, gasoline, and carbon tetrachloride. They are also available for ammonia and carbon monoxide, as well as dusts, fumes, fog and smokes in combination with these. A gas mask commonly comprises a facepiece connected via a flexible tube to a canister containing the active agents. Both multipurpose and single purpose masks are available.

Air suppliers

Air suppliers do more than merely remove contaminants; they supply clean air or oxygen from some uncontaminated source. There are two types of air suppliers, namely air line respirators, and self-contained breathing apparatus.

Air line respirator: The air line respirator comprises a mask or facepiece through which the worker can breathe and see. It is connected by a long hose to a compressor which incorporates a regulator to control the flow.

The compressor can be a hand-operated pump, although more commonly it is power operated with a blower.

The length of the hose for a mask without a blower is normally limited to 75 feet; with a blower it may be 100 feet. To ensure adequate flow at high breathing rates it should deliver 50 litres per minute. A blower is normally considered to be mandatory in atmospheres where the hazard is immediate, that is where escape or failure of equipment is hazardous because of toxic, inflammable, or oxygen deficient atmospheres.

Because of the weight of the hose, and the difficulty in manoeuvering it in through cluttered areas, the hose requires a body harness to support it. This should be certified for a strain of up to 250 lb.

The air line respirator provides protection against any type of atmosphere. Although it may be awkward to manipulate because of the attached air hose, it provides less resistance to breathing than does the self-contained breathing apparatus to be considered next. It is again emphasized, however, that the air supply must be free of contaminants, including, for example, carbon monoxide generated from burning oil in the compressor.

Self-contained breathing apparatus (SCBA): SCBA equipment is essentially similar to the equipment used in underwater swimming. It comprises a mask, connecting hose, and a supply of compressed air or oxygen carried in a portable tank on the back. The system may be open circuit, that is where respiratory air is exhaled directly to the exterior, or it may be closed circuit, where the expired air is processed and re-circulated.

The open circuit system is simpler and cheaper, but is wasteful of air or oxygen which is inevitably exhaled with the unwanted carbon dioxide. The closed circuit system is more sophisticated and incorporates devices for the removal of the carbon dioxide and water vapour before the exhaled air is returned for inhalation.

Still another system avoids the cumbersome oxygen or compressed air backpack by chemically generating oxygen as required. The oxygen-generating chemicals are carried in a small canister which is connected by tubes to the facepiece through a regulator and breathing bag. The breathing bag acts as a reservoir for adjustment of breath volume. The canister contains superoxide or peroxide chemicals which evolve oxygen on contact with carbon dioxide and water vapour, absorbing the latter. The supply is good for several hours.

Other protective equipment

Eye and face protection is provided by goggles and face shields which act to protect against corrosive solids, liquids, vapours and the impact of dust and particulates. As with respiratory protection the requirements are defined by national standards such as the *Practice for Occupational and Educational Eye and Face Protection,* Z87.1 (ANSI).

Standards also exist for protective clothing, such as gloves, helmets, aprons, coveralls, or overalls, boots of leather, rubber and plastic, and so on.

Protective creams and lotions

Protective creams and lotions are intended to provide a barrier between the skin and the offending chemical or chemicals. Oily creams are used to protect against watery solutions. They should not, however, be used as a substitute for protective clothing, gloves, or face covering. They are best used as a lubricant for gloves, or where gloves are impracticable, where the material is of low toxicity, and where there is minimal skin contact.

Chapter 24
A review of the principles of ventilation for contaminant control

Provision of adequate ventilation is one of the keys to development and maintenance of a contaminant-free workplace. Design and installation of a proper ventilation system requires both skill and knowledge. It is normally a task undertaken by a professional engineer or industrial hygienist. This chapter then is not intended for the engineer or hygienist. They should consult the sources from which it is derived. It is the objective of this chapter to provide those other than engineers and hygienists, with a broad general understanding of the nature of ventilation systems and the processes used in their design. While there are several popular texts on the subject, the reference used by most professionals, and from which much of this material is derived, is *Industrial Ventilation, A Manual of Recommended Practice*, published by the American Conference of Government Industrial Hygienists, which defines in detail the theory, the necessary calculations, and the practical requirements for design and installation of a ventilation system. Useful and pertinent data are also found in the *Engineering Field Reference Manual* published by the American Industrial Hygiene Association, which also contains other valuable material useful to someone concerned with the problems of man in the workplace. Other institutions have also published relevant texts.

Basic principles

Ventilation serves two major purposes, namely, contaminant control and thermal control. Provision of adequate fresh air and removal of contaminants assists in maintaining a clean workplace, while provision of cooled or warm air, as required, assists in maintaining thermal comfort.

There are two basic types of ventilation, namely, *dilution ventilation*, or general ventilation, and *exhaust ventilation*, or local ventilation. By introducing large quantities of clean air, a dilution ventilation system attempts to reduce the concentration of sundry contaminants to an acceptable level. A local ventilation system attempts to remove the contaminants at the source and transport them to a special area for further disposal. It must be recognized that a ventilation system removes the problem from one area and deposits it in another. In particular, a ventilation system may remove the problem outside of the plant, where, apart from the fact that dumping of contaminants into the general environment is undesirable, it may contravene environmental regulations.

Dilution ventilation

A dilution ventilation system has certain advantages. The cost of equipment, installation, and maintenance is relatively low in comparison with local exhaust systems, and at the same time it can be effective in the control of small quantities of mild to moderately toxic gases, mists, and vapours. However, it also presents significant disadvantages. For example, and most importantly, it does not eliminate exposure, and in particular it does not eliminate local exposure of the worker at the source. Consequently it should not be used where the quantities of contaminant are large or where the toxicity is high. Furthermore, because of their relative density, it is not effective against heavy particulates or high density metal fumes, and because it is generalized it cannot compensate for sudden local increases in outflow of contaminant materials.

Design of a dilution system

Should it be considered that a dilution system would be adequate, it can be relatively simple to determine how much dilution is required. It is first necessary to determine the rate at which air must be moved to ensure that the contaminant is removed, assuming it is evenly and broadly distributed throughout the work space. Table 24.1, derived from the AIHA Field Manual, lists some of the air volumes required to dilute contaminants to a level equivalent to their TLV. The air volumes are calculated at standard temperature and pressure, dry, (25°C, and 760 mm Hg) and presented in terms of the volume required per pint or per pound evaporation.

For solvents not included in the above list the dilution air volumes required can be calculated by applying the following formula:

$$V = \frac{403 \times \text{sp.gr. liquid} \times 1\,000\,000 \times K}{\text{M.W. liquid} \times \text{TLV}}$$

where,
V = volume of air required
sp.gr. = specific gravity of liquid
M.W. = molecular weight of the liquid
K = constant (to be defined)

Table 24.1. Dilution air volumes for solvents

Liquid	Cu. ft. air (STP) required for dilution to TLV per pint evaporation	per pound evaporation
acetone	7 300	8 850
n-amylacetate	27 200	29 800
isoamyl alcohol	37 200	43 900
benzene	not recommended	
n-butanol (butyl alcohol)	88 000	104 000
n-butyl acetate	20 400	22 200
butyl cellosolve	61 600	65 600
carbon tetrachloride	not recommended	
chloroform	not recommended	
ethylene dichloride	not recommended	
1,-dichloroethylene	not recommended	
ethyl acetate	10 300	11 000
ethyl alcohol	6 900	8 400
ethyl ether	9 630	13 100
isopropyl alcohol	13 200	16 100
isopropyl ether	11 400	15 140
methyl acetate	25 000	26 100
methyl alcohol	49 100	60 500
methyl butyl ketone	not recommended	
methyl cellosolve	not recommended	
methyl cellosolve acetate	not recommended	
methyl ethyl ketone	22 500	26 900
methyl propyl ketone	19 000	22 400
naptha	special consideration	
nitrobenzene	not recommended	
n-propyl acetate	17 500	18 900
Stoddard solvent	30 000–35 000	20 000–50 000
1,1,2,2-tetrachloroethane	not recommended	
toluene	38 000	42 000
xylol	33 000	36 400

Having decided on the volume of air required for evaporation, it is desirable to multiply that volume by a safety factor, K. The K value varies from a minimum of three to a maximum of ten, and is determined on personal judgment on consideration of several factors, as follows:

Toxicity of material

The more toxic the material the greater is the K value. The actual value is determined from the TLV as follows:

Slightly toxic: TLV 500 ppm or greater
Moderately toxic: TLV from 100 to 500 ppm
Highly toxic: TLV 100 or less

Dilution ventilation should be used only for materials where the TLV is 100 or less, and only when the quantities are small.

Rate of evaporation of material

The higher the evaporation rate the greater is the K value.

Proximity to the operation

The closer the worker is to the operation during his normal day's work the greater is the K value.

Effectiveness of clearance

The effectiveness of the clearance is determined by the location of the source of contamination with respect to the air flow. Where the general level of air flow in the area is poor, as when the work is in a corner, or shielded, the K factor is high.

The final step in determining the air capacity required in a given period of time is to divide the calculated volume by the number of minutes during which the exposure under consideration is occurring. This calculation will provide the necessary volume in cubic feet per minute, which in turn will determine the size of fan required to move that air. If the volume is large and the room is small then additional make-up air will be required, which may also need to be heated in cold weather.

Should more than one contaminant be present in significant concentration it is necessary to consider their effects as additive, and calculate for each separately, adding the calculated volumes to give a total required volume.

Inlets and outlets

The effectiveness of dilution ventilation depends to a large extent on the positioning of inlets and outlets. In this connection it is important to ensure that only clean air is moved through a worker's breathing zone while at the same time draughts of cold or hot air should be avoided. The positioning of the inlets and outlets should be such that the air is drawn from a clean source, across, and away from, the breathing zone.

Exhaust ventilation

An exhaust ventilation system is designed to capture contaminants at the source and remove them from the workplace for disposal. It can be made capable of handling all types of aerosol using relatively small amounts of make-up air and relatively low exhaust volumes. However, although the air flow may be smaller than required in a dilution system, the cost of design and installation can be expensive. It also has the disadvantage of requiring regular cleaning and maintenance.

An exhaust system normally has four major components, namely, a hood or hoods, air ducting, an air cleaner and a fan. An air cleaner is not always required. The contaminants are collected by the hood and transported via the ducts to the air cleaner, where present, by the fan. Replacement air is required to compensate for the air lost in the exhaust.

Design factors

There are three factors, or sets of factors, used in calculations for the design of exhaust systems. These are the air capacity, the air pressure, and the air resistance. Details of the calculations belong in the design reference manuals already mentioned. Some of the concepts underlying the use of these calculations are discussed here. In addition it is necessary to ensure that air exhausted from the building is replaced with make-up air.

Air capacity

The air capacity (Q) is the volume of air that can be moved per unit time. Using the concepts discussed previously in determining the K factor, a capture velocity and an associated air capacity can be chosen to handle the processes and types of contaminants under consideration. This flow rate is given by the formula below, and in turn defines the necessary duct dimensions to achieve that flow:

$$Q = AV$$

where
 Q = volume of air flow (cfm)
 A = cross-sectional area (ft^2) through which air flows
 V = velocity of air flow (fpm)

Air pressure

The air pressure for ventilation purposes is the difference as measured in inches of water between the gauge or measured pressure and the atmospheric pressure.

Three types of pressure should be considered, namely, static pressure, velocity pressure, and total pressure. Static pressure is the pressure exerted by the air at rest. Velocity pressure is the pressure required to move air at a given velocity. It is related to air velocity under standard conditions by the equation:

$$v = 4005\sqrt{VP}$$

where
 v = velocity (fpm)
 VP = velocity pressure (ins water)

Air pressure may be negative if it is below atmospheric pressure and positive if it is above it. Static pressure can be positive or negative. On the suction side of a fan it is always negative. Velocity pressure is always positive. The total pressure is the algebraic sum of the static pressure and the velocity pressure.

Air resistance

The third factor is air resistance which is the sum of the resistances exerted by the entrance to the hood, the friction of the ducts, the resistance of the cleaner, and the resistance of the discharge from the system. Hood resistance occurs because of turbulence and acceleration as it enters the hood. Duct friction occurs from friction between the air and the walls of the ducts, and also by obstruction from change in duct direction and duct diameter. Air cleaner resistance occurs as the air moves through a filter, and discharge resistance occurs as the air moving from the system meets still air or air moving in the outside atmosphere. Resistance is measured as static pressure in inches of water.

Make-up air

Replacement air will enter the building to equal the volume exhausted whether provision is made for this or not. Where the volume of air is small, and where the building is open with various sources of incoming air from windows, doors, and leaks, no specific replacement system will be needed. In a large building, however, particularly where it is reasonably airtight, the building may become 'air-starved' if there is no appreciable exhaust ventilation. To ensure balance of the system it is necessary to provide make-up air such that the volume of the air coming in equals the volume of the air going out, otherwise hoods may cease to work, high velocity cross drafts may occur via windows and doors, the operation of natural draught stacks such as combustion flues may be impaired, and doors may be difficult to open or shut. In cold climates, or cold conditions it may also be necessary to heat the incoming air, which unfortunately adds to the expense of the system.

Hood design

The selection of a hood depends on several different factors, such as the toxicity of the material, and the nature of the material (for example, dust, fume). Although different types of hood are required for different purposes, certain general principles can be defined, as follows:

Air currents: Before installing a hood ensure that air currents and draughts that might disturb the air flow around the hood are regulated.

Enclosure: Ensure that the hood is enclosed to the extent possible, for example by a flange, which is a broad flat rim around the edge of the open hood, by baffles, or even by curtains if baffles are not feasible. These reduce the air currents and concentrate the exhaust.

Proximity: Place the hood as close to the source of the contaminant as feasible, since the velocity of air moving towards the hood opening at a point distant from the hood is inversely proportional to the square of that distance.

Breathing zones: Locate the hood to ensure that the breathing zone of the worker is maintained clear of contaminant.

Natural motion: Use natural motion of air to advantage. For example, place a canopy hood over a hot process to trap the upwardly moving contaminants, but ensure at the same time that the location of the hood, for example over a hot tank or vat, does not allow the worker to imperil his breathing zone by being forced to place his head under the canopy.

Capture velocities: Provide capture velocities which draw contaminants into the hood. The expected range of capture velocities is presented in Table 24.2 below (ACGIH, 1978):

Table 24.2. Range of capture velocities (after ACGIH, 1978)

Condition of dispersion	Example	Capture velocity (fpm)
Practically no velocity; quiet air	Evaporation from tanks; degreasing	50–100
Low velocity; moderately still air	Spray booths; intermittent container filling; low speed conveyor transfers; welding; plating; pickling	100–200
Active generation into zone of rapid air	Spray painting in shallow booths; barrel filling; conveyor loading; crushers	200–500
High initial velocity into zone of very rapid air motion	Grinding; abrasive blasting; tumbling	500–2000

In each category above, a range of capture velocity is shown. The proper choice of values depends on several factors, namely:

Lower end of range
1. Room air currents minimal or favourable to capture.
2. Contaminants of low velocity or nuisance value only.
3. Intermittent low production.
4. Large hood, large air mass in motion.

Upper end of range
1. Disturbing room air currents.
2. Contaminants of high velocity.
3. High production, heavy use.
4. Small hood, local control only.

Plain or flanged hood

A plain hood is normally the open end of a duct, commonly flared as shown in Figure 24.1 below. As already noted it should be placed as close to the source of the contaminant as possible. Surrounding the end of the duct with a flange increases the capture efficiency of the hood, as does forming the hood in the shape of a cone. Plain hoods may be mounted on the end of a flexible duct so that they can be moved to the site of operations as required.

without flange   with flange

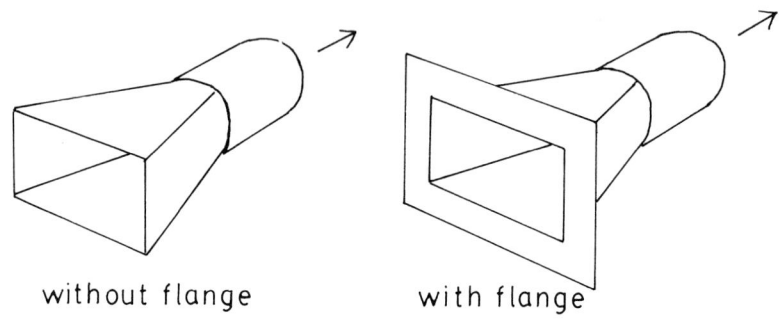

Figure 24.1. Plain hoods with and without flanges.

The air capacity required in the hood is determined by the capture velocity needed at the source. For solvents and gases that capture velocity is normally set at 100 fpm, while for welding fumes it is 200 fpm. The air capacity is calculated from the equation:

$$Q = V(10X^2 + A)$$

where
Q = air capacity (cfm)
V = capture velocity (fpm)
X = distance from source to hood opening (ft)
A = cross-sectional area of hood opening (ft^2)

Since the cross-sectional area is normally much smaller than 10X, it is usually ignored and the equation becomes:

$$Q = V(10X^2)$$

Canopy hood

A canopy hood is commonly installed permanently above a specific process. To increase efficiency it may be equipped with side curtains of such materials as fibreglass cloth, sheet metal, metal or cloth strips, and so on, depending on circumstances. It should normally overlap in area the sides of the source operation and be located as close to the operator as is feasible without interference.

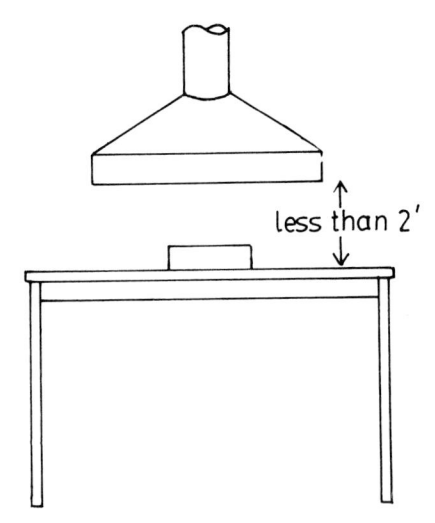

Figure 24.2. Canopy hood.

Calculation of flow requirements is complex and is presented in the reference manuals. As a rule of thumb it is desirable to provide an air capacity of 100 cfm for every square foot of hood face.

Slot hood

A slot hood comprises a plenum chamber, commonly placed at the back of a work bench so that the contaminants source lies between the plenum chamber and the worker. The front of the chamber is equipped with one or more slots, each the length of the chamber with the width of each being not more than one-fifth of the length. The plenum chamber tapers up to a duct.

Grinder hood

A grinder hood encloses most of the grinder, leaving only the operating portion open. The hood design should take advantage of the movement of

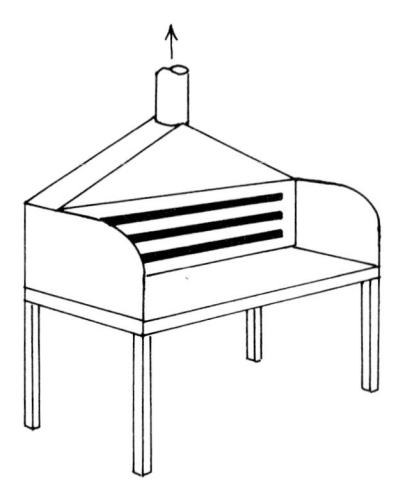

Figure 24.3. Slot hood.

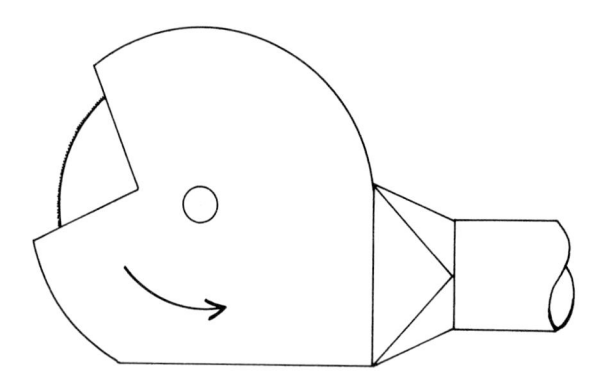

Figure 24.4. Grinder hood.

the grindstone to encourage the natural movement of air into the receiving duct.

Calculation of required duct velocities is presented in the appropriate ventilation manuals. Table 24.3 below, derived from the ACGIH Industrial Ventilation manual, indicates the range of design velocities required to prevent settling and plugging of ductwork.

Ducts

Ducts transport the air and associated aerosols to the air cleaner, or the exterior; consequently one of the major principles in duct design is to

Table 24.3. *Duct Velocities Required to Prevent Settling (ACGIH, 1978)*

Dust	Examples	Velocity (fpm)
Very fine, light	cotton lint, wood flour, litho powder	2000–2500
Dry dusts, powders	fine rubber dust, jute, lint, cotton dust, light shavings, soap dust, leather shavings	2500–3500
Average industrial	sawdust (heavy and wet), grinding dust, buffing lint, wool, jute dust, coffee beans shoe dust, granite dust, silica flour, general materials handling, brick cutting, clay dust, foundry dust, limestone dust	3500–4000
Heavy dusts	metal turnings, foundry tumbling barrels and shakeout, sandblast dust, wood blocks, hog waste, brass turnings, cast iron boring dust, lead dust	4000–4500
Heavy or moist	lead dust with small chips moist cement dust, asbestos chunks from transit pipe cutting machines, buffing lint (sticky), quicklime dust	4500 and up

reduce the resistance to air flow within the duct. Resistance occurs from friction with the wall and is aggravated by curves, bends (particularly right-angle bends), junctions, aggregations of dust, and changes in duct calibre.

Duct size

The required size of the duct depends on the air velocity necessary to ensure flow at a nominal 2000 fpm for solvents and gases, and at least 3500 fpm for dusts. Assuming the required velocity and the necessary air capacity in the light of previous considerations, the cross-sectional area is given by the relationship $A = Q/V$. Standard duct sizes vary from four to twelve inches in diameter, with one inch or half inch increments. Larger diameters come in one inch or three inch increments depending on size.

Duct materials

Ducts are commonly made in galvanized steel, twenty-gauge for the smaller and eighteen-gauge for the larger, with elbows and junctions being made of thicker steel. Ducting can also be made, although more expensively, of plastic.

Flexible ducts

In some situations, where for example it is desirable to have a movable hood, the duct must be made of flexible material, commonly wire-reinforced fabric such as cotton with a neoprene coating. Flexible metal hose is occasionally necessary for operations where the cotton material would be damaged or destroyed by the contaminants within.

General design principles

As noted, one of the main principles in good duct design is to minimize the internal resistance. Certain guidelines in achieving this objective can be stated as follows:

Duct shape: Circular ducts are preferable to rectangular, since they reduce the internal surface area. The interior surface should be smooth to avoid accumulations of dust.

Duct length: Duct length should be kept as short as possible. The frictional resistance is directly proportional to the length of the duct.

Curves, bends, and junctions: Ductwork should be straight, with no curves. When bends are necessary, by way of elbows in the duct, these should be at angles no less than 135 degrees, if possible; junctions should be made at no less than 45 degrees. Right angled bends should be avoided since they increase resistance and encourage accumulations of material. Changes in duct diameter should be as few as possible. When they are necessary they should be gradual rather than sudden.

Leakage: Leakage, at junctions or from holes in the ductwork, should be eliminated. Not only may the leak drastically alter system pressure, but it may also allow contaminants to be dispersed into the plant.

Balancing: Where more than one hood is connected to a duct the air flow through each branch should be balanced by an appropriate combination of duct diameters and duct lengths, or by installing suitable baffles, known as blast gates, into one or more ducts to control the air flow. A disadvantage of blast gates is that they will encourage deposit of dust at the gates, and they may be misused.

Maintenance: Regular, routine monitoring and maintenance, on an established schedule, should be conducted on hoods and ductwork, to ensure freedom from clogging, and also to ensure the integrity of the system.

Fans

The purpose of a fan is to move the air through the hoods, ducts, and air cleaner, where present, against the combined resistance of the system.

General considerations

Selection of a fan for anything other than a small local system is a matter for an engineer or industrial hygienist. Some general factors, however, can be considered here.

Before selecting a fan, consideration must be given to the volume of air which must be moved, as previously discussed, and also the fan static pressure, or the amount of resistance in the system that the fan will be required to work against. To ensure selection of the proper type of fan it is also necessary to determine the type of substance to be exhausted, for example, clean air, fibre, dust, vapour, and so on.

These considerations lead to further considerations of efficiency, in terms of the energy required to operate the fan, versus the losses in the fan operation. Different types of fans have different efficiencies. For example, a radial fan (see later) tends to be less efficient than a forward curved fan, although it has other advantages which will be considered in due course. Efficiency information is available from the manufacturers.

Other considerations include the availability of space for installation, and the amount of noise generated. Forward curved fans (see later) tend to have low space requirements and are relatively quiet in comparison with others, but as will be seen they have certain disadvantages.

The nature of the drive is also of significance. Fans that are directly driven off the motor such as the common household propeller fan, are generally limited to one or two speeds from the motor itself. They take up less space, and require less maintenance than belt driven fans, but the latter allow for a relatively easy change in speed as required.

All operating fans generate heat at the bearings in use. Since they are normally in constant use, the type of bearing becomes important. Sleeve bearings can operate up to temperatures of 250°F, and ball bearings up to 550°F. Above those temperatures special cooling devices are required. Recommendations for specific fans are available from the manufacturers.

Information on the applicability of different fans in explosive, inflammable, or corrosive atmospheres is also available from the manufacturers.

Types of fan

There are two basic types of fan, namely, the *axial fan* in which the air flows through the fan past the impeller, and the *centrifugal fan*, where the air is pulled by the blades into the centre of the fan and discharged perpendicularly. Different types of fans are illustrated in Figure 24.5 below (ANSI Z9.2-1979).

Axial fan

The propeller fan, as illustrated in Figure 24.5 is the most common axial fan. It is best used for dilution ventilation, and for cooling. The capacity of a propeller fan for air movement is good but it is inefficient against resistance. As a rule of thumb it can be assumed that a propeller fan will deliver 1000 cfm of air for each square foot of surface area.

Centrifugal fan

There are three basic types of centrifugal fan, as illustrated in Figure 24.5 namely the forward curved, or 'squirrel-cage', the backward curved, and the radial or straight blade.

> *Forward curved:* In the forward curved fan, the blades are tilted forward in the direction of rotation. This fan can be used in dilution ventilation, cooling, and for the exhaustion of air with no

AXIAL FLOW

Propeller fan   Tube-axial fan   Vane-axial fan

CENTRIFUGAL

Backward curved blades   Straight or radial blades   Forward curved blades

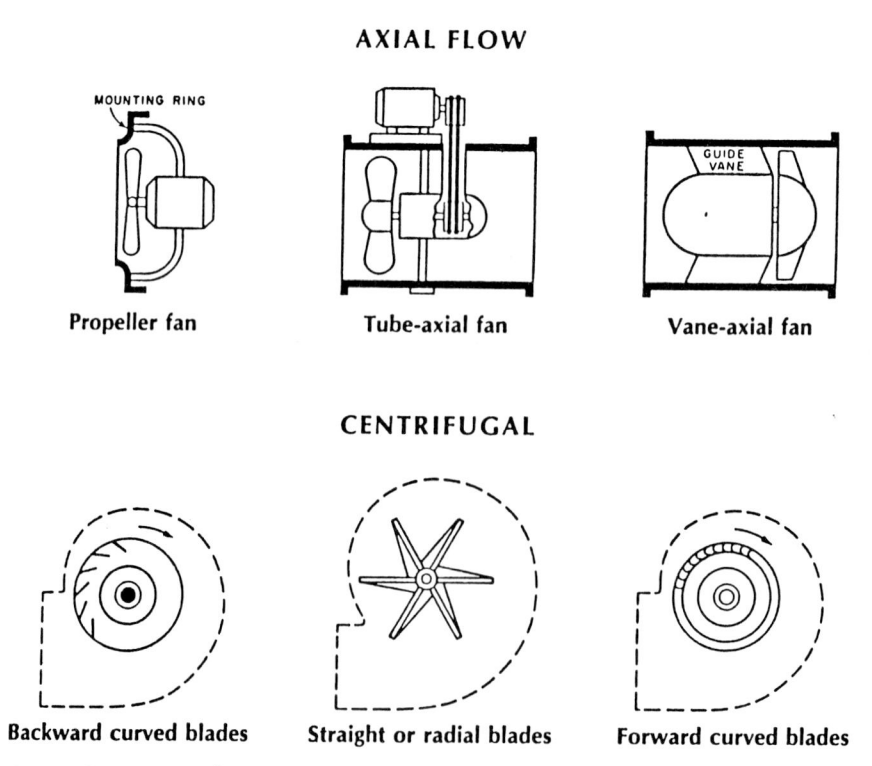

Figure 24.5. Basic fan types (after ANSI, 1979).

particulates. It should not be used for removing dusts, fumes, or vapours which can condense on the blades, since these materials will tend to accumulate on the blades and increase the resistance. The forward curved fan has low space requirements, and low tip speeds, and hence will be less noisy and have low running costs for the same efficiency than say the propeller fan.

Backward curved: The backward curved fan has blades tilted against the direction of rotation. Like the squirrel cage fan it should normally be used with clean air, although it can be used with light dust, fumes, or condensable vapour loads. Standard backward curved fans rotate at higher tip speeds and have a higher efficiency than forward curved fans but they are noisier. Backward curved fans have a built-in non-overloading, or power limiting, characteristic which makes them more useful than forward curved fans in situations of varying resistance, since forward curved fans do not have this feature.

Radial: The radial fan, or paddle-wheel fan, has blades at right angles to the fan shaft. It is commonly used in exhaust systems where there is a requirement to handle dusts and other solid or condensable particles. It has a medium tip speed and a medium noise factor, but does not tend to become clogged to the same extent as other types of fan. Consequently they have wide use in conditons where there is a heavy dust or sticky solvent load.

Air cleaning devices

Air cleaning devices remove the collected materials from the ventilation system for subsequent disposal. There are many different types available to meet different requirements, with different efficiencies, different space, installation, energy, and maintenance requirements, and, of course, different costs. The evaluation of different types for selection purposes is complex and normally needs the skill and knowledge of an engineer or industrial hygienist. The specifications, capacities, and limitations of different categories and types of air cleaner are outlined in detail in the previously noted ACGIH publication, *Industrial Ventilation*, which should be consulted for further information. This section will describe some of the principles involved.

Equipment selection

There are four basic types of dust collectors, namely, electrostatic precipitators, fabric collectors, wet collectors, and centrifugal collectors, each of which has several sub-types which will be discussed in due course. There are, however, certain general principles for selection that should be considered, as follows:

Extent of collection

The extent of collection required depends on such factors as the location of the plant, for example, whether it is in a remote area, an industrialized area, or a residential area; the nature of the contaminant, for example, whether it is a health hazard, a public nuisance, or it has salvage value; and the requirements of regulations. A general recommendation is that a collector should be selected which will contain as much contaminant as possible, meet the required regulations, while meeting reasonable cost and maintenance requirements.

Characteristics of carrier gas

The characteristics of the gas stream need consideration. Gas streams exceeding 180°F will prevent the use of standard cloth media in fabric collectors; the presence of steam or condensed water vapour will cause packing and plugging; corrosive gases, particularly in the presence of water, may attack system components.

Characteristics of contaminants

Reactive contaminants may also attack the system, while sticky materials or lints may adhere to collector surfaces, and abrasive materials may cause rapid wear.

Types of collector

Electrostatic precipitators

Electrostatic precipitators are high efficiency, high cost devices most efficient for fine particles and fumes. They are unsuitable for dusty operations since they clog easily, and they should not be used with highly toxic non-particulate materials since these can be recirculated into the atmosphere. Nor should they be used in the presence of flammables or explosives since electric arcing can occur with the danger of fire or explosion. The efficiency is improved when the air is humidified, but reduced when it is wet.

They operate by imposing an electronegative charge on particles in the airstream. These particles are then attracted to positively charged collecting plates from which the collected material is then removed by vibration, or 'rapping', at regular intervals. The collection efficiency is high but the space requirements are also relatively large and the cost is high.

Fabric collectors

Fabric collectors are high efficiency, medium cost collectors. They are used extensively in industry in a wide range of applications. They require more space than many other devices and are commonly installed outside the plant. In operation, the air is directed into a box with an inverted conical base. Within the box are tubular-shaped, or envelope-shaped, filters hung in parallel through which the air is passed. The fabric is often woven cotton, but wool, paper, glass cloth and other synthetics are also used. Silicone-treated glass cloth is becoming more common. During the passage of air, dust is built up on the filters and the dust itself becomes the primary filter. When the subsequent resistance to flow becomes unacceptable the fabric is restored by shaking, vibrating, or reverse flow which causes the dust to drop into the conical hopper from whence it is removed for disposal.

Use of the collectors is limited to conditions where there is no condensation or deposit of free moisture which will tend to cause bonding between the particles and the fabric. The rate of flow varies from one collector to another as well as with the type and concentration of dust. It is usually selected such that the pressure drop will not exceed five inches of water.

Wet collectors

There are several types of wet collectors each of which has the ability to collect not only ordinary dust, but also some inflammable, or potentially explosive dust, as well as high temperature and moisture-laden gases. The presence of water, however, may encourage corrosion, and, if the device is outside, may make it subject to freezing.

Chamber or spray towers

The chamber, or spray tower, comprises a rectangular or cylindrical chamber through which air is passed and into which water is introduced by way of spray nozzles. The dust in the air is impacted on the water droplets created by the nozzles. These droplets are subsequently separated from the air stream by centrifugal force or impingement on water eliminators.

While the air pressure drop is low, about a half to one and a half inches, the water pressure is of the order of 10 to 400 psi, although commonly found at the low pressure end. High pressure devices, using fogging rather than spraying, operate at the high end of the scale.

Packed towers

A packed tower collector is commonly a cylindrical container within which water flows over a series of ceramic or plastic weirs, or beds. The water flows at rates of 5–10 gallons per minute per cfm of air. The pressure drop for four feet of packing is in the range of one and a half to three and a half inches of water. The contaminated gases pass through the beds concurrently, counter-currently, or in cross flow. The device is used primarily for gas, vapour, and mist removal. While it will extract solid particulates, the wetted dust tends to clog the packing.

Wet centrifugal collectors

Wet centrifugal collectors are the most common of the wet collectors. They can be used with dust, and utilize centrifugal force to accelerate dust particles on to a wetted collector surface. The water may be provided by nozzles, or gravity flow. Water rates are in the order of 2–5 gallons per minute per 1000 cfm of gas, and pressure drop is some two to six inches of water.

Wet centrifugal collectors are more efficient than chamber collectors. They can be made available with different numbers of impinger sections, which, although less efficient, offer lower cost, less pressure drop, and smaller space requirements.

Several other kinds of wet collector are in use, such as wet dynamic precipitators, orifice type collectors, and venturi collectors. Information on these is available in source references.

Dry centrifugal collectors

There are two basic types of dry centrifugal collector, namely the cyclone collector, and the high efficiency centrifugal collector.

> *Cyclone collector:* The cyclone collector comprises an inverted cone into which is directed air containing medium to coarse particles. The air in the cyclone is spun, and the particles are thrown to the edge of the airstream where they settle out into the apex of the cone for disposal. A cyclone is not effective in removing fine particles, and is best used for the removal of coarse dusts, or as a pre-cleaner for more efficient dry or wet collectors, or as a separator in an airstream which is being used for the transportation of particles. Some cyclones have integral fans. Cyclones are very efficient when used for the proper purpose; the presssure drop is low (three-quarters to one and a half inches of water), and they require little maintenance. They are low in cost, but can produce excessive noise.
>
> *High efficiency centrifugal collector:* In high efficiency centrifugal collectors the airstream is accelerated to high velocities. To achieve this a number of small diameter units may be mounted in parallel, or sometimes in series. A skimmer may be added to increase the efficiency. These units tend to be more efficient than conventional

cyclones, but less so for small particles than the conventional electrostatic, fabric, or wet-type units. Other centrifugal types are also in use. Their specifications are presented in the source references.

Recirculation of air

As previously noted, the needs of air balance require that outgoing air be replaced with incoming air. Frequently this air must be heated. Where the quantities are large this becomes an expensive process. To reduce the costs it may be feasible to clean and recirculate the contaminated air. Some authorities do not permit recirculation of air that has been contaminated with hazardous materials, while others permit it under stringent control. Control requirements vary from one jurisdiction to another; the relevant regulations should be consulted.

The ACGIH (1978) makes recommendations in this regard, stating as follows:

1. An air cleaner or air cleaning system must be furnished of adequate efficiency to provide an exit concentration not more than the allowable C value (see below). This will be called the primary air cleaning system.

2. A secondary air cleaning system of equal or greater efficiency than the primary system shall be installed in series with the primary system; or a reliable monitoring device shall be installed to provide a representative sample of the recirculated air. The monitoring system shall be fail-safe with respect to failure of power supply, environmental contamination, or typical results of poor maintenance.

3. Provision shall be made for a warning signal indicating the need for attention to the secondary air cleaning system or above-limits concentrations detected by the monitor.

4. Provision shall be made for immediate by-pass of recirculated air to the outdoors or complete shut down of the contaminant-generating process if conditions occur which activate the warning device.

The permissible concentration of contaminant in recirculated air under equilibrium conditions may be calculated by the following equation:

$$C_R = \tfrac{1}{2}(TLV - C_O) \times \frac{Q_T}{Q_R} \cdot \frac{1}{K}$$

where,

C_R = concentration of contaminants in exit air from the collector before mixing, in consistent units

Q_T = total ventilation flow through affected space, cfm
Q_R = recirculated air flow, cfm
K = an 'effectiveness of mixing' factor usually varying from 3 to 10, with 3 = good mixing
TLV = threshold limit value of contaminant
C_O = concentration of contaminant in worker's breathing zone with local exhaust discharged outside

The use of recirculation precludes the recirculation of substances which can cause permanent damage or significant physiological harm from a short exposure. It is usually considered necessary also to provide general ventilation air in addition to that recirculated so that there is, in effect, continuous dilution of any recirculated contaminants.

Recirculated air procedures can be applied also to the control of 'nuisance' contaminants and odours as well as to potentially hazardous contaminants. When applied to nuisance contaminants and odours, the constant, ½, preceding the variables, is replaced by 0.9.

Appendix.
Walk-through survey:
A checklist questionnaire

Demographic and operational information

What is the Company name and address?
What division (if applicable)?
What specific operation(s) are to be examined (if applicable)?
Who is responsible for operation(s) to be investigated?
Who are significant contact persons?
Who is the safety/health officer (if any)?
Who is medical officer, nurse (if any)?
Who are worker representatives (if any)?
How many shop floor employees in plant?
    Male? Female?
How many office employees, laboratory employees, etc?
    Male? Female?
How many employees in the operation of interest?
What trade unions or other workers' organizations are involved (if any)?
What is the incidence of sickness and other absenteeism?
What is the incidence and nature of work-related injury and medical treatment?
Are there observable patterns in absenteeism, injury, and need for medical care?
What is the incidence and nature of cases reported to workers' compensation or insurance agencies?
What do management perceive as the main health and safety problems?
What do the worker representatives perceive as the main health and safety problems?

Work organization pertinent to occupational health and safety

What are the Company operational procedures and policies with respect to health and safety?

What are the laws and regulations pertinent to the operation(s) of concern?
What, if any, is the system of health protection and surveillance, with respect in particular to:
> pre-employment, pre-placement, and routine physical (medical) examination?
> special medical (biological) testing?

What is the supervisory organization? Shop floor supervisors? Lead hands, etc?
Are they familiar with, and do they implement good health and safety practices?
What is the shift organization?
Are there supervisory persons knowledgable in health and safety on all shifts?
What allowance is made for rest pauses?
Is there any system of formal job rotation? If so specify?
Is work organized to minimize isolation, boredom, fatigue, or emotional stress?
Are injuries and complaints pertinent to health and safety adequately investigated?
After investigation, are corrective procedures applied?
Is there formal and repeated training in occupational health and safety?
Are there written instructions pertaining to safe and healthy work practices?
Is there adequate feedback to workers on the results of their performance with respect to health and safety?

Workstation layout

Are work benches, equipment, and/or machines suitably designed to meet anthropometric requirements? If not, specify.
Is work done seated? Standing? Sit/stand? If standing or sit/standing can more be done seated?
Is there a requirement for working in awkward postures? Upraised arms? Standing unbalanced?
Is there a requirement for extended reach? If so, how much and how frequent?
Is there adequate clearance for handling and maintenance tasks?
Is special handling equipment (hoists, scissors, tables, drum carts, etc.) available? In use? Specify.
Are seats (if any) ergonomically suitable, easy to adjust, with adequate footrest, back rest, and arm rest if feasible?
Is workplace layout suitable for efficient motions?
Is storage for parts, components, and completed products within easy reach and adequate in available space?

Are work benches, seats, etc. adjustable?
Are arm and wrist supports provided where appropriate?

Perceptual and mental demands

Do the activities require complex mental processing?
Is there a need to keep track of multiple factors simultaneously?
Is there a demand for paced performance as in externally controlled machine operations?
Is the work highly repetitive, monotonous, and/or unvarying in its demand?
Are there critical tasks demanding high accountability where errors are not tolerated?
Is there a requirement for a heavy short-term memory load?
Is there an excessive amount of information to be handled in a short time?
Is the design of individual displays and controls ergonomically suitable?
Are controls and displays laid out systematically and consistently?
Are displays difficult to read from the worker's usual position?
Does the reading of displays require strained head and neck motions?
Are displays and controls adequately lit?
Do displays and controls meet the requirements of population stereotypes?
Is display and control labelling appropriate?
Are controls easy to distinguish one from another?
Are controls easily reached?
Is the number of controls required to perform the job excessive?

Physical demands

Have the physical dimensions of the workers been considered?
Have the actions been planned for average capabilities?
Are the required actions compatible with human limitations of strength, speed, accuracy, and reach?
Can females cope with loads designed for males?
Are the capacities of the workers fully utilized?
Is there need for:
   Frequent heavy lifting (e.g. greater that 18 kg for two hours or more per day?
   Occasional very heavy lifting (greater than 23 kg), or force exertion (225 N)?
   Short duration heavy effort?
   Moderate to heavy effort sustained throughout shift?
   Handling difficult-to-grasp or bulky items?
   Handling oversized objects requiring two-person lift and carry?
   Awkward lifts or carries, near the floor, above shoulders, or far in front of the body?

Exertion of forces in awkward positions, to the side, overhead, at extended reaches?
Machine-pacing of handling (e.g. movement on and off conveyors)
Unrelieved standing?
Frequent daily stair, ramp, or ladder climbing?
Sudden movements during manual handling?
Lifting with a twisting motion?
Static muscle loading?
High pressure on hands from thin edges (bucket handles, sheet metal edges)
Use of hand tools or equipment difficult to grasp?
High precision movements?
Are passages, work spaces adequate for dimensions of workers?

Physical environment

Does noise level appear to exceed permissible intensity?
Is a noise survey needed? If so has it been done?
Is there need for a hearing protection programme? Does one exist? If so, is it implemented?
Does noise affect communication? Comfort?
Is there a vibration problem? Whole body? Segmental?
Are anti-vibration techniques needed? In use?
Is lighting adequate? General? Local?
Does glare interfere with efficient and comfortable work?
Is the thermal environment satisfactory? If not, does it affect health? Well-being? Comfort?

Chemical environment

What is (are) the end-product(s) of the operation?
    e.g. chemicals, manufactured materials, hardware, etc.?
What are the primary materials?
    e.g. chemicals (gas, liquid, solid) part finished products, etc.?
What is (are) the process(es)?
    i.e. describe what happens, prepare flow chart
What equipment is used in processing?
    e.g. reactors, presses, grinders, furnaces, hand tools, etc?
What materials are added during the process(es)?
    e.g. special chemicals, solvents, degreasers, etc.
What are the by-products?
    e.g. gas effluents, mists, fumes, dusts, residues, etc.
Is there potential for:
- toxicity (corrosivity)?
- flammability?

- explosivity?
- inadvertent reactivity?

Does that potential (if any) occur among:
- primary materials?
- added materials?
- end-products
- by-products?

Does any potential hazard occur in connection with:
- storage of raw materials?
- adding or mixing of raw materials?
- transport of raw materials?
- processing materials?
- transport and distribution of by-products?
- storage and distribution of end-products?

What is (are) the specific hazards?
What is (are) the probability(ies) of exposure?
Is the exposure local at the site, or general through some area of the plant? Specify.
How many employees are exposed?
    Job category? Number? (name?)
What sampling, monitoring and/or measurement procedures are in effect at the site(s) or in the area?
Is there need for further sampling, monitoring and/or measurement?
What control procedures are in effect at the site(s)?
- administrative?
- engineering?
- personal protective?

Specify.
What, if any, ventilation system is used?
- general?
- local?

Is the ventilation system, if any, adequate with respect to design and use of:
- hoods?
- ducts?
- fans?
- air cleaners?
- recirculation systems?
- replacement air?

Is there need for new or additional control procedures, including ventilation? Specify.
Are any processes or operations enclosed? Specify.
Is there need for new or additional enclosure? Specify.
Are hazardous materials, by-products, etc., stored according to best practice?
Are containers (materials, waste, etc.) covered when not in use?

Is there vacuuming or other cleaning of dust?
Is the floor area clean and free from clutter?
Are separate eating facilities provided?
Are separate washing facilities provided?
Are eating and smoking prohibited on the job?
Are clean work clothes provided?
If not, are work clothes kept free from impregnation or soaking with hazardous materials?
Is it desirable and feasible to change a hazardous process for one less hazardous?
Is it desirable and feasible to introduce wet methods for dust control?
Is it desirable and feasible to improve the general level of housekeeping and hygiene? If so, how?

References

Aaonsen, A., (1964), *Shiftwork and Health.* Norwegian Monographs on Medical Science. (Oslo: Universitets forlaget).

ACGIH, (1978), *Industrial Ventilation, A Manual of Recommended Practice.* American Conference of Government Industrial Hygienists, Committee on Industrial Ventilation, Lansing, MI.

ACGIH, (1986–87), *Threshold Limit Values for Chemical Substances in the Work Environment Adopted by ACGIH.* American Conference of Government Industrial Hygienists, Cincinnati, OH.

AIHA, (1980), *Engineering Field Reference Manual.* American Industrial Hygiene Association, Akron, OH.

Akerstedt, T., (1976), Interindividual differences in adjustment to shiftwork. *Proceedings of the 6th Congress of the International Ergonomics Association,* Santa Monica, CA.

Altman, S., (1976), When heat is the hazard. *Job Safety and Health,* **4**, 5–10.

Andersen, E. J., (1957), The main results of the Danish medico-psycho-social investigations of shift workers. *Proceedings of the International Congress on Occupational Health,* Helsinki, Finland.

Anon., (1973), American National Standard Practice for Office Lighting. *Journal of the Illuminating Engineering Society,* **3**, 3–27.

Anon., (1976), Asbestos in the office air. *Job safety and Health,* **4**, 13–14.

Anon., (1980), Regulation respecting asbestos made under the Occupational Health and Safety Act, (1978). *Revised Statutes of Ontario, Chapter 321.*

Anon., (1982), Policy letter, Radiation Protection Branch, Health and Welfare, Canada, Ottawa, Canada.

ANSI, (1979), *American National Standard Fundamentals Governing the Design and Operation of Local Exhaust Systems,* American National Standards Institute, New York, ANSI Z9.2-1979.

ANSI, (1967), *Practices for Respiratory Protection,* American National Standards Institute, New York, ANSI Z88.2.

ANSI, (1967), *Practice for Occupational and Educational Eye and Face Protection,* Z87.1 American National Standards Institute, New York.

Anticaglia, J. R. and Cohen, A., (1969), Extra-auditory effects of noise as a health hazard. *Proceedings of the Conference of the American Industrial Hygiene Association,* Denver, CO.

Armonstrong, H. G., (1939), *Principles and Practice of Aviation Medicine,* (Baltimore, MD: The Williams and Wilkins Company).

Armstrong, T. J. and Chaffin, D. B., (1979), Carpal tunnel syndrome and selected personal attributes. *Journal of Occupational Medicine,* **21**, 481–486.

Armstrong, T. J., Foulke, J. A., Joseph, B. S. and Goldstein, S. A., (1982), Investigation of cumulative trauma disorders in a poultry processing plant. *American Industrial Hygiene Association Journal,* **43**, 103–115.

Arndt, R., (1981), The development of chronic trauma disorders among letter sorting machine operators. *Proceedings of the Conference of the American Industrial Hygiene Association,* Portland, OR.

ASHRAE, (1965), *Guide and Data Book for 1965–66,* American Society of Heating, Refrigerating, and Air-Conditioning Engineers, Inc., New York, NY.

ASHRAE, (1977), *Handbook and Product Directory, (Fundamentals),* American Society of Heating, Refrigerating, and Air-Conditioning Engineers, Inc., New York.

Astrand, P-O. and Rodahl, K., (1970), *Textbook of Work Physiology,* (New York, NY: McGraw Hill Book Company).

Ayoub, M. and LoPresti, P., (1971), The determination of an optimum size cylindrical handle by use of electromyography. *Ergonomics,* **14**, 509–518.

Azer, N. Z. and Hsu, S., (1977), The use of modelling human responses in the analysis of thermal comfort of indoor environments. In *Thermal Analysis — Human Comfort Indoor Environments,* National Bureau of Standards, Washington, DC.

Bailey, R. W., (1982), *Human Performance Engineering: A Guide for System Designers,* (Englewood, NJ: Prentice-Hall, Inc.)

Baleshta, M. M. and Fraser, T. M., (1986), An arm movement notation system in investigation of cumulative trauma disorder. In *Trends in Ergonomics/Human Factors* III, edited by W. Karwowski, (Amsterdam: North Holland).

Barbash, J., (1974), *Job satisfaction and attitude surveys.* Organization for Economic Cooperation and Development, OECD-MS/IR/74.31, Geneva.

Bartlett, F. C., (1953), Psychological criteria for fatigue. In *Symposium on Fatigue,* edited W. F. Floyd and A. T. Welford, (London: H. K. Lewis).

Belding, H. S. and Hatch, T. F., (1955), Index for evaluating heat stress in terms of resulting physiological strains. *Heating, Piping, and Air Conditioning,* **27**, 129–136.

Bell Canada, (1986), *Ergonomic Guidelines for Visual Display Terminals,* Info Pro, A division of Bell Canada, Toronto, Canada.

Bennett, C., Chitanglia, A. and Pringekar, A., (1977), Illumination levels and performance of practical visual tasks. *Proceedings of the 21st Annual Meeting of the Human Factors Society,* Santa Monica, CA.

Bennett, P. B. and Elliot, D. H. (editors), (1969). *The Physiology of Diving and Compressed Air Work,* (Baltimore, MD: The Williams and Wilkins Company).

Beranek, L. L., (1947), The design of speech communicaction systems. *Proceedings of the Institute of Radio Engineers,* **35**, 880–890.

Beranek, L. L., (1960), *Noise Reduction.* (New York, NY: McGraw-Hill Book Company).

Berenson, P. J. and Robertson, W. G., (1973), Temperature. In *Bioastronautics Data Book,* 2nd edition, edited by James F. Parker Jr, and Vita R. West,

BioTechnology Inc., Scientific and Technical Information Office, National Aeronautics and Space Administration, Washington, DC.

Berry, C. A., Catterson, D. A., (1967), Pre-Gemini medical predictions versus Gemini flight results. *Report SP-138, Gemini Summary Conference,* National Aeronautics and Space Administration, Washington, DC.

Berry, C. M., (1983), Occupational hygiene. In *Encyclopedia of Occupational Health and Safety,* p. 1511, edited by L. Parmeggiani, (Geneva: International Labour Office).

Billings, C. E., (1973), Barometric pressure. In *Bioastronautics Data Book,* 2nd edition, edited by J. F. Parker and Vita R. West, BioTechnology Inc., Scientific and Technical Information Office, National Aeronautics and Space Administration, Washington, DC.

Blackwell, H. R., (1963), Visual benefits of polarized light. *American Institute of Architects Journal,* November, 87–92.

Blackwell, H. R. and Blackwell, O. M., (1968), The effect of illumination quantity upon the performance of different visual tasks. *Illumination Engineering,* **63**, 142–152.

Blanchard, F., (1975), Annual Report to the International Labour Office, Geneva.

Blockley, W. V., (1964), Temperature. In *Bioastronautics Data Book,* 1st edition, editor W. V. Blockley, Scientific and Technical Information Office, National Aeronautics and Space Administration, Washington, DC.

Blockley, W. V., (1965), *Human sweat response to activity and environment in the compensable zone of thermal stress,* SA-CR-25682, Scientific and Technical Information Office, National Aeronautics and Space Administration, Washington, DC.

Bobbert, A. C., (1960), Optimal form and dimensions on certain concrete building blocks. *Ergonomics,* **3**, 141–147.

Brandt, A. D., (1947), *Industrial Health Engineering,* (New York: John Wiley and Sons).

British Standards Institute, (1975), *Guide to the evaluation of exposure of the hand arm system to vibration,* Draft document DD43, London.

Brouha, L., (1960), *Physiology in Industry,* (New York: Pergamon Press).

Brown, M. C. and Dwyer, J. M., (1984), Repetition strain injury, *Proceedings of the Conference on Repetition Strain Injury,* New South Wales, Australia.

Brown, C. D., Nolan, B. M. and Faithfull, D. K., (1984), Occupational repetition strain injury — guidelines for diagnosis and management. *Medical Journal of Australia,* **140**, 329–332.

Browning, E., (1969), *Toxicity of Industrial Metals,* (London: Butterworth).

Buckhout, R., (1964), Effect of whole body vibration on human performance. *Human Factors,* **6**, 157–163.

Burrows, A. A., (1960), Acoustic noise, an informational definition. *Human Factors,* **2**, 163–168.

Burton, A. C., (1946), Clothing and heat exchanges. *Federal Proceedings,* **5**, 344–351.

Cakir, A., Hart, D. J. and Stewart, T. F. M., (1980), *Visual Display Terminals,* (New York: John Wiley and Sons).

Casarett, L. J. and Doull, J., (editors), (1975), *Toxicology: The Basic Science of Poisons,* (New York: Macmillan).

Catterson, A. D., Hoover, G. N. and Ashe, W. F., (1962), Human psychomotor performance during prolonged vertical vibration. *Journal of Aviation Medicine*, **33**, 598-602.

Chaffin, D. B., (1974), Human strength capability and low back pain. *Journal of Occupational Medicine*, **16**, 248-254.

Chaffin, D. B., Herrin, G. D., Keyserling, W. M. and Garg, A., (1977), A method for evaluating the biomechanical stresses resulting from manual materials handling jobs. *American Industrial Hygiene Association Journal*, **38**, 662-675.

Chaffin, D. B. and Park, K. S., (1973), A longitudinal study of low back pain as associated with occupational lifting factors. *American Industrial Hygiene Association Journal*, **34**, 513-525.

Chaney, R. E. and Parks, D. L., (1964), *Tracking performance during whole-body vibration*. Report D3-3512-6, Boeing Company, Wichita, KS.

Chapanis, A. and Kincade, R. G., (1972), Design of controls. In *Engineering Guide to Equipment Design*, edited by H. P. Van Cott, and R. G. Kincade, (Washington, DC: American Institute for Research).

Christensen, E. H., (1953), Fisiologiska synpunkter pa arbetskrav och arbetsplacering. *Nord. Med*, **50**, 1380.

Clarke, N. P., Taub, H., Scherer, H. F., Temple, W. E., Vykukal, H. E. and Matter, M., (1965), *Preliminary study of dial reading performance during sustained acceleration and vibration*. Document AMRL-TR-65-110, Wright-Patterson Air Force Base, OH.

Cleary, P. J., (1974), Life events and disease. A review of the methodology and findings. *Report No. 37*. Laboratory for Clinical Stress Research, Karolinska Sjukhuset.

Coermann, R. R., Magid, E. B. and Lang, H. O., (1962), Human performance under vibrational stress. *Human Factors*, **4**, 315-324.

Conover, D. W. and Kraft, C. L., (1958), *The use of color in coding displays*. Document TR55-471, Wright-Patterson Air Force Base, OH.

Corlett, E., Madely, S. and Manenica, I., (1979), Posture targetting: a technique for recording working posture. *Ergonomics*, **22**, 357-366.

Crouch, C. L. and Buttolph, L. J., (1973), Visual relationships in office tasks. *Lighting Design and Application*, May, 23-25.

Damon, A., Stoudt, R. W. and McFarland, R. A., (1966), *The Human Body in Equipment Design*. (Cambridge, MA: Harvard University Press).

Dashevsky, S. G., (1964), Check-reading accuracy as a function of pointer adjustment, patterning, and viewing angle. *Journal of Applied Psychology*, **48**, 344-347.

Datta, S. R. and Ramanathan, N. L., (1971), Ergonomics comparison of seven modes of carrying loads on the horizontal plane. *Ergonomics*, **14**, 269-278.

Dean, R. D., Farrell, R. J. and Hitt, J. D., (1967), Effect of vibration on the operation of decimal input devices. *Proceedings of the 11th Annual Conference of the Human Factors Society*, Boston, MA.

Dennis, J. P., (1965), Some effects of vibration upon visual performance. *Journal of Applied Psychology*, **8**, 193-205.

Drillis, R. W., (1963), Folk norms and biomechanics. *Human Factors*, **5**, 427-441.

Ducharme, R. E., (1975), Problem tools for women. *Industrial Engineering*, September, 46-50.

Duncan, J. and Ferguson, D., (1974), Keyboard operating posture and symptoms in operating. *Ergonomics,* **17**, 651–662.

Durnin, J. V. G. A. and Passmore, R., (1967), *Energy, Work and Leisure.* (London: William Heinemann).

Edholm, O. G. and Murrell, K. F. H., (1973), *The Ergonomics Research Society, A History.* (London: Ergonomics Research Society).

Endo, S. and Kogi, K., (1975), Monotony effects of the work of motor men during high speed train operation. *Journal of Human Ergology,* **4**, 129–140.

Feallock, J. B., Southard, J. F., Kobayashi, M. and Howell, W. C., Absolute judgments of colors in the Federal Standards System. *Journal of Applied Psychology,* **50**, 266–272.

Ferguson, D. A. and Duncan, J., (1974), Keyboard design and operating posture. *Ergonomics,* **17**, 731–744.

Ferguson, D. A., Major, G. and Keldoulis, T., (1974), Vision at work. Visual demands of tasks. *Applied Ergonomics,* **5**, 84–93.

Fletcher, H. and Munson, W. A., (1933), Loudness, its definition, measurement, and calculation. *Journal of the Acoustical Society of America,* **5**, 82–108.

Fox, G. A., (1971), Personnel selection, vocational guidance, and job satisfaction. *Proceedings of a Seminar on Job Satisfaction,* Division of Occupational Health and Pollution Control, Department of Health, NSW, Australia.

Fraser, T. M., (1960), Aspects of the physiological response to whole body vibration. *Physiology,* **3**, 60.

Fraser, T. M., (1964), Aspects of the human response to high-speed, low-level flight. *Aerospace Medicine,* **35**, 365–368.

Fraser, T. M., (1964), Reliability and quality assurance of man in a man-machine system. *Journal of Environmental Sciences,* **7**, 18–22.

Fraser, T. M., (1966), *Human response to sustained acceleration.* Document NASA SP-103, Scientific and Technical Information Office, National Aeronautics and Space Administration, Washington, DC.

Fraser, T. M., (1974), Human stress, work, and job satisfaction. *Occupational Safety and Health Series,* **50**, International Labour Office, Geneva.

Fraser, T. M., (1978), Job satisfaction and work humanization: an expanding role for ergonomics. *Ergonomics,* **21**, 11–19.

Fraser, T. M., (1979), Human quality assurance in a man-machine-environment system. *Proceedings of the 7th International Ergonomics Association Congress,* Warsaw, Poland.

Fraser, T. M., (1980), Ergonomic principles in the design of hand tools. *Occupational Safety and Health Series,* **44**, International Labour Office, Geneva.

Fraser, T. M. (1983a), Human Stress, work, and job satisfaction: a critical approach, *Occupational Safety and Health Series,* **50**, International Labour Office, Geneva.

Fraser, T. M., (1983b), Ergonomics and the office. *Journal of the Royal Society of Health,* **103**, 196–200.

Fraser, T. M., (1984), Ergonomics and industrial hygiene: a complementary relationship. *American Industrial Hygiene Association Journal,* **45**, B5–B8.

Fraser, T. M., (1985), How to conduct a walk-through survey. *Occupational Health and Safety, Canada,* **1**, 26–31 (64).

Fraser, T. M., Hoover, G. N. and Ashe, W. F., (1961), Tracking performance during low frequency vibration. *Aerospace Medicine,* **32**, 829-833.

Fry, H. J. H., (1986), Overuse syndrome of the upper limb in musicians. *Medical Journal of Australia,* **144**, 182-185.

Galitz, W. O., (1980), *Human Factors in Office Automation,* (Atlanta, GA: Life Office Management Association).

Gell, D. F., (1961), Table of equivalents for acceleration termination: recommended for general use by the Acceleration Committee of the Aerospace Medical Panel, AGARD, *Aerospace Medicine,* **32**, 1109-1111.

Goldthorpe, J. H., (1968), *The Affluent Worker: Industrial Attitudes and Behaviour.* (Cambridge: The University Press).

Goodman, M. W. and Workman, R. D., (1965), *Manual recompression, oxygen-breathing approach to treatment of decompression sickness in divers and aviators.* Research report 5-65 BuShips Project SFO 22 06 05, Task 11513-12, United States Navy, Bureau of Medicine and Surgery, Washington, DC.

Grandjean, E., (1980), *Fitting the Task to the Man: An Ergonomic Approach,* 3rd edition (London: Taylor and Francis).

Grandjean, E. and Vigliani, E., (editors), (1980), *Ergonomic Aspects of Visual Display Terminals.* (London: Taylor and Francis).

Greenberg, L. and Chaffin, D. B., (1976), *Workers and their Tools. A Guide to the Ergonomic Design of Hand Tools and Small Presses.* (Grand Rapids, MI: Pendal Press).

Grether, W. F. and Baker, C. A., (1972), Visual presentation of information. In *Human Engineering Guide to Equipment Design,* edited by H. P. Van Cott and R. G. Kincade, (Washington, DC: American Institute for Research).

Grounds, M. D., (1964), Raynaud phenomenon in users of chain saws. *Medical Journal of Australia,* **1**, 270-272.

Guyton, A. C., (1984), *Physiology of the Human Body,* 6th edition, (Philadelphia: Saunders College Publishing).

Hall, J. F., Jr., Polte, J. W. and Kelly, R. L., (1953), *Cooling of clothed subjects in cold water.* Report WADC-TR-323, Wright Air Development Center, Wright-Patterson Air Force Base, United States Air Force, Dayton, OH.

Hamilton, H., (1966), A study of spastic anemia in the hands of stone cutters. *Bulletin of the United States Bureau of Labor Statistics No. 236, Industrial Series No. 19,* (Washington DC: Government Printing Office).

Harris Associates, (1980), *Comfort and Productivity in the Office of the '80s.* Steelcase National Study of Office Environments.

Harris, C. M., *Handbook of Noise Control,* (1957), (New York: McGraw Hill).

Harris, J. C., (1981), Toxicology of urea formaldehyde and polyurethane foam insulation. *Journal of the American Medical Association,* **245**, 243-246.

Heglin, H. J., (1973), *NAVSHIPS display illumination guide: II Human Factors,* Document NELG/TD 223, United States Navy, Naval Electronics Center, San Diego, CA.

Hempstock, T. and O'Connor, D., (1977), Assessment of human exposure to hand-transmitted vibration. *Proceedings of a symposium on human factors and industrial design in consumer products,* (Medford, MA: Tufts University).

Hertzberg, H. T. C., (1972), Engineering anthropology. In *Human Engineering Guide to Equipment Design,* edited by H. P. Van Cott and R. G. Kincade, (Washington DC: American Institute for Research).

Herzberg, F., (1960), *Work and the Nature of Man*, (New York: World Publishing).
Hildebrand, G. and Rohmer. W., (1974), 12 and 24 hour rhythms in error frequency of locomotive drivers and the influence of tiredness. *International Journal of Chronobiology*, **2**, 175–180.
HMSO, (1969), Powered hand tools, 1. Electric tools. *Advisory leaflet no. 18*, Ministry of Public Buildings and Works, (London: Her Majesty's Stationery Office).
Holland, C. L., (1967), Performance effects of long term vibration. *Human Factors*, **9**, 93–104.
Hopkinson, R., (1972), Glare from daylighting in buildings. *Applied Ergonomics*, **3**, 206–215.
Hornick, R. J., (1973), Vibration. In *Bioastronautics Data Book*, 2nd edition, edited by J. F. Parker and V. R. West, Scientific and Technical Information Office, National Aeronautics and Space Administration, Washington, DC.
Hornick, R. J. and Lefritz, N. M., (1966), A study and review of human response to prolonged random vibration. *Human Factors*, **8**, 481–492.
Hornick, R. J., Boetcher, C. A. and Simons, A. K., (1961), The effects of low frequency, high amplitude, whole body, longitudinal and transverse vibration upon human performance. *Final Report, Contract DA-11-022-509-ORD-3300*, Bostrom Research Laboratories, Milwaukee, WI.
Hultgren, G. V. and Knave, B., (1974), Discomfort glare and disturbance from light reflections in an office landscape with CRT display terminals. *Applied Ergonomics*, **5**, 194–200.
Humantech Inc, (1986), Queen Street, Cookstown, Ontario, (president: Franz Schneider), Personal communication.
Hunt, J. W., (1971), Job satisfaction and the organization. *Proceedings of a Seminar on Job Satisfaction*, Division of Occupational Health and Pollution Control, Department of Health, NSW, Australia.
Hunting, W., Grandjean, E. and Maeda, K., (1980), Constrained postures in accounting machine operators. *Applied Ergonomics*, **11**, 145–149.
Hymovitch, L. and Lindholm, M., (1966), Hand, wrist, and forearm injuries as a result of repetitive motions. *Journal of Occupational Medicine*, **8**, 573–577.
IBM Corporation, (1979), *Human Factors of Work Stations with Display Terminals*, Human Factors Center, San Jose, CA.
IES, (1972), *IES Lighting Handbook*, 2nd edition, (New York: Illuminating Engineering Society).
IES Nomenclature Committee, (1979), Proposed American national standard nomenclature and definitions for illuminating engineers. *Journal of the Illuminating Engineering Society*, **9**, 2–46.
ILO, (1962), *Recommended weight limits for lifting tasks*. CIS Information Sheet, 3, International Labour Office, Geneva.
ISO, (1974), *Guide for the evaluation of human exposure to whole-body vibration*. International Standard 2631, International Organization for Standardization, Geneva.
ISO, (1975), *Draft proposal for guide for the measurement and evaluation of human exposure to vibration transmitted to the hand*. Document ISO/TC 108/SC4, International Organization for Standardization, Geneva.
ISO, (1978), *Ergonomic principles in the design of work systems*. Draft International

Standard ISO/DIS 6385, International Organization for Standardization, Geneva.

ISO, (1984), *Acoustics—determination of occupational noise exposure and estimation of noise-induced hearing impairment.* International Standard ISO/DIS 1999, International Organization for Standardization, Geneva.

Jamal, M., (1987), 'Rotating Shifts Wreak Havoc with Families'. Quoted in *Toronto Star*, Ontario, April 24, 1987.

Jenkins, W. and Connor, M. B., (1949), Some design factors in making settings on a linear scale. *Journal of Applied Psychology*, **33**, 395–409.

Jones, M. R., (1962), Colour coding. *Human Factors*, **4**, 355–365.

Kamon, E., (1980), Ergonomic aspects of exposure to thermal and other environmental stresses. In *Environment and Health*, edited by N. Triefs, (Ann Arbor, MI: Ann Arbor Science).

Kane, J. M., (1945), Design of exhaust systems. *Heating and Ventilating*, **42**, 68.

Kerr, W. A., (1950), Accident proneness of factory departments. *Journal of Applied Psychology*, **34**, 167–170.

Key, M. M., (1977), *Occupational diseases: a guide to their recognition.* Publication 77-181, National Institute of Occupational Health and Safety, Cincinnati, OH.

Keyserling, M. M., Herrin, G. D. and Chaffin, D. B., (1978), An analysis of selected work muscle strength. *Proceedings of the 22nd Annual Meeting of the Human Factors Society*, Detroit, MI.

Kindwall, E. P., (1975), Medical aspects of commercial diving and compressed air work. In *Occupational Medicine*, edited by C. Zenz, (Chicago: Year Book Publishers Inc.)

Kivi, P., (1984), Rheumatic disorders of the upper limbs associated with repetitive occupational tasks in Finland in 1975–1979. *Scandinavian Journal of Work, Environment, and Health*, Volume **5**, (Supplement 3).

Kjellberg, A. and Wikstrom, B-O., (1985), Whole-body vibration: exposure time and acute effects—a review. *Ergonomics*, **28**, 535–544.

Kjellberg, A., Wikstrom, B-O. and Dunberg, U., (1985), Whole-body vibration: exposure time and acute effects—experimental assessment of discomfort. *Ergonomics*, **28**, 545–554.

Klemmer, E. T., (1971), Keyboard entry. *Applied Ergonomics*, **2**, 2–6.

Knauth, P. and Rutenfranz, J., (1976), Experimental shiftwork studies of permanent night, and rapidly rotating, shift systems. *International Archives of Occupational and Environmental Health*, **37**, 125–137.

Kodak, (1983), *Ergonomic Design for People at Work*, Volume 1. Edited by Human Factors Section, Health, Safety and Human Factors Laboratory, Eastman Kodak Co, Rochester, NY, (Belmont CA: Lifetime Learning Publications, a division of Wadsworth, Inc.).

Komoike, Y. and Horiguchi, S., (1971), Fatigue assessment on key punch operators, typists, and others. *Ergonomics*, **14**, 101–109.

Konz, S., (1974), Design of hand tools. *Proceedings of the 18th Annual Meeting of the Human Factors Society*, Huntsville, AL.

Konz, S., (1979), *Work Design*, (Columbus, OH: Grid Publishing Inc.).

Konz, S., Jeans, C. and Rathmore, R., (1969), Arm motion in the horizontal plane. *American Institute of Industrial Engineers Transactions*, **1**, 359–370.

Kroemer, K. H. E., (1971), Foot operation of controls. *Ergonomics*, **14**, 333–361.

Kroemer, K. H. E., (1974), Horizontal push and pull forces. *Applied Ergonomics,* **5**, 94–102.

Kroemer, K. H. E. and Robinette, J., (1969), Ergonomics in the design of office furniture: a review of European literature. *Journal of Industrial Medicine and Surgery,* **38**, 115–125.

Kryter, K. D., Ward, J. D., Miller, J. D. and Eldredge, D. H., (1966), Hazardous exposure to intermittent and steady state noise. *Journal of the Acoustical Society of America,* **39**, 451–464.

Kuorinka, I. and Koskenin, P., (1979), Occupational and rheumatic diseases and upper limb strain in manual jobs in a light mechanical industry. *Scandinavian Journal of Work, Environment and Health,* Volume **5**, (Supplement 3).

Lamb, T. W. and Tenney, S. M., (1966), Nature of vibration hyperventilation. *Journal of Applied Physiology,* **21**, 404–410.

Laubach, L. L., (1976), Comparative muscular strength of men and women: a review of the literature. *Aerospace Medicine,* **43**, 738–742.

Lees, R. E. M., Workman, D. G. and Laundry, B. R., (1987), *Influence of Shift Duration on Injuries and Morbidity: An Analysis of Experience on Eight- and Twelve-hour Shift Systems,* (Kingston, Ontario: Queens University, Department of Community Health).

Lehmann, G., (1962), *Praktische Arbeitsphysiologie,* 2 Auflage. (Stuttgart: Thieme Verlag).

Leithead, C. S. and Lind, A. R., (1964), *Heat Stress and Heat Disorders.* (London: Cassell and Company, Ltd.).

Lille, F., (1967), Le sommeil du jour d'un groupe de travailleurs de nuit. *Le Travail Humain,* **30**, 85–97.

Luckiesh, H. and Moss, F. K., (1927–1932), The new science of seeing. In *Interpreting the Science of Seeing into Lighting Practice,* Volume **1**. (Cleveland, OH: General Electric Company).

Luxon, S. G., (1984), A history of industrial hygiene. *American Industrial Hygiene Association Journal,* **45**, 731–739.

Maas, J. B., (1974), Effects of spectral difference in illumination and fatigue. *Journal of Applied Psychology,* **59**, 524–526.

McCormick, E. J. and Sanders, M. S., (1982), *Human Factors in Engineering and Design,* 5th edition, (New York: McGraw-Hill Book Company).

McDermott, F. T., (1986), Repetition strain injury: a review of current understanding. *Medical Journal of Australia,* **144**, 196–200.

Mackie, R. R. and Miller, J. C., (1978), *Effects of hours of service, regularity of schedules, and cargo loading on truck and bus drivers' fatigue.* Report DOT HS-5-01142, National Technical Information Service, United States Department of Commerce, Washington, DC.

MacPherson, R. K., (1960), *Physiologic responses to hot environments.* Document MRC-SRS-298, Medical Research Council, (London: Her Majesty's Stationery Office).

Magid, E. B., Coermann, R. R. and Ziegenruecker, G. H., (1960), Human tolerance to whole body sinusoidal vibration: short time, one-minute and three-minute studies. *Journal of Aviation Medicine,* **31**, 915–924.

Maslow, A. H., (1954), *Motivation and Personality,* (New York: Harper and Row).

Mathews, J. and Calabrese, N., (1982), *Health and safety bulletin,* ACTU-VTHC,

Document D129-22, Occupational Health and Safety Unit, Victoria, NSW, Australia.

Miwa, J., (1967), Evaluation methods for vibration effects, 1. *Industrial Health*, **5**, 103-120.

Miwa, J., (1968), Evaluation methods for vibration effects, 2. *Industrial Health*, **6**, 1-27.

Moruzzi, G. and Magoun, H. W., (1949), Brain stem reticular formation and activation of the EEG. *Electroencephalography and Clinical Neurophysiology*, **1**, 455-573.

Mozell, M. M. and White, D. C., (1958), Behavioral effects of whole-body vibration. *Journal of Aviation Medicine*, **29**, 716-724.

Muc, L., (1981), *Radiation from VDT Terminals.* Symposium on hazards of video display terminals, Centre for Occupational Health and Safety, University of Waterloo, Ontario.

Muller, P. F., Jr., Sidorsky, R. C., Slivinske, A. J., Alluisi, E. A. and Fitts, P. M., (1955), *The symbolic coding of information on cathode ray tubes and similar displays.* Document WADC TR 55-375, Wright Air Development Center, Wright-Patterson Air Force Base, United States Air Force, Dayton, OH.

Munsell Company, (1929), *Munsell Book of Colors*, (Baltimore, MD: Munsell Color Company).

Murray, W. E., Cox, C., Smith, M. J. and Stammerjohn, L. W., (1981), *Potential Hazards of Video Display Terminals*, Division of Biomedical and Behavioral Sciences, National Institute of Occupational Safety and Health, Cincinnati, OH.

Murrell, K. F. H., *Ergonomics*, (1969) (London: Chapman and Hall).

Napier, J. R., (1956), The prehensile movements of the human hand. *Journal of Bone and Joint Surgery*, **38B**, 902-913.

Nemeck, J. and Grandjean, E., (1973), Results of an ergonomic investigation of large-space offices. *Human Factors*, **15**, 111-124.

Neuloh, O., Ruke, H. and Graf, O., (1957), *Der Arbeitsunfall und seine Ursachen*, (Stuttgart: Ring Verlag).

NIOSH, (1961), *Work Practices Guide for Manual Lifting*, United States Department of Health and Human Services, Public Health Service, Centers for Disease Control, National Institute for Occupational Safety and Health, Division of Biomedical and Behavioral Science, Cincinnati, OH.

Nixon, C. W., (1962), Influence of selected vibration on speech: range of 10 cps to 50 cps. *Journal of Auditory Research*, **2**, 247-260.

NSC, (1971), *Fundamentals of Industrial Hygiene*, edited by B. Olishfiski, (Chicago: United States National Safety Council).

Oborne, D. J., (1983), Whole-body vibration and International Standard ISO 2631: a critique. *Human Factors*, **25**, 55-69.

Ott, J., (1976), *Health and Light*, (New York: Pocket Books).

Passmore, R. and Durnin, J. V. G. A., (1955), Human energy expenditure. *Physiological Reviews*, **35**, 801.

Patty, F. A., (editor), (1963), *Industrial Hygiene and Toxicology*, Volume 2, (New York: Wiley Interscience).

Pfeffer, H., (1971), Pneumatic tools. In *Encyclopedia of Occupational Health and Safety*, Volume **1**, pp. 1083-1085, edited by L. Parmeggiani, (Geneva: International Labour Office).

Phalen, G. S., (1966), The carpal tunnel syndrome. *Journal of Bone and Joint Surgery*, **48A**, 211–228.
Pheasant, S. and O'Neill, D., (1975), Performance in gripping and turning. A study in hand/handle effectiveness. *Applied Ergonomics*, **6**, 205–208.
Poulton, E., (1970), *Environment and Human Efficiency*, (Springfield, IL: C. T. Thomas).
Pradko, F., (1964), Human vibration response. *Proceedings of the 10th Annual Conference on Human Factors Research and Development*, Fort Rucker, AL.
Proctor, N. H. and Hughes, J. P., (1978), *Chemical Hazards in the Workplace*, (Philadelphia: J. B. Lippincott).
Ramazzini, B., (1713), *De Morbis Artificum*, translated by W. C. Wright, 1940, (Chicago: University of Chicago Press).
Ramsay, J. D., (1980), Occupational vibration. In *Occupational Medicine*, edited by C. Zenz, (Chicago: Year Book Publishers, Inc.).
Raven, P. B., Dodson, A. and Davis, T. O., (1979), Stresses involved in wearing PVC supplied-air suits: a review. *American Industrial Hygiene Association Journal*, **40**, 592–599.
Reynolds, R. E., White, R. M. and Hilgendorf, R. L., (1972), Detection and recognition of colored signal lights. *Human Factors*, **14**, 227–236.
Robinson, D. and Dodson, R., (1957), Threshold of hearing and equal-loudness relations for pure tones, and the loudness function. *Journal of the Acoustical Society of America*, **29**, 1284–1288.
Roebuck, J. A., Kroemer, K. H. E. and Thomson, W. C., (1975), *Engineering Anthropometry Methods*, (New York; John Wiley and Sons).
Roethlisberger, F. J. and Dickson, W. J., (1939), *Management and the Worker*, Western Electric Company, Hawthorne Works. (Cambridge, MA: Harvard University Press).
Rohmert, W., (1960), *Die Grundlagen der Beurteilung statischer Arbeit, Forschungsberichte des Landes Nordrhein-Westphalen Nr 938*, (Koln: Westdeutscher Verlag).
Rosner, J., (1979), *President's Address*, 7th Triennial Congress of the International Ergonomics Association, Warsaw, Poland.
Roth, E. M., (1967), Selection of space-cabin atmospheres. *Space Science Reviews*, **6**, 452–492.
Rothstein, T., (1918), *Report of the physical findings in eight stonecutters from the limestone region of Indiana*. Bulletin of the United States Bureau of Statistics, Industrial Accidents and Hygiene Series, No. 19. (Washington DC: Government Printing Office).
Salvendy, G. and Seymour, W., (1973), *Production and Development of Industrial Work Performance*, (New York: John Wiley and Sons).
Schmitz, M. A., Simon, A. K. and Boettcher, C. A., (1960), *The effect of low frequency, high amplitude, whole-body vertical vibration on human performance*. Report 130, Contract DA-49-007-MD-797, Bostrom Research Laboratories, Milwaukee, WI.
Schohan, B., Ranson, H. E. and Soliday, S. M., (1965), Pilot and observer performance in simulated low altitude high speed flight. *Human Factors*, **7**, 257–265.
Selye, H., (1950), *The Physiology and Pathology of Exposure to Stress*, (Montreal, Quebec: ACTA Inc., Medical Publishers).

Selye, H., (1960), The concept of stress in experimental physiology. In *Stress and Psychiatric Disorder,* edited by J. M. Tanner, (Oxford, England: Blackwell).

Selye, H., (1973), Stress and aerospace medicine. *Aerospace Medicine,* **44,** 190-193.

Selye, H., (1974), Stress without distress. In *World Health,* World Health Organization, Geneva, December.

Shaw, E. A. C., (1985), *Occupational noise exposure and noise-induced hearing loss: scientific issues, technical arguments, and practical recommendations.* Report prepared for the Special Advisory Committee on the Ontario Noise Regulation, Ministry of Health, Ontario, Canada.

Shoskes, L., (1976), Space planning—designing the office environment. *Architectural Record,* **84.**

Siple, P. A. and Passell, C. F., (1945), Measurement of dry atmospheric cooling in subfreezing temperatures. *Proceedings of the American Philosophical Society,* **89,** 177-199.

Smith, F. E., (1955), *Indices of heat stress.* MRC Memo-29, Medical Research Council, (London: Her Majesty's Stationery Office).

Smith, M. J., (1980) *An Investigation of Health Complaints and Job Stress in Video Display Operators,* National Institute for Occupational Safety and Health, Division of Biomedical and Behavioral Science, Cincinnati, OH.

Snook S. H., (1978), The design of manual tasks. *Ergonomics,* **21,** 963-985.

Soliday, S. M. and Schohan, B., (1965), Task loading of pilots in simulated low-altitude, high-speed flight. *Human Factors,* **7,** 45-53.

Spitzer, H., Hettinger, T., (1958), *Tafeln fur Kaloriennmsatz bei Korperlilcher Arbeit,* (Darmstadt: REFA publication).

Stammerjohn, L. W., Smith, M. J. and Cohen, B. G. F., (1981), Evaluation of workstation design factors in VDT operations. *Human Factors,* **23,** 401-412.

Stewart, T. F. M., Ostberg, O. and MacKay, C. J., (1974), *Computer terminal ergonomics, a review of recent human factors literature,* (Loughborough, England: University of Loughborough).

Strydom, N. B., (1971), Age as a causal factor in heat stroke. *Journal (S. Africa) of Industrial Mining and Metallurgy,* **72,** 112-118.

Strydom, N. B., (1975), Physical work and heat stress. In *Occupational Medicine,* edited by C. Zenz, (Chicago: Year Book Publishers Inc.).

Strydom, N. B. and Wyndham, C. H., (1963), Natural state of heat acclimatization of different ethnic groups. *Federal Proceedings,* **22,** 801.

Tamburi, G., (1983), Social security. In *Encyclopedia of Occupational Health and Safety,* edited by L. Parmeggiani, p. 2073, (Geneva: International Labour Office).

Tasto, D. L., Colligan, M. J. and Skjli, E. W., (1978), *Health consequences of shift work.* NIOSH Technical Report, SRI Project URU-4426, Contract No. 210-75-0072, National Institute for Occupational Safety and Health, United States Department of Health, Education and Welfare, Public Health Service, Center for Disease Control, Behavioral and Motivational Factors Branch, Cincinnati.

Taub, H. A., (1964), Dial reading performance as a function of frequency of vibration and head restraint system. *Report AMRL-TR-66-57,* Wright-Patterson Air Force Base, OH.

Taylor, W., Pearson, J., Kell, R. L. and Keighly, L., (1971), Vibration syndrome

in Forestry Commission chain saw operators. *British Journal of Industrial Medicine,* **28**, 83–89.
Taylor, W. and Pelmear, P. L., (1975), *Vibration White Finger in Industry*, (London: Academic Press).
Taylor, W., Pelmear, P. L. and Pearson, J., (1974), Raynaud's phenomenon in Forestry Commission chain saw operators. In *The Vibration Syndrome*, edited by W. Taylor, (London: Academic Press).
Thiis-Evansen, E., (1958), Shiftwork and health. *Industrial Medicine,* **27**, 493–497.
Tichauer, E. R., (1978), *The Biomechanical Basis of Ergonomics. Anatomy Applied to the Design of Work Situations.* (New York: John Wiley and Sons).
Trumball, R., (1956), *Environmental modification for human performance.* Report ONR-ACR-105-11, United States Navy, Office of Naval Research, Washington, DC.
Ulich, E., (1964), *Schicht und Nachtarbeit im Betrieb.* (Koln: RKW publication, Westdeutscher Verlag).
U.S. General Accounting Office, (1980), *Indoor air pollution: an emergency health problem.* Document CED-88-111, Washington, DC.
U.S. Navy, 1970, *Diving manual.* Document NAVSHIPS 0994-001-9010, United States Navy, Government Printing Office, Washington, DC.
Walker, C. A. and Guest, R. H., (1952). *The Man on the Assembly Line,* (Cambridge, MA: Harvard University Press).
Webster, J. C., (1969), SIL—past, present, and future. *Sound and Vibration,* **3**, 22–26.
W.H.O., 1979, Health aspects related to indoor air quality. *EURO Reports and Studies,* **21**, (Copenhagen: World Health Organization).
W.H.O., (1986), *Early Detection of Occupational Diseases,* (Geneva: World Health Organization).
Wikstrom, K., (1969), Allergic contact dermatitis caused by paper. *Acta Dermato-Venereologica,* **49**, 547–551.
Wolff, H. S., (1970), *Biomedical Engineering.* (New York: World University Library, McGraw-Hill Book Company).
Woodson, W. E. and Conover, D. W., (1964), *Human Engineering Guide for Equipment Designers,* (Berkely, CA: University of California Press).
Wyndham, C. H. and Strydom, N. B., (1967), An examination of certain factors affecting the heat tolerance of mine workers. *Journal (S. Africa) of Mining and Metallurgy,* **68**, 79.
Zeff, C., (1965), Comparison of conventional and digital time displays. *Ergonomics,* **8**, 339–345.
Zenz, C., (1975), *Occupational Medicine. Principles and Practical Applications.* (Chicago: Year Book Publishers Inc.).
Ziegenruecker, G. H. and Magid, E. B., (1959), Short time tolerance to sinusoidal vibrations. *Report WADC-TR-59-391,* Wright-Patterson Air Force Base, OH.

Index

abrasives, dust from 380
absenteeism 56, 58, 185
acetyl choline 107
Action Limit for lifting 87–8
administrative factors
  and hazard control 392
  and RSI 99–100
air
  cleaning devices 413–4
  collectors 414–7
  ducts 408–10
  purifiers 396
  quality of office 198
  recirculation 417–18
  replacement 403–4
  sampling methods 389–91
  suppliers 396–7
alcohols as hydrocarbons 339
aldehydes as hydrocarbons 339
alkenes as hydrocarbons 340–1
allergies to cotton 381
Alveolar Equation 308
ammonia as an irritant 358
anatomy, musculo-skeletal 80–1
aniline, toxicity of 373
anthropometry 64–5
antimony 370
apneustic centre and
  respiration 305
aerosols, types of 344–5
arm reach 76
asbestos 382–3
asbestosis 383
asceptic necrosis of bone 332

atmosphere 300
  see also barometric pressure
audiometry 243
axes 168

barium dust 380
barometric pressure
  evolved gas effects 324–9
  partial effects 312–21
  total effects 321–7
  trapped air squeeze 324–9
basal metabolic rate
  (BMR) 52–4, 257
'bends' *see* dysbarism
benzene 372–3
  ring 341–3
beryllium, toxic effects of 366
biological agents in environment
  20, 381
biomechanics of lifting 83–4
blood system 308–10
  and core temperature 255
  toxic materials in 371–3
Bohr effect 310–11
bone marrow and toxic materials
  371, 372–3
boredom at work 48–50
byssinosis 381

cadmium oxide, toxicity of 366
calcium dust 380
carbohydrates 37
carbon 337–8
  see also hydrocarbons

carcinogens 374
carpal tunnel syndrome 92, 93
cervico-brachial syndrome 93
chain saws 176-7
chair design 189-9
chemical industry, hazards of 372-3
chemicals
  agent 20
  control of 384-5, 391
  corrosive 298-9
  toxic
    accumulation 349-51
    blood and 371-3
    as carcinogens 374
    and digestive system 367
    elimination of 351
    exposure limits 353
    and respiration 358
chemoreceptors and respiration 305-6
chisels 169-70
chlorine, toxic effects of 359
circadian rhythms 59, 112
  extended working hours and 114-15
  night work 113
  shift work 115
  sleep and 113-14
clothing against cold 277-9
cold
  performance in 277
  protection from 277-9
colour
  blindness 286-7
  contrast and 288-9
  Helmholtz theory of 285-6
  perception of 285-7
communication and noise interference 234-6
compensation
  schemes 11, 12
  suits 10
controls
  display ratio 150-1
  function of 146-9
  identification of 148-9
  layout 152
  requirements for 149-51

stereotypes 151
conveyors 77-8
cotton dust 381
creams, protective 398
creativity, human 26-7

data processing 186-7
deafness 237-41
decompression tables 329-30
design 26-7
  materials 162-4
  percentile values in 67
  tool 153
    grips 154-8
    handles 158-63
desks, design of 189
detoxification 350-1
Devonshire colic 8
dinitrobenzene 373
displays
  analogue 138-9
  categories of 135-8
  criteria for 134-5
  design of 139-43
  electronic 144-5
  organization of 141-2
  pseudoquantitative 137-8
  qualitative 138-9
  representational 139
  selecting 145-6
  types of 135-6
divers and barometric pressure 322
diving, underwater
  ascent 331
  effects of cold 328-9
  recompression 332-3
  saturation 330-1
ducts 408-10
dusts
  concentration of 376
  dispersion of 377-8
  effects of 378-83
  inhaled 375
  nuisance 379-80
  proliferative 381-3
  respiratory system and 377
  size 376
  toxic 379
dysbarism 326-32

ear, human 224-5
  deafness 237-41
  pressure and 322-3
  range of hearing 226
ecosphere 17-18
education for health and safety 101
effluents in the workplace 385
electric power tools 174-5
electrostatic collectors of
    contaminants 414
enclosure of hazards 394
endocrine system 255
  neuro- 105-6
energy
  cellular 38-9
  consumption 50
  cost of work 50-4
  expenditure 51-4
  measurements of 50
  metabolism 36-42
  sources of 36-40
environment
  cultural 21
  factors in RSI 99
  human 28
  office 195, 199
  operational 18
  person-machine- 17-23
  physical 19-20
  psychosocial 20-1
  workers 60-1
enzymes 38-41
epicondylitis 93
epidemiological approach to manual
    work 81-2
ergonomics
  background to 5-8
  occupational hygiene and 13-14
  office 184-5
  RSI and 98-9
ergotropic adjustment 50

fabric collectors for
    contaminants 415
fans 411-13
fatigue
  physical 46
  skill 46-8
files 169

fluorides, toxic effects of 366-7
formaldehyde 359
fructose 37

ganglions 106-7
gases 358-63
  asphyxiant 363
  evolved gas effects 324-9
  exchange in respiration 306-7
  masks 396
  trapped 322
General Adaptation
    Syndrome 109-11
glands
  adrenal 109
  hypophysis 108-9
  hypothalamus 107, 254
  lymph 377-8
glare 295-6
grasp of hand tools 154-5, 172
  handedness and 155-6
grip, hand tool 154-5, 168

H-point in seat design 65
haemoglobin 309-11
  see also blood system
Haldane Tables for decompression
    329-30
hammers 165-8
harmonic motion and
    vibration 203
Hawthorne Study of
    motivation 60
hands
  -edness 76-7
  strength of 156
  tools and 156-7
hazards in workplace
  chemical 384-5, 391-8
  control of
    administrative 392-4
    engineering 394-5
  evaluation of 391-2
  others 387-9
health surveillance 59, 100-1
hearing
  conservation 241-3
  Damage Risk Criteria 239-40
  deafness 237-41

evaluation of 243-4
loss 237
measuring 226-8, 243-4
range of 226
status 243-4
threshold shifts 238-9
heat
  adaptation to 267-8
  balance 259-60
  body 252-3, 257-8
  controlling 254-7, 270-4
  exhaustion 268
  loss 275-7
  metabolic 253-4
  protective clothing against 274
  pulses 269-70
  storage 268-9
  stress 261, 270-4
  stroke 269
heavy metals, toxic effects of 367-70
homeostasis 105, 127-8
hoods, ventilation 403-8
hormones 108, 112, 254
human body
  dimensions of 64-5
  factors engineering and 5
  heat and 252-4
  vibration and 205-8, 209-16
    prevention of 218
    segmental 216-22
hydrocarbons 338-43, 360
hydrogen sulphide 360
hygiene
  industrial 6
  occupational 3-8, 13-14, 385
hyperoxia 315
  occurrence of 318-9
hypoxia 312-17, 371

'induced colour' phenomena 288-9
illumination
  office 289-94
  VDTs and 195-8
industrialization
  effects of 120-1
  and job satisfaction 125-6
  RSI and 94-6
  toxicology and 344, 345-6

insulation from cold 278-9
International Labour Organization (ILO) 12-13
ionization and VDTs 180-2
isocyanates 362
isolation as hazard control 394

job
  satisfaction 12, 121-4, 126-9
  dissatisfaction 125-6, 128

kaolin dust 381
ketones 339

labour unions and workers' safety 385
lead
  ingestion of 367
  toxic effects 368-9
learning and systems viewpoint 25-6
Lens Makers' Equation 282
lighting, industrial 290-5
limbic system 107-8
limitations, human 22
  behavioural 27
  environmental 28
  occupational 28
  structural 27
load carrying 74-5
luminance ratio 297
lungs 302-5, 309-10, 323-4, 376-9
lymph system 377-8

machines in systems 16-31
management 23-5, 60, 251, 385-6
manual lifting
  age and 82
  biomechanical approach to 83-4
  gender and 81-2
  guidelines on 85-90
Maximum Permissible Limit for lifting 87-88
mechanical
  impedance 205-8
  limitations 22-3
mercury 369
metabolism 52-4, 253-7

metal
  carbonyls   270
  fume fever and   365-7
  heavy   267-70
methane, toxic effects of   363
mica dust   381
muscles, human   40-2

National Institute of Occupational
    Safety and Health
    (NIOSH)   13, 115-16
nervous system   49, 105-7
  autonomic   107-8
nickel carbonyl   370
nitrogen
  narcosis   263, 319-21
  oxides of   361
noise   224-6, 242-6, 255
  control   244-50
  as hazard   236-41, 251
  as interference   234-6
  loudness   229-30
  measuring   226-8
  office   199
  personal protection from   248-51
  prevention of   244-8
  sound power levels   228-9
nuisance dusts   379-80

occupational
  health   12-13
  hygiene   3-6, 8, 13-14, 385
  medicine   8
Occupational Safety and Health
    Administration (OSHA)
office
  equipment   191-5
  system   178-9, 184-200
  work station   178-9, 189
    air quality   198
    equipment   191-4
    ergonomics   184-5
    furniture   189-90
    layout   199-200
    work surfaces   194
overuse injury and RSI   92
oxygen   50-2, 307-11
ozone   361-2

particulates   344-5
  see also dusts
person-machine-environment
    17-23
  capacities   23-7
  interface   133-4
  limitations in   27-8
  systems control   146-7
  and tool systems   153-4
  VDTs and   186
petroleum industry and toxicity
    358-60
physiology   5-6, 44, 76, 254
  job satisfaction and   128-9
  lifting and   84-5
  stress and   36, 46, 104-5
  ventilation and   418
  vibration and   212
planes   170
plastics industry and toxic
    effects   358-9, 362
pneumonicosis   357, 380-1, 382-3
polychlorinated biphenyl (PCB)   362
psychology   5
  job satisfaction and   128-9
  manual lifting   85-6
  stress and   104-5
psychophysical approach and
    lifting   85-6
psychosocial effects   20-1, 115-16

radiation
  exposure limits   183-4
  ionization   180-2
  office   179-81
  VDTs and   179, 180-4
Raynaud's disease   216-22, 374
recompression   332-3
reflectance   294
refrigeration and toxicity   358
Repetitive Strain Injury (RSI)   91-2
  ergonomics and   98
  in industry   94-6
  pathology of   96-7
  prevention of   100-2
respiration   301-11, 395-8
  and asphyxiants   358-62
  and dust   377-83

heavy metals and 367, 370–1
solvents and 363–7
retina 281

safety 72, 125, 385
 legislation for 8–9
 RSI and 101
 shift work and 112–17
salt and heat control 273–4
saws 168
scraping tools 168–9
screwdrivers 170–1
shift work 58–9, 112–17
silica dust 382
sinuses 322
sinusoidal vibration 203
skills and job satisfaction 118–19
sleep 113–14
smoke dust 279–80
social security 11–12
solvents 363–5
soot dust 379–80
sound
 parameters of 223
 sound power level (SPL) 228–30
 pressure level 228–30
Speech Interference Level (SIL) 234
starches and energy 37–8
stimuli, lack of 50
stress at work 35–6, 103–5, 109–11, 126–7
sweat production 255–6, 261–3
 P4SR index 261–2
synapse 106
systems approach 15–19, 23
 function 28–31
 human limits in 27
 mechanical limitations 22–3
 person-machine-environment 17–19, 23, 29–31

talc dust 383
task design and RSI 99
temperature
 core 255, 257
 measurement of 260

Oxford index 266
 scale 263
 skin 257–9
 wet-bulb globe 266–7
tennis elbow 93
tenosynovitis 92–3
thermal indices 260–7
tin oxide dust 380
Threshold Limit Values (TLV) 253–6
toluene 273
tools
 boring 170
 compressed air 176
 explosive drive 177
 generic 165–73
 handles, design of 154–61
 internal combustion 176–7
 manually driven 165–73
 materials for 162–4
 percussive 165–8
 power driven 173–7
 RSI and 98–9
 size of 164
 weight of 161
toxic materials
 carcinogens 374
 concentration 352–3
 effects of 346–7, 349–51, 371–3
 human body and 347–9, 351, 353
 gases 358–63
 heavy metals 367–70
 metal carbonyls 370
 solvents 363–7
 in workplace 357–8

ventilation 408–10, 411–13, 403–8
 dilution 399, 400–2
 exhaust 399, 402–4
vibration 203–8
 mechanical impedance and 205–8
 prevention of 218
 protection against 215–6
 Raynaud's disease 216–22
 safety guide for 213–4
 segmental 216–22

sinusoidal 203
whole body response to 208–16
video display terminals
    (VDTs) 179, 185, 192–3
  and fatigue 187–8
  guide to exposure 183
  keyboards 191–4
  management of 184
  office furniture 189
  radiation and 179–84
  screens 192–4, 195–8
vinyl chloride 374
vision
  adaptation of 284–5
  depth of 280, 287–8
  stereoscopic 287
visual
  hazards 296–9
  purple 284–5

walk-through survey 385–9
water sterilization and toxic
    chemicals 273–4, 359
welding 360–1, 365–7
wet collectors of contaminants
    415–17
wetting down as heat control 395
windchill 275–7
word processors 178–9, 186–7
  see also visual display terminals

work
  breaks 57–8
  environment 21–3
  fatigue 42–6, 55–9
  load 273
  motivation 21–3, 48–50
  practices 273
  shift 58–9, 112–17
  station analysis 62–3, 100
    area 67
    conveyors 77–8
    design 66–7, 74–9
    dimensions 64–5
    hazards 384–5
  types of 68–73
workers
  compensation 11–12, 384
  environment 60–1
  health 59–60
  job satisfaction 118–19, 125–7
  noise and 248
  office 187–8
  RSI and 97–8
  stress and 119–20, 126–9
  vibration and 205–8, 211–15
  word processors and 187, 195–8
working
  height 73–4
  hours 55–7

xylene 341–2